GLOBAL ENVIRONMENTAL CHALLENGES

GLOBAL ENVIRONMENTAL CHALLENGES

PERSPECTIVES *from the* SOUTH

EDITED BY JORDI DÍEZ & O.P. DWIVEDI

BROADVIEW PRESS

LIBRARY AND ARCHIVES CANADA CATALOGUING IN PUBLICATION

Global environmental challenges : perspectives from the South / edited by Jordi Díez and O.P. Dwivedi.

Includes bibliographical references and index.

ISBN 978-1-55111-820-8

1. Environmental management—Developing countries. 2. Environmental policy—Developing countries. 3. Globalization—Environmental aspects. 4. Environmental degradation. 5. Global environmental change. I. Díez, Jordi, 1970- II. Dwivedi, O.P., 1937-

GE149.G56 2007 363.7009172'4 C2007-906336-5

BROADVIEW PRESS is an independent, international publishing house, incorporated in 1985. Broadview believes in shared ownership, both with its employees and with the general public; since the year 2000 Broadview shares have traded publicly on the Toronto Venture Exchange under the symbol BDP.

We welcome comments and suggestions regarding any aspect of our publications—please feel free to contact us at the addresses below or at: broadview@broadviewpress.com / www.broadviewpress.com.

North America
Post Office Box 1243,
Peterborough, Ontario, Canada K9J 7H5

2215 Kenmore Ave.,
Buffalo, New York, USA 14207
tel: (705) 743-8990; fax: (705) 743-8353

customerservice@broadviewpress.com

UK, Ireland and continental Europe
NBN International
Estover Road, Plymouth, UK, PL6 7PY
tel: 44 (0) 1752 202300;
fax: 44 (0) 1752 202330;
enquiries@nbninternational.com

Australia and New Zealand
UNIREPS, University of New South Wales
Sydney, NSW, Australia, 2052
tel: 61 2 96640999; fax: 61 2 96645420
infopress@unsw.edu.au

Broadview Press acknowledges the financial support of the Government of Canada through the Book Publishing Industry Development Program (BPIDP) for our publishing activities.

Book Design by Michel Vrána, Black Eye Design.
Printed in Canada

This book is printed on paper containing 100% post-consumer fibre.

10 9 8 7 6 5 4 3 2 1

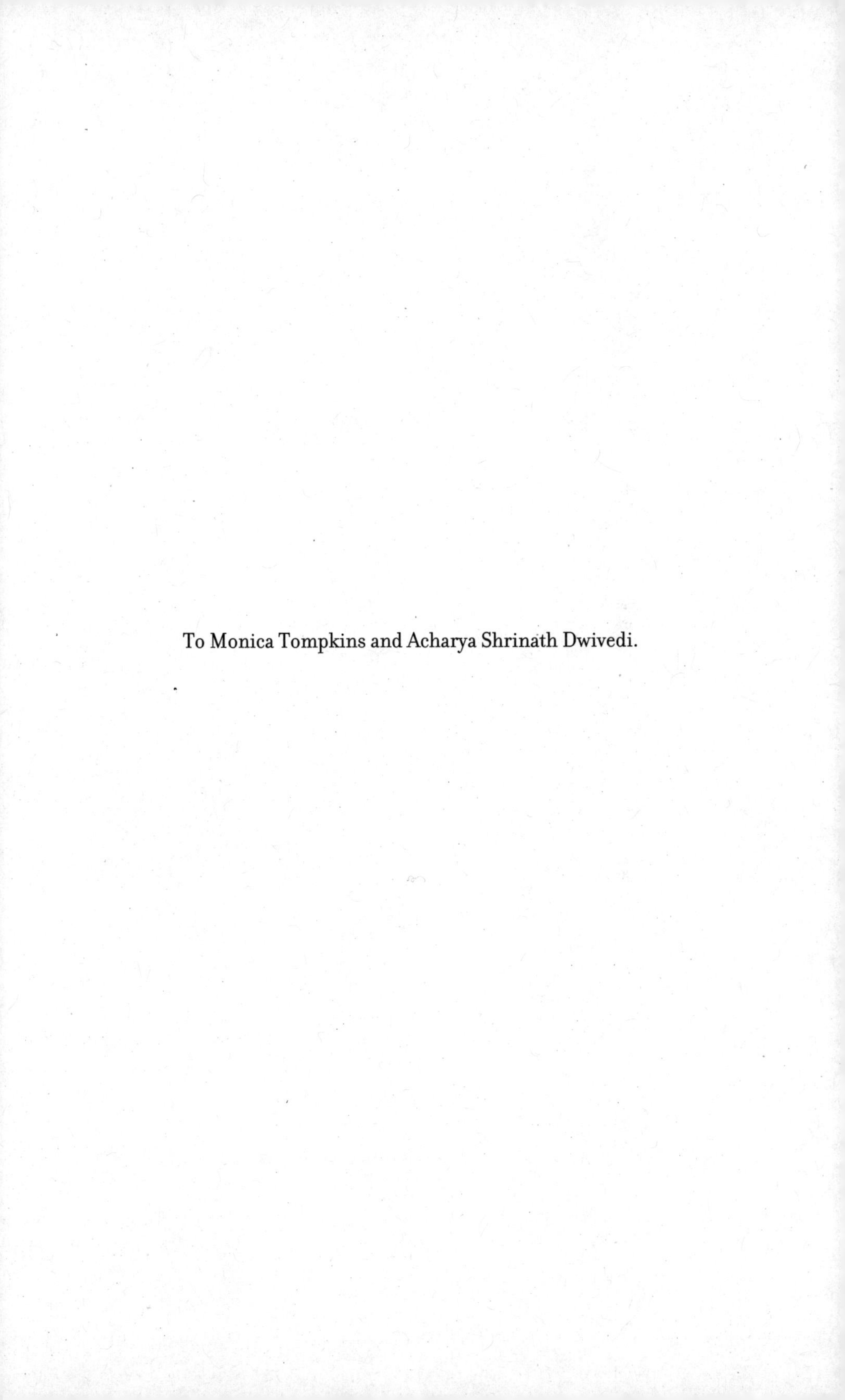

To Monica Tompkins and Acharya Shrinath Dwivedi.

CONTENTS

ACKNOWLEDGMENTS

In our attempt to offer a uniquely southern perspective to the numerous environmental challenges facing our planet, we present here a collection of essays from an array of junior and senior scholars with expertise on environmental issues in countries of the Global South. Our biggest debt, therefore, is to them, for this volume would not be possible without their truly excellent contributions. We thank them for having agreed to participate in this project, but especially for having so graciously accepted to revise their chapters as requested. We would also like to acknowledge and thank the anonymous referees for their insightful and very helpful comments. Very special thanks must go to Greg Yantz at Broadview, who helped us shepherd this project from its very conception to its completion with professionalism, graciousness, and an unbeatable sense of humor.

Jordi Díez
O.P. Dwivedi

INTRODUCTION

Jordi Díez and O.P. Dwivedi

GLOBALIZATION AND THE ENVIRONMENT

Our world faces significant environmental challenges. Over the last half-century, world population has doubled and the size of the global economy has expanded from less than $5 trillion to more than $35 trillion. The consequences of such increased pressure on our natural environment have become all too obvious. Only about 20 per cent of the world's original forests remain in their natural state, having made tens of thousands of plant and animal species extinct. Overfishing has already caused the collapse of approximately 30 per cent of fisheries and, if current trends continue, it is estimated that all fisheries will have collapsed by 2050. Soil is being degraded on a large scale throughout the world, and approximately 80 per cent of agricultural land in dry regions suffers from severe to modest desertification. Vast amounts of waste, hazardous materials, and heavy metals have been dumped into our oceans and rivers, causing severe and irreversible changes to sea environments. Increased pollution is responsible for the deaths of hundreds of thousands of people around the world every year, and, despite some skepticism by a relatively small number of people, it is now clear that pollution is responsible for the warming of the Earth, a process that will have a direct and significant impact on our planet. The demands placed on natural resources by our current development course are simply not sustainable. Given the prevailing levels of population growth, the Earth will have to accommodate and provide for an additional three billion people over the next 50 years, placing added pressure on our natural environment. The continuing deterioration of our planet's ecology poses a major threat to the viability of our societies.

The fact that the degradation of our natural environment has accelerated over the last 30 years should also be viewed against the backdrop of increased global economic interconnectedness, a process known as globalization. The lowering of trade barriers, which has been part of processes of economic reform introduced by numerous countries around the world, has resulted in the integration of economies as goods and services flow more freely across borders. Such integration has contributed to the emergence of a global economic and trade system to which an increasing number of national economies are now more directly linked. This process of globalization has been unfolding for quite some time, but its latest phase appears to be different in that it has been characterized by deeper and faster global interconnections that go beyond economic and trade relationships. Largely fuelled by significant advancements in information technology, global integration appears to have reached more places in a much deeper manner.

Debates on the nature, significance, and effects of globalization have preoccupied many social scientists over the last decade, and little agreement appears to have been reached. Most of these debates, however, have focused on the social, political, and economic effects of globalization; much less attention has been paid to its environmental repercussions. As the global economy continues to expand, and as populations continue to grow, it becomes increasingly more important to look at the extent to which the acceleration of global integration has affected the environment. That is the goal of this collection of essays. There is no doubt that economic integration has promoted economic prosperity to some parts of the world, such as many Asian countries. But the acceleration of economic activity prompted by globalization can also potentially bring further deterioration to our natural environment. As economic activity accelerates, the demand for natural resources and pollution levels necessarily increases. In this book, we look at whether globalization has brought about new environmental challenges to countries of the Global South by looking at the effects of global integration on the natural environments of eleven developing countries selected for study in this volume.

Most environmental problems are global in nature. However, there are vast differences between the industrialized countries and countries of the Global South in both their contribution to these problems and their effects. The Global North generates and consumes most of the world's wealth. By the turn of the century, people living in the world's richest 20 per cent of countries were consuming 86 per cent of the world's Gross National Product (GNP). As a result, people in industrialized countries also contribute more inordinately to environmental degradation. Industrial countries account for 70 per cent of carbon dioxide emissions, about 3.3 tons per capita, while developing countries emit half a ton per capita. The United States alone has roughly the same amount of greenhouse gas emissions as 150 developing countries. However, while the Global South consumes less and pollutes less globally, the effects of environmental degradation tend to be felt more severely by its peoples; environmental problems are often manifested far more intensely because survival margins are much smaller. Water pollution, for example, has a more direct and severe effect on people in developing countries; each year about 5 million people die from diseases caused by unsafe drinking water and lack of sanitation and hygiene. The effects of globalization on the environment can therefore be felt much differently in developing countries than in industrialized ones. It is because of this difference that we present in this volume a distinctly "southern" perspective to the link between globalization and the environment. The contributors to this book tell the stories of how countries of the Global South have been affected by problems of environmental degradation and how they have responded. What specific environmental challenges do these countries face? What new challenges has globalization brought about? What have been the effects? How have they dealt with them? Through analyses of eleven case studies selected from across the Global South, the various contributors to this book answer these questions. We believe that it is important to present the stories of how the people, institutions, and leaders of countries of the Global South have been affected by environmental challenges in an increasingly globalized world, and how they have responded.

GLOBALIZATION AND THE ENVIRONMENTAL MOVEMENT

The process of globalization has not been limited to global economic integration. It has also been characterized by increased global social interaction. Economic globalization has been accompanied by a revolution in information technology that has greatly facilitated cross-border interaction among non-state actors. Such interaction has in turn contributed to the transnational organization and mobilization of individuals in pursuit of numerous common objectives, such as better protection of human rights. We have therefore witnessed a thickening of cross-national links among individuals, and we are quite possibly experiencing the emergence of a global civil society. Environmental groups have been at the heart of this phenomenon. Over the last several decades, the number of Environmental Non-Governmental Organizations (ENGOs) that operate across borders has markedly increased, as has their international visibility. ENGOs have thus become very active players in international politics. Indeed, the elevation of environmental issues to the international agenda has in large part been the result of increased environmental mobilization. Beyond their ability in influencing the international environmental agenda, however, ENGOs have also been influential in the construction of international environmental regimes; ENGOs have been active players in the negotiation of the numerous environmental treaties and agreements that have been established in recent years. Perhaps nowhere was such influence more evident than the Rio Summit, where ENGOs were important actors who had an unprecedented impact on the conference negotiations and on drafting the principles included in the Rio Declaration and elements of *Agenda 21*.

The growth in the number of international ENGOs has also been characterized by an increase in their operations worldwide, especially in developing countries. As they have become more organized and better funded, these ENGOs have expanded their efforts in the Global South. Over the past two decades, organizations such as Greenpeace, Conservation International, and the World Wildlife Fund have increased the number of projects and programs they launch and manage in developing nations. Indeed, not only are most of these organizations visible at international fora, but they have also become important players in domestic environmental issues. The expansion of the operations of these organizations has often contributed to the transfer of resources and expertise to individuals and institutions in developing countries, thereby strengthening environmental mobilization in the Global South. Frequently this has occurred through close collaboration between northern and southern Non-Governmental

Organizations (NGOs) in a variety of environmental projects. As ENGOs become more visible and active within developing countries, it seems warranted to look at the extent to which they have influenced and contributed to environmental management in the Global South. Consequently, in addition to looking at the effects of economic globalization on countries of the Global South, the contributors to this volume explore the role, influence, and impact that ENGOs have had in environmental management in developing countries.

THE INTERNATIONAL ENVIRONMENTAL AGENDA

Increased global interaction and the acceleration of environmental degradation over the last 30 years have also occurred against a backdrop of collective efforts at the international level to deal with environmental problems, as we have witnessed increased environmental awareness across the world. The link between globalization and environmental degradation must therefore be studied in relation to the evolution of the international environmental agenda. The rise of concern and consciousness about environmental matters in the Global South can be understood only in the broader context of international efforts in general, and the domestic pressures that force countries to develop environmental policies and programs in particular. The history of such concerns and consciousness can be traced back to the call by industrialized nations for a world assembly in 1972. This call was actually spearheaded by the Club of Rome report, *The Limits to Growth*, which argued effectively that there was an urgent need to control the present growth trends in world population, industrialization and pollution, food production, and resource depletion if we wished to save our earth from a disaster of planetary proportions. To that end, a meeting of world leaders took place in Stockholm in June 1972. Issues raised in that meeting included whether poverty in the South or affluence in the North was the main cause of pollution, and what efforts should be undertaken by both international organizations and international development agencies to alleviate this problem. Developing nations argued that just as the pollution of air and water, accumulation of wastes, urban blight, the loss of wildlife, and the dereliction of land and forests were examples of environmental degradation in the industrialized nations, so were the various companions of poverty such as disease, malnutrition, hunger, and squalor in the developing nations. Despite differences among the nations of North and South, environmental awareness started seeping though the consciousnesses of developing countries as they started to create environmental policy and institutions. But the main impetus came from the Brundtland Commission (1984-87), whose report, *Our Common Future*, stated that no country can develop in isolation from others, and that the unity of human needs requires the pursuit of sustainable development with changes in the domestic and international policies of every nation. The Commission's report also set the stage for the 1992 Earth Summit (also called the Rio Meeting).

The Earth Summit took place in Rio de Janeiro, Brazil, in June 1992, and it demonstrated that environmental protection could no longer be regarded as a luxury of the rich alone. Rather, environmental factors would have to be integrated with economic and social issues, which then would have to become a central part of the policy-making process in all countries. To this end, the Summit adopted *Agenda 21*, a blueprint for sustainable development, with a common vision for growth, equity, and nature conservation for future generations. There was no doubt that by 1992 the issues of the South had gained legitimacy, as indicated by the keen participation by the leaders, activists, and NGOs from those nations. Finally, ten years after the Rio meeting, the United Nations organized another assembly, the World Summit on Sustainable Development (WSSD), in Johannesburg, South Africa, in June 2002. Issues discussed in Johannesburg included the state of the world's environment, which still remained in a very fragile state: the North had not fulfilled its promises made through *Agenda 21*, whereas the South continued to pursue reckless practices of development while paying only lip-service to environmental protection; and the failure of the North could not discourage multinational corporations from exploiting the resources of poor countries. Nevertheless, by this time, all developing nations had national policies and institutional mechanisms in place to control pollution and usher in sustainable development strategies, although the effectiveness of such institutional mechanisms remained questionable.

How have developing countries responded to the international environmental agenda? The Stockholm Conference of 1972 inspired nations and green activists everywhere. It also led to the establishment of environment ministries, departments, and agencies worldwide, and put the environment on the international political agenda. The result was that many developing nations, particularly in Asia and Latin America, started taking a greater interest by creating the necessary management tools for environmental protection in their countries. However, the biggest impetus came around 1990–92, when all developing nations were encouraged (and in many cases conditionalities were imposed by international development agencies such as the World Bank, the IMF, and other regional organizations) to create their own national environmental policy document and to enact environmental protection legislation. Further impetus came after the Earth Summit and its *Agenda 21*, which strengthened the resolve in many developing nations to enact laws and regulations suited to country-specific conditions. As various chapters in this book illustrate, international pressures and the growth of ENGOs in the South enabled most of the developing nations to respond in earnest to environmental concerns, and also to give effect to international environmental norms. Nevertheless, developing nations suffer from the accelerating rate of environmental deterioration because either their laws and institutions remain largely diffuse and sectoral, or their management tools and regulatory mechanisms remain ineffective, thereby rendering the implementation of international environmental norms an elusive goal. It is from this perspective, exploring the link between globalization and the environment, that the essays commissioned for this volume analyze both the various environmental management tools that have been adopted in the Global South to deal with environmental degradation and the extent to which they have been effective. Let us look at what lies in the pages ahead.

THE STRUCTURE OF THE BOOK

This book is divided into twelve chapters. In Chapter 1, Jordi Díez provides an analysis of the relationship between globalization and environmental issues in very general terms as context for the eleven country case studies that follow. Díez begins his chapter by presenting an overview of the debate surrounding globalization. He argues that while there is little agreement on the effects of the latest phase of the process of globalization, its environmental effects on the Global South have become more than obvious. Díez suggests, for example, that global economic integration has increased the demand for natural resources that many countries in the Global South have in abundance. This has in turn resulted in increased levels of deforestation and overfishing. He argues, however, that while globalization has brought about new environmental challenges in developing countries, it has also brought about new opportunities to tackle environmental problems. He suggests that because globalization has facilitated communication among people around the world, it has contributed to the communication among non-state actors across borders, helping them to organize and mobilize in the pursuit of better environmental protection.

In Chapters 2 to 12, scholars with expertise in environmental issues present their country case studies. In Chapter 2, Stephen Ma explores the effects of the relationship between globalization and the environment in China. Ma argues that globalization has presented China with numerous new challenges, and its integration as a new member of the "global village" has had a strong impact on the nation's environmental management. Ma shows that China faced environmental challenges as early as the 1970s, long before it achieved the astonishing levels of economic growth that it has been experiencing more recently. However, China's integration into the global economy has brought with it an acceleration of environmental degradation, as it has attracted multinational corporations that seek softer environmental legislation and which, in collusion with corrupt government officials, have relocated their heavily polluting factories to China. Global integration, Ma shows, has also resulted in increased desertification, exacerbating sandstorms that occur in the country's capital city, Beijing; an increase in urban population along coastal areas, which has created increased pollution and sanitation problems; and a dramatic increase in the demand for natural resources to satisfy its "neck-breaking industrialization." Ma argues that China's opening to the outside world, which was one of the core modernization plans of its Cultural Revolution, has contributed to the emergence of environmental consciousness in China as well as to an increase in the number of NGOs that operate in the country. Having benefited from the transfer of funding from international organizations, ENGOs have not only had an impact on environmental management in China, but they actually function as conduits between governments and citizens, providing an avenue for motivating and mobilizing public participation in environmental protection.

In Chapter 3, Craig Johnson presents Thailand's recent experience with environmental management. Johnson shows that the export-oriented economic policies that Thailand introduced in the latter part of the twentieth century, and which led to a "neo-liberal boom," provided strong incentives for Thailand to "mine" its natural resources, policies that have resulted in an array of environmental problems, such as land and water degradation, very high levels of deforestation, and the depletion of the country's coastal fisheries. Johnson also shows that, along with the neo-liberal boom, the 1980s and 1990s were characterized by important environmental mobilization that was responsible for several environmental victories, a phenomenon that represents a watershed moment for Thai politics. These environmental victories translated into the establishment of several pieces of environmental legislation. In his chapter he argues that cycles of economic crises and political infighting within the country's elite provided the political opportunity structures for Thailand's environmental movement to emerge. Johnson shows that the emergence of Thai environmentalism was facilitated by a new transnational discourse that encompassed the concept of sustainability, thereby affording environmental concerns legitimacy, through growing interest by national and international ENGOs in the country's environment, and through the 1992 Rio Summit. The movement in turn provided an important vehicle through which previously excluded groups could advance a new political agenda highlighting the environmental costs of Thailand's neo-liberal success.

In the next chapter, Chapter 4, Bruce Mitchell explores the relationship between environmental management and globalization in Indonesia through an analysis of two case studies in which he was personally involved: the development of a sustainable development strategy for the province of Bali during the 1990s, and the process that led to the enhancement of environmental management tools in the province of Sulawesi. Based on these two cases, Mitchell argues that development and environmental decisions and their subsequent impact on the environment are significantly influenced by external forces. This, he argues, can be seen through the impact that tourism has on Indonesia. Mitchell's analysis also shows that culture plays an important role in the formulation and implementation of environmental policies, that ENGOs play an important role in these processes, and that they, in fact, act more quickly than governments in addressing environmental problems.

In the last Asian case, O.P. Dwivedi presents an analysis of India's experience with environmental management. Dwivedi shows that India's environmental challenges have increased significantly over the last thirty years, and argues that the main factors behind the deterioration of the country's natural environment are poverty and the government's inability to enforce environmental legislation and regulation. Despite India's extensive environmental and sustainable strategies, Dwivedi argues that substantial change is neither noticeable nor liable to emerge soon, given that the factors that contribute to environmental degradation are likely to continue. On a more positive note, Dwivedi shows in his chapter that environmentalism has strengthened in India, beginning in the late 1980s, and that NGOs have been able to contribute to raising awareness among the population and to aid the government in monitoring policy implementation.

In Chapter 6, Paul Kingston presents an analysis of the relationship between globalization and the environment in the Middle East by focusing on the case study of Lebanon. In his chapter, Kingston argues that environmental politics in Lebanon are strongly influenced by agents of global capitalism. In addition to multinational corporations, governments of industrialized countries, and international economic institutions, Kingston argues, international NGOs are important agents of "this global environmental civilization project." By looking at Lebanon, he argues that international NGOs and scientific experts, operating under the banner of promoting sustainable development and global biodiversity, have penetrated deep into the local political ecologies of developing countries, setting off processes that have led to significant, and often counterproductive, restructurings of local environmental governance in the Global South.

The next three chapters present a uniquely African perspective. In Chapter 7, Christopher Gore looks at the case of Uganda and argues that, in Africa, the process of globalization has had a definite impact on environmental issues and resources. With reference to Uganda, Gore argues that nowhere does this impact play out more poignantly than in national environmental decision- and policy-making processes. He suggests that it is in this political milieu that tensions between NGOs and governments as a result of calls made by international donors come to the fore: African governments are antagonistic to these calls because of their support for policies that rely on the extraction of natural resources, as they are a source of foreign exchange. Focusing on policy- and decision-making processes in Uganda's energy and electricity sector, Gore argues that globalization and the attendant actors, ideas, processes, and demands it carries place an enormous strain on systems of environmental management in Africa—and, in certain cases, they undermine them. Finally, Gore shows that NGOs have had limited influence on environmental affairs due to the power of the World Bank's policy agenda and the commitment of the Ugandan government to that agenda.

In Chapter 8, Godwin Onu looks at the case of Nigeria. In his chapter, Onu reviews four areas of environmental degradation in Nigeria—desertification, pollution, soil erosion, and flooding—and points to an acceleration of this degradation in recent years. In his contribution, Onu argues that, despite the increased role of civil-society actors and the establishment of numerous environmental management tools in the country, better environmental protection in Nigeria will depend entirely on increased political will by the government to implement environmental regulations, more participatory decision- and policy-making processes, and a change in the citizenry's attitudes toward environmental damage.

In the last African case, Rebecca Tiessen looks at South Africa. Analyzing the country's experience with environmental management over the last decade, Tiessen argues that, in South Africa, environmental management is the result of an array of pressures on the government to adopt sustainable development practices coming from within the government, civil society groups, the corporate sector, and international commitments the country has undertaken. Tiessen argues that these cross-pressures must be examined in the context of globalization and changing global trends, as well as the emergence of new political and environmental actors both within South Africa and internationally. She suggests that the international environmental agenda has had a significant effect on environmental politics and policy in South Africa.

The last chapters present three Latin American cases. In Chapter 10, Jordi Díez analyzes Mexico's performance in the area of environmental management within the country's experience of economic reform. Similar to most other Latin American countries, Mexico has undergone significant economic reforms, and its economy has become directly linked with global economic and trade forces. Such economic transformation, Díez shows, has brought about significant challenges to the country's natural environment, such as increased pollution, increased pressure on the country's natural resources, and soil erosion due to the expansion of agricultural land. Díez also shows, however, that Mexico's increased interaction with outside actors and institutions, especially within the North American Free Trade Agreement, has resulted in an unprecedented increase in the number of ENGOs in the country, and that these organizations have not only become important domestic political actors, but they have achieved numerous environmental triumphs.

Chile is the next Latin American country analyzed in this collection. In Chapter 11, Jorge Nef looks at the Chilean case and argues that, although the country's environmental problems may not be new, they have in many cases worsened and accelerated since Chile's adoption of a neo-liberal model in the mid 1970s. Air, water and marine pollution, deforestation, and soil degradation have resulted from the country's export-led policies that concentrate upon non-traditional exports (e.g., fruits). Despite the country's eager adoption of the model, Nef argues that the public sector has remained one of the most important actors in national life in Chile, and it has been through state intervention that environmental problems have been tackled. However, given the country's highly formalistic and legalistic bureaucratic systems and its fragmented and weak environmental management system, Nef argues that the environmental agenda in post-coup Chile remains largely an unfulfilled promise. In his chapter Nef also shows that an "environmental alliance" made up of domestic and international scientists, environmentalists, and activists, has emerged in Chile but, despite their organization and mobilization, they have been unable to influence important environmental policy decisions.

In the last chapter, Angus Wright looks at the case of Brazil. In his contribution, Wright argues that while most Brazilians care about the country's truly vast natural wealth, and while people and organizations have been vigorous in attempting to protect the natural environment, environmental management by society and the country remains disappointing. Wright argues that this is due to profound problems caused by deep social inequalities and a closely related inability or unwillingness of public institutions to govern effectively. Importantly, Wright shows in his chapter that, despite the great hopes with which environmentalists greeted the election of reformist Ignacio (Lula) da Silva in 2001, environmental degradation has continued apace.

Let us begin.

CHAPTER 1

GLOBALIZATION AND ENVIRONMENTAL CHALLENGES CONFRONTING THE SOUTH

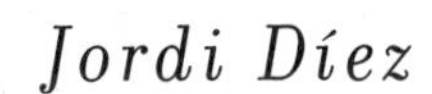

Jordi Díez

INTRODUCTION

It would not be an exaggeration to state that our world has undergone some rather profound changes over the last two decades. The collapse of the Soviet Union gave way to a more relaxed international system as the ideological confrontation that dominated the Cold War dissipated and a new spirit of international cooperation emerged during the 1990s. Politically, the so-called Third Wave of democratization swept the world and numerous countries moved away from authoritarian regimes and into democratic rule. Today, more people around the world are ruled by democratically elected governments than ever before. On the economic front, the inward-oriented development models that characterized the post-war international system have been dismantled and most countries have adopted market-friendly economic policies. At the same time, advances in communication technologies have propelled a true information revolution, the consequences of which will likely be as transformative of our society as the Industrial Revolution of the mid-nineteenth century was of its. All these developments have contributed to the process that we have now come to know as globalization. As economic and political barriers have come down, and as technology has increased interaction among people around the world, our societies have become more interconnected socially, economically, and politically. Within the context of increased global interconnectedness, we have increasingly become aware of the environmental damage to which our planet has been subjected. Our world faces several environmental challenges, and our current development paths are simply not sustainable.

What exactly is globalization? Is this process of increased global interaction new or simply a continuation of a process that started a long time ago? Who has benefited from this process and who has not? Has globalization contributed to further environmental degradation? In this chapter, I answer these questions and explore the relationship between globalization and the environment and its effects on the Global South in an attempt to provide the context for the eleven case studies that follow in this collection. This may help the reader follow the stories that the various contributors are about to tell. In this chapter, I argue that the latest phase of globalization—which, while not new, has been characterized by its scope and speed—has brought about numerous environmental challenges to countries of the Global South. However, globalization has also provided important opportunities to tackle environmental problems. Because globalization does not strictly refer to growing economic integration, increased global interconnectedness has facilitated communication among people around the world, and this has contributed to both the strengthening of transnational links and interaction among non-state actors as well as the emergence of a system of environmental regimes. The chapter is divided into three main sections. The first reviews the debate over globalization and argues that its latest phase, given its extent and scope, appears to be distinct: while globalization has been unfolding for a while, several factors have made the latest phase different from previous ones. The second section explores the various environmental challenges that globalization has brought about for countries of the Global South. The last section looks at the effects globalization has had on the strengthening of a global civil society and the emergence and evolution of our system of global environmental governance.

THE GLOBALIZATION DEBATE

What is Globalization?

In very general terms, globalization is usually understood as the process of growing social and economic interconnectedness among people around the world. Among scholars, however, there are various conceptualizations of this process that can be grouped into four general categories. For a first group, globalization refers merely to an increase in economic interaction among countries in terms of trade and investment. Paul Hirst and Grahame Thompson have defined globalization in terms of "large and growing flows of trade and capital investment between countries" (1999: 48). Seen from this economic perspective, globalization is strictly measured by the total amount of world production that is traded. For a second group, globalization equates economic liberalization, and the process primarily refers to the removal of state restrictions to movements across countries in order to create an "open" world economy. Scholars who adhere to this view use globalization to describe the general process of international economic integration that has resulted from the lifting of restrictions to economic activity (Sander 1996: 27). A third conception of globalization sees the process as the spread of Western values. According to this view, globalization refers to the "westernization" of values, which has been responsible for the destruction of pre-existing cultures around the globe. Adherents to this definition readily refer to the process as the "Americanization" of the world and often describe it as the "imperialism" of culture (Barber 1996; Gowan 1999).

For a fourth group, to which this volume belongs, globalization is seen as growing connections of people around the world, connections that go beyond simply economic integration and affect the manner in which we look at the world. According to this conception, globalization has provoked a "respatialization" of social relations that entails the reconfiguration of social geography, making the world smaller (Held et al. 1999; Giddens 1990; Robertson 1992). That is, beyond growing economic interconnectedness, and greatly aided by the revolution in information and communication technologies, people across the globe are able to interact with one another more frequently and faster, thereby creating closer social relations despite geographical distances. These social relations have in turn helped create a sentiment of global belongingness because of the global consciousness that has recently developed. Roland Robertson argues that there has been a "compression of the world" and an emergence of a "global consciousness" (1992). Anthony Giddens, for his part, characterizes globalization as "the intensification of world-wide social relations, which link distant realities in such a way that local happenings are shaped by events that take shape miles away" (1990: 64). According to this view, the world has become a smaller place, which has had an effect on the way we look at it and how we see ourselves in it.

Although there exists some overlap among these definitions of globalization, there is little agreement on the extent and significance of the process, something which has created a great deal of academic debate. For some people, commonly referred to as the "globalization sceptics," there is nothing new about this process because, they argue, it has been unfolding for centuries, if not millennia. Observers who focus on the economic aspects of globalization, for example, argue that, in proportional terms, levels of trade and investment were higher toward the end of the nineteenth century, when the world experienced an expansion of the capitalist system fuelled by the Industrial Revolution (Hirst and Thompson 1999; O'Rourke and Williamson 1999). Others have argued that, while there may have been an acceleration of the process in recent years, global integration can be traced back 500 years, when the current international economic system was established with the colonization of the Americas (Robertson 1992).

On the other side of the debate there are those who argue that the process of globalization is not only real and new, but is also the most important feature of contemporary history. Adherents to this view are commonly referred to as the "hyper-globalizers." For them, social and economic relations around the world have become thoroughly globalized, and the process is similar in scope and importance to some previous watershed developments in human history. For example, for former Brazilian president Fernando Henrique Cardoso (1996), the emergence of global consciousness as a result of globalization is as important as Copernicus's discovery five centuries ago that the world is not flat.

The debate over the significance of globalization is undoubtedly influenced by political persuasions. But while analyses and interpretations of social phenomena are generally influenced by the ideological bent of the observer, there is no question that those within the hyper-globalizer camp have taken a decidedly ideological position. For many among them not only argue that globalization is a phenomenon of paramount historical importance, but they also take a normative position, arguing that globalization is a "good" thing. Their arguments have therefore taken on discursive characteristics, as they have strongly defended global integration. Such discourse has come to be known as "globalism." Whereas globalization refers to the process of global social and economic integration, globalism entails the discourse that emerged toward the end of the 1980s, the goal of which is to further the process of globalization.

Globalism has stemmed mostly from actors and institutions in the Global North and has been actively pursed by international financial institutions such as the International Monetary Fund (IMF), the World Bank, the US Treasury Board, executives of large transnational corporations, and investment bankers. The discourse is associated with the so-called "Washington Consensus" of the 1980s, which was firmly grounded in the tenets of classical economic theory.[1] Perhaps nothing exhibits more the ideological component of the globalist argument than the idea that globalization is inevitable, a notion best encapsulated by the concept of TINA (There Is No Alternative). Proponents of this view argue that globalization reflects the irreversible spread of market forces, driven by technological innovations, which makes the march toward integrated market economics completely inevitable (Steger 2002). A central component of this argument is the gradual and inevitable weakening of the nation-state; as economies around the world become more integrated and more dependent on trade and investment, the view holds, decisions made by states will become less and less relevant as they are replaced by the rationales dictated by market forces.[2] One of the main weaknesses of the globalist position is the paradox that exists between seeing market forces as the amalgamation of individuals' rational choices and the view that these are in turn inevitable. It is ironic that globalists adopt the determinist historical narrative that many Marxist theorists, whom they have fiercely criticized, also adopt. Clearly, given that humans are rational, decision-making individuals, the only phenomena that are inevitable are natural disasters.[3]

As with most debated subjects in the social sciences, agreement on the significance of globalization is unlikely to transpire any time soon, and the debate will no doubt continue. Some observations can nevertheless be made. There is no doubt that global economic and social integration has been taking place for centuries, as the globalization sceptics maintain. Indeed, it can be argued that the process began some 12,000 years ago, when a group of hunters and gatherers reached South America, and accelerated when the Europeans reached the Americas in 1492, thereby establishing the first truly international economic system through mercantilist policies that were central to colonialism. Since this international system was established by the European colonial powers, the process of interconnectedness has gradually intensified. However, it appears that such a process has accelerated over the last few decades and that it has entered a new phase, one that exhibits characteristics that are different on many levels from those of the past and that has been driven by three main factors. First, the dismantling of trade barriers during the 1980s and 1990s has increased trade and commerce to unprecedented levels. While the post-war Bretton Woods international system was characterized by Keynesian interventionism and protectionism—the so-called "golden age of controlled capitalism"—the arrival to power of Margaret Thatcher and Ronald Reagan in the early 1980s spurred a neo-liberal revolution that saw free markets as the main promoters of economic growth. This revolution gained legitimacy with the collapse of the Soviet Union in the mid-1980s, as it took a paradigmatic advantage, and it became the dominant economic model as international financial institutions, heavily influenced by Washington and London, acquired an enhanced economic role internationally. The ascension of neo-liberalism during the 1980s resulted in the introduction of economic reform programs throughout the world, as barriers to trade were taken down. These reforms have increased trade and investment to unprecedented levels and have allowed for the emergence of truly global markets for a myriad of goods and services.

Globalization sceptics also argue that the total amount of cross-national trade and investment has been higher before, especially toward the end of the nineteenth century, the classical Gold Standard era. However, as one of the most comprehensive studies on globalization suggests, there is higher trade intensity among countries today than there has been ever before (Held et al. 1999: 169).[4] When the trade of services is taken into account, trade intensity is even higher.[5] Cross-national trade has not only increased, but it has also linked national economies at a deeper level than in the past; the division of trade and domestic activity has been blurred as countries have become increasingly more dependent on components from overseas for their production processes (Held et al. 1999). What we are witnessing is not only an increase in international trade, but a deeper connection among world economies as national economies become ensnared with each other in such a way that trade becomes integral to national economic prosperity (Held et al. 1999: 170).

The current phase of globalization has also been propelled by the rather dramatic drop in the costs of transportation of people and goods. While the transportation of goods, services, and people is nothing new, the cost of doing so has steadily decreased over the last few decades as a result of technical progress, containerization, and cheaper fuel, thereby increasing its volume. For example, the average freight ocean cost to transport a ton of import and export cargo was $95 (US) in 1920. That cost had been reduced to $63 by 1940, and it dropped to $29 by 1990 (Hufbauer 1991). The cost of air transportation, as measured by the average air transport revenue per passenger mile, was $0.68 in 1930, $0.24 in 1960, and $0.11 in 1990 (Hufbauer 1991). As these costs have decreased, they have become negligible components of the price of traded goods; this has fuelled an increase not only in the volume, but also in the number, of goods and services that are traded. What is distinctive, therefore, about the new phase of globalization is the emergence of global markets, helped by the reduction in barriers and transportation costs, of a much wider array of goods and services than during the Gold Standard era (Held et al. 1999: 171).

The third, and perhaps most important, factor that has contributed to the distinctive characteristics of the current phase of globalization has been the revolution in information technologies. The advancements made in communications over the last three decades have been dramatic and unprecedented and have facilitated communications among people around the world. The cost of communication has steadily decreased during the twentieth century. For example, the cost of a three-minute telephone call from New York to London was US$244.65 in 1930. The same call cost $45.86 in 1960 and $3.32 in 1990 (Hufbauer 1991). The reduction of telephone costs has contributed to an increase in trade and finance. More importantly, it has been accompanied by the advent of technologies such as fibre optic cables, satellites, and digitization, which have contributed to the emergence of a truly global telecommunication system. Such a system has made it increasingly cheaper and hence easier for people to communicate with others around the world and has allowed people to receive information from faraway places instantly. Globalization has therefore not been limited to increased economic integration but has encompassed an increase in social interaction. Growing communication has had an impact on the human consciousness: people have become more conscious of the world around them and an ever larger number of them are connected to the rest of the world through all forms of communications. Globalization is thus more than economic integration; it is a multidimensional set of social processes that involves the intensification of human contact on a variety of areas, social, cultural and economic. The latest phase of contemporary globalization is then characterized by its scope and speed.

The Effects of Globalization

If there exists little agreement on the extent and significance of globalization, there appears to be even less when the debate focuses on its effects. For its implacable defenders, globalization has mainly had positive effects. Taking a merely economic perspective, it has been argued that increased trade has unambiguously been good in reducing poverty and fostering economic development (Bhagwati 2004). According to this line of thinking, everyone will eventually be better off in a globally integrated economy. Francis Fukuyama (1992) famously argued that, with the collapse of the Soviet Union, the virtues of global capitalism were confirmed, proving that further global integration was needed as the world reached the "end of history." Others have highlighted the economic efficiencies gained in a globalized world and the growth, usually measured in terms of Gross Domestic Product (GDP), that it has fuelled.[6] Proponents of globalization admit that the process does carry with it some shortcomings. However, they argue that, overall, and in the long run, everyone will be better off. Indeed, Jagdish Bhagwati has argued that those parts of the world that have lagged behind economically, such as sub-Saharan Africa, need more globalization, not less (2004). It has also been argued that globalization has fostered a sense of global community, which has in turn been responsible for a decrease in violent conflict (Holm and Sørensen 1995), and that globalization, partly through the expansion of information technologies, has been responsible for the unprecedented spread of democracy in the world (Diamond and Plattner 1996; Hill and Hughes 1998).

To fierce opponents of globalization, increased global interconnectedness has had mostly negative effects. Opponents argue that global economic integration has given too much power to multinational corporations and that these now "rule" the world (Barlow and Clarke 2001; Korten 1995). A common criticism made by these observers is that despite their increased power, multinational corporations have become less accountable to national governments given their ability to move their operations across borders, an ability that has increased with economic liberalization. Others have contended that the manner in which globalization has unfolded has resulted in a "usurpation" of power from national governments by international financial institutions, such as the International Monetary Fund (IMF) and the World Bank, taking away, irreversibly, their autonomy to formulate national policy (Burbach and Danaher 2000). Along this vein, some have gone even further to argue that national states have become "obsolete" or in crisis (Horsman and Marshall 1994; Bamyeh 2000; Ohmae 1991).

From a Marxist perspective, it has also been argued that globalization is but the continuation of the relentless drive to subordinate all forms of social organization to the directives of the market, and so, according to this view, globalization has further weakened the working classes (Gills 2000). Central to this approach is the idea that globalization is, more than a process, a new phase of a project of capitalist subjugation and domination.[7] Others have suggested that globalization has brought about a new global "disorder," which has been manifested by an increase in violence, religious fundamentalism, intolerance, and terrorism (Kaplan 2000; Brzezinski 1993). Finally, in terms of the effects of globalization on social equality, critics of globalization maintain that it has had negative effects, as disparities between rich and poor in almost every country have increased, as have disparities between the North and the South (Gray 1998; Mishra 1999).

Between these two starkly different positions, some scholars have more recently taken a more intermediate position and have argued that, while globalization has been responsible for some positive outcomes, the process needs to be "civilized" so as to make market forces serve society, and not the other way around. For example, Richard Sandbrook, influenced by the writings of the great twentieth-century economic historian Karl Polanyi, has argued that the current version of globalization, which is neo-liberal at its core, must be harnessed "so that the economy serves society and not vice versa" (2003: 2).

Who is right? There is no doubt that, similarly to what occurs with every significant process of social change, there are people who are negatively affected by globalization and people who have benefited from it. Clearly, the effects of such profound social change are felt very differently by a wealthy, middle-aged investment banker who lives in New York and for whom the lowering of investment barriers represents new opportunities than they are by a poor, older Latin American farmer who has been put out of business because of a reduction in subsidies—a staple of rural living in the Global South under globalization—and who has no social security. However, it appears that assessments of the process of globalization will need to be based on the evidence that emerges over the next few decades, as the process is quite new and its social, political, and economic effects are still unfolding. It would seem that additional time is needed in order to obtain a fuller and more definitive assessment of the effects of globalization. In terms of the Global South, the evidence appears to be mixed at best, although it begins to tilt over the negative side of the balance sheet. Much has certainly been achieved in the last two decades as globalization has accelerated. On average, people in the Global South are healthier, better educated, and generally less impoverished, and they are more likely to live in a multiparty democracy (UNDP 2005: 3). Life expectancy has increased by two years since 1990, there are three million fewer annual child deaths per year, and more than 30 million have escaped extreme poverty. Many of these accomplishments have been the result of increasingly global and prosperous economies, especially in Asia. But progress in these areas is overshadowed by severe regressions in others. For example, 18 countries of the Global South have registered a decrease in human development, as measured by the human development index, since 1990. Overall, however, it appears that the disparities between the North and the South have increased during the latest phase of globalization. The debates on the issue continue to rage in full force, but evidence points to growing inequality; there are seven different ways in which inequality has been measured, and all but one of the formulae used in these calculations suggest that the gap grew between 1980 and 2000 (Wade 2004).

What are the effects of globalization on the environment? The debates on the relationship between globalization and the environment have focused mainly on the relationship between free trade, a staple of globalization, and the environment, debates that appear to have gained force with the introduction of the North American Free Trade Agreement (NAFTA) in 1994.[8] Within this debate there are those who argue that free trade helps in expanding the level of domestic economic activity, thereby alleviating poverty and providing citizens and their governments with greater resources to spend on protecting the environment. They believe that environmental regulations may hinder trade (Bhagwati 1993, 1978; Lash 1994; Anderson and Leal 1992). Free trade is then seen as a creator of wealth, which can then get taxed, resulting in more resources to protect the environment. A second group includes those who see free trade as a threat to the environment. According to their argument, the expansion of trade is seen as resulting in higher pollution levels. They believe that free trade has the potential to apply downward pressure on environmental standards as a result of increased competition, the so-called "race to the bottom." Free trade's focus on economic production, they argue, fails to internalize properly the cost of resource use and degradation and the social costs of job loss and of economic disruptions, while ignoring the proper limits on activity dictated by the environment. Increased volume of trade, then, undermines both environmental standards and incentives for investing in clean technology (Anderson 1992; Arrow 1995; Charnovitz 1993, 1994; Chichilinsky 1994; Daly 1991; Sanderson 1992; O'Brien 1993; Ferguson 1995). Finally, there is a third group that includes those who see trade and the environment as compatible. According to this view, international trade is essential to the well-being of humankind, and it is not irreconcilable with environmental protection; they believe it is possible to increase trade while respecting high environmental standards (Esty 1993; Schoenenbaum 1997; Grossman and Kruger 1992; Stern 1996; Muñoz and Rosenberg

1993). This encapsulates the broader debate of the relationship between globalization and environmental degradation, and there does not seem to be a consensus. However, while the debate will likely continue, there is no doubt that the process of globalization has thus far had an effect on our natural environment and on the manner in which humans interact with it. The environmental effects of increased global interaction and the acceleration of economic activity can already be felt. Let us now turn to these environmental challenges.

GLOBALIZATION AND ENVIRONMENTAL CHALLENGES

The degradation of the environment is not a new phenomenon. Because many types of economic activity necessarily require the use of natural resources, the economic development of societies has resulted in the degradation of the environment in various periods of human history. Indeed, in some cases, the rapid and unsustainable use of natural resources has been a primary cause for the decline and collapse of some classic civilizations, such as the Sumerians and the Maya (Wright 2004). In more recent times, the economic transformations that took place in Europe from the seventeenth to the late nineteenth centuries also witnessed enormous ecological disruptions.[9] However, while human and economic activity have always had an effect on the environment, the extent of environmental damage was limited to certain regions of the world, and it did not reach global proportions until the establishment of the current international economic system in 1492, with the arrival of European conquerors to the Americas. The emergence of the new colonial system saw the most devastating effects on the world's natural environment, which was particularly felt in the Western Hemisphere. The conquest of the Americas brought with it numerous micro-organisms, flora, and fauna that had a destructive effect on indigenous species. The immediate effect was felt through the spread of disease, which killed millions of people on the continent, leaving approximately only 10 per cent. But colonization would start a long and steady process of environmental transformation through the expansion of grasslands and the felling of vast forest areas. The process that was then unleashed has not yet been arrested.

Even though the process of global integration that began through colonialism has had an effect on the environment since the beginning, it was not until the latter part of the twentieth century that environmental degradation accelerated and that its effects became apparent. Over the last thirty years it has become more than clear that most of our seas and oceans are overfished, that soil is being degraded and eroded on a larger scale in many parts of the world, that numerous natural habitats have been destroyed, resulting in the extinction of hundreds of species every year, and that air and water pollution have reached unprecedented levels. Environmental degradation may not be new, but the extent and speed with which it has taken place have increased over the last three decades, and this process has been largely due to the acceleration of global integration. Because global social and economic interactions are increasingly more intense, they have had a more significant impact on the environment. The latest phase of globalization has therefore exacerbated many environmental problems.

An aspect of environmental degradation that has become apparent over the last thirty years has been its global dimension. Although many environmental problems have transnational repercussions, it was not until the late twentieth century that their global dimension became more evident. This has been particularly the case for some environmental problems such as global warming and the depletion of the ozone layer. The release of chlorofluorocarbons (CFCs) and greenhouse gases into the atmosphere, for example, contributes to the depletion of stratospheric ozone and to global warming, which are inherently global problems irrespective of who is responsible for the emissions. The effects of such emissions are inherently global. As we shall later see, it has been the realization over the last several decades that environmental problems are global in nature which has prompted a generation of political leaders, activists, and scientists to pursue coordinated efforts to tackle these problems.

Nevertheless, while many environmental problems are global, their effects tend to be felt more inordinately by countries of the Global South. Because there is a direct connection between environmental degradation and poverty, poorer countries tend to be most severely affected by environmental degradation and its effects. Consider pollution. All environmental problems are directly connected with population growth, and higher levels of population mean more consumption, which in turns means more pollution. Given that almost all contemporary population growth takes place in the developing world, it means that the environment is at a higher risk in the Global South. The world's population grew in the twentieth century from 1.5 billion to six billion, and it is projected to grow by another 25 per cent in the next twenty years. The bulk of this growth has taken place in the Global South, and the trend is expected to continue: at current levels, the population in developing countries will double in 36 years (Speidel 2003: 4).

What are the environmental challenges that the latest phase of globalization has brought about? As we have seen, one of the factors behind the acceleration of globalization has been the opening up of domestic economies and their integration with global market forces. Since the onset of the Debt Crisis in the early 1980s, countries have introduced a series of economic reform programs that have dismantled the economic models that characterized the post-war era and that were based on the protection of domestic economies, generally known as Import Substitution Industrialization (ISI). Barriers to trade were thus lifted, allowing for a freer flow of goods across borders. Economic reform has been based on the concept of comparative advantages of classic economic theory. According to this economic notion, countries would benefit the most from exporting products that they have an advantage to produce more cheaply given their endowment of the various factors of production. Because levels of industrialization in the Global South are lower than in the North, the concentration of economic activity on the production of goods on which they have a comparative advantage has meant that developing countries have increasingly focused on exporting commodities and agricultural goods. As a result, the demand for these goods has increased, placing greater pressure on the natural environment. Countries that are endowed with natural resources have increased their extraction for export markets, and in many cases such extraction has been done in highly unsustainable ways (Altieri and Rojas 1999; Figueroa 2001).[10]

Two natural resources that have suffered the most from this increased pressure are forests and fisheries. The Global South is home to two thirds of the world's animal and plant species, and most of the world's temperate and tropical forests are located between the Tropics of Cancer and Capricorn. Brazil, Indonesia, and the Democratic Republic of Congo alone contain half of the world's tropical forests, approximately one billion acres. It is estimated that about half of the world's forests have been cleared and that only 20 per cent of them remain in a wild state. Deforestation rates have increased over the last two decades, to a rate of one acre per second. The net loss of forest area during the 1990s was approximately 94 million hectares, the equivalent of 2.4 per cent (UNEP 2002). Although the factors behind deforestation are multiple, the increased demand for forest products, which has been largely fuelled by economic liberalization, has been a main contributing factor (UNEP 2002; Roberts and Thanos 2003; Dauvergne 2005).

Deforestation is also the result of agricultural expansion. Because developing countries generally enjoy a comparative advantage in the production of agricultural goods, the integration of their economies into global markets has also increased the demand for these products. Encouraged by international financial institutions such as the World Bank, many developing countries have promoted agricultural development and cattle ranching, often with government subsidies, in previously forested areas. The pressure on forests has therefore increased and deforestation accelerated. During the 1990s alone, almost 70 per cent of deforested areas were changed to agricultural land, predominantly under permanent rather than shifting systems (UNEP 2002). Beyond the loss of diversity, deforestation has an impact on local indigenous communities that depend on the forests. Forest clearance leaves human forest dwellers without food or shelter. Moreover, the clearing of forests contributes to climate change; trees and plants absorb carbon dioxide, a main greenhouse gas, and transform it into oxygen through the process of photosynthesis.[11] The clearing of forests then eliminates one of the ways in which the environment can cope with this greenhouse gas in a natural manner. The effects of global warming are expected to have severe effects for people in the Global South, such as an increase in the geographic spread and incidence of insect-borne diseases, particularly malaria and dengue;[12] the displacement by rising sea levels of tens of millions of people from small island states and the low-lying delta areas of China, Egypt, and Bangladesh;

and a decrease in the amounts of precipitation in many arid and semi-arid areas, most of which are located in developing countries.

Fish stocks have also been brought under increased pressure within the new global economy, and their depletion has accelerated over the last decade. The lowering trade barriers, combined with the drop in transportation costs, have resulted in an increased demand for fish products. The total global catch of fish has gone down steadily since 1988 as demand has accelerated on a global scale. Current trends are simply not sustainable; whereas in 1960 only five per cent of marine fisheries were fished to capacity or overfished, the number of collapsed fisheries had increased to 29 per cent by 2003, and estimates predict that, by 2048, all marine fisheries will have collapsed (Worm et al. 2006: 790).[13] The decline of fish stocks poses a serious challenge to the Global South. A fifth of the world's population get a fifth or more of their protein from fish and, in Asia, fish is the main protein source for approximately half of the population. Fishery exports are an important economic asset for developing countries; they are responsible for half of the world's exports. As stocks decline, economies of developing countries will be severely affected.

Globalization has also increased air pollution. Increased economic activity and transportation have resulted in unprecedented levels of air pollution, particularly carbon dioxide, the principal greenhouse gas. Global carbon-dioxide emissions climbed by 22 per cent between 1980 and 2000, and the concentration of this pollutant is now at its highest level in over 420,000 years. The increase is largely due to anthropogenic emissions from fossil-fuel combustion. Consider the increase in air travel, a staple of our globalized world. As the global economy has grown, and as the technology for air transport has been significantly improved over the last several decades, the cost of air travel has been reduced in real terms, stimulating traffic growth. As a result, domestic and international air traffic increased from 9 million passengers in 1945 to over 1.5 billion in 1999 (ICAO 1999). On average, passenger travel has grown at approximately ten per cent annually, and it is expected to grow at an average rate of 4.5 per cent between 1997 and 2010. When aircraft engines burn fuel, they produce emissions that are similar to those resulting from fossil-fuel combustion. However, aircraft emissions are unusual in that a significant portion is released at very high altitudes, making it difficult for natural mechanisms to absorb and transform them into oxygen, such as the natural process of photosynthesis by plants and trees. One of the most comprehensive studies on the subject estimates that 3.5 per cent of global emissions of pollutants that have an effect on climate change are released by aircraft (ICAO 1999). While the accumulation of CO_2 is a global environmental problem, industrialized countries are responsible for 70 per cent of its emissions; industrialized countries emit 3.3 tons per capita, whereas developing ones emit only half a ton. However, developing countries are a lot more susceptible to climate change. People in the Global South are a lot more dependent on their natural base for food and shelter, they are more exposed to extreme weather conditions, and they are less capable of adopting the necessary technology to lessen its impact. For example, air pollution also affects forests and it has been implicated in large-scale forest die-offs in China; the World Bank estimates that the cost of air pollution in China's forests and crops exceeds $5 billion a year (Speth and Haas 2006: 21).

The lowering of barriers to trade and commerce has also facilitated the relocation of companies to some countries in the Global South in which environmental regulations are not as stringent as they have "gone global." In what has been commonly referred to as a "race to the bottom," companies have increasingly tended to move their operations in search of the so-called "pollution havens." There is an important debate regarding the extent of this phenomenon,[14] and while some people have certainly exaggerated it, there is mounting evidence suggesting that, in many cases, less rigid environmental standards are taken into account when companies decide where to relocate (Mani and Wheeler 1998; Clapp 2001). This has been particularly the case for polluting-intensive industries. Less rigid environmental standards are not the only consideration for the relocation of firms; as it has been argued, access to large markets, such as India and China, as well as cheaper labour are also factors considered by companies in deciding where to locate their operations, but "softer" environmental regulation has certainly also been a factor (Barton and Jenkins 2002). Global economic competition may not have applied pressure on developing countries to lower their environmental standards, but competition among themselves results in incentives not to increase them, and they have been "stuck at the bottom" (Porter 1999). The result has been increased pollution, and its associated health consequences, in many countries of the Global South.

The lowering of trade barriers has also been partly responsible for what some people call "garbage imperialism." Global trade in toxic and hazardous waste, the so-called "pollution flight," has steadily increased from countries in the north to the Global South as globalization has facilitated the cheap transportation network and economic incentives to create the modern hazardous waste trade (Clapp 2001). As Jennifer Clapp demonstrates, the new waste trade has brought about numerous environmental challenges to countries of the Global South, since they lack the expertise and technology to deal with industrial wastes that have been imported into those countries (2001). The result is increased stress on the health of local populations. Steps have been taken to deal with this problem, such as the Basel Convention, which banned the export of certain hazardous products. However, the Basel ban is not always effective: developing countries do not always have the appropriate regulation, and when they do it is not readily enforced. One study calculated that firms made over 500 attempts between 1989 and 1994 to export a total of more than 200 million tons of waste from countries belonging to the Organization of Economic Cooperation and Development (OECD) to the Global South (Bellamy Foster 1994, as cited in Scholte 2005: 286). Ironically, while the flow of this type of north-south trade has increased, industrialized countries, despite their calls for greater trade liberalization through their voices at international financial institutions, have continued to subsidize their agricultural sectors, the very sectors in which the Global South has a comparative advantage. It is estimated that the developing world's agricultural producers lose about US $24 billion a year due to industrial countries' trade barriers and subsidies. The inability to place agricultural goods in northern markets pushes millions of people into fragile and marginal lands to make a living. The problem is especially acute in Africa, which accounts for approximately 70 per cent of the world's drylands, which become increasingly degraded and undergo desertification as a result of overgrazing, overcultivation, and poor irrigation. It is estimated that 80 per cent of the world's drylands suffer from moderate to severe desertification.

The globally interconnected market for agricultural goods has also increased the unregulated entry of Genetically-Modified Organisms (GMOs) into developing countries. The trade pressures that the globalized international system places on countries result in the unregulated entry of these goods (Clapp 2006). This is not to say that some developing countries do not regulate the flow of these goods. Clapp, for example, shows that several African countries have had strong and rapid reactions to this type of trade. But others, such as Mexico, have not (2006). GMOs, on their own, do not necessarily have negative environmental consequences. Indeed, GMOs may actually be the solution to the environmentally unsustainable agricultural methods currently in use in the Global South as their development requires less irrigation, fewer fertilizers, and limited pesticides (Meninato 2001). However, the problem arises when the trade in the products is unregulated, as it can irreversibly contaminate the genetic make-up of some aboriginal species. GMOs may not be a problem, but the consequences of their unregulated spread are yet to be known.

While the social and economic effects of globalization can be debated, there is no question that its acceleration over the last two decades has brought about these environmental challenges to developing countries, as the foregoing discussion suggests. However, globalization is not limited to economic integration; the process has also facilitated increased interaction among individuals, organizations, and institutions across the globe, thereby bringing about opportunities to deal collectively with these environmental challenges. As the process of globalization has intensified, a new system of environmental governance has also begun to take shape, a system shaped by an increasingly active and more global civil society. Let us turn to this aspect of globalization.

GLOBAL ENVIRONMENTAL MOBILIZATION AND INTERNATIONAL GOVERNANCE

Globalization and Social Movements

While the process of globalization has brought about several environmental challenges to countries of the Global South, it has also brought about opportunities for non-state actors, from both industrialized and developing countries, to organize and address environmental issues affecting developing countries. Beyond economic integration, the process of globalization has also been characterized by the internationalization of social movements, which have also "gone global." Important policy decisions have traditionally been made by states, and social mobilization—around women's issues, human rights, and other areas—has tended to focus their attention on raising awareness and influencing policy at the national level. However, the process of globalization has, in turn, propelled the globalization of these movements. Increasingly, decisions are made by international and supranational institutions, such as the World Trade Organization and the European Union, and social movements have thus been forced to look beyond their borders to influence policy. As a result, they have more actively sought to cooperate and coordinate efforts with like-minded actors across borders in their mobilization efforts. Such cooperation has been greatly facilitated by the development of cheap communication networks (telephone, fax, and the Internet) that have allowed for faster, cheaper, and denser links among people to organize movements on a social scale (Cohen and Shirin 2000: 8). Moreover, the development of mass, and sometimes instant, forms of communication has contributed to a growing consciousness among people across the globe that many issues are global in nature. Perhaps no other environmental challenge reflects this phenomenon better than climate change. Coverage of freak weather occurrences and natural disasters caused by climate change are transmitted instantly across the world, contributing to the realization that these issues have global repercussions regardless of the source.

The result has been that social movements have increasingly become globalized. The various anti-globalization protests that are now a fixture of the meetings held at some of the main international financial institutions attest to it; members of the movement not only have international institutions and fora as their targets, but they have relied heavily on communication technologies to organize their mobilization and have used the media to gain worldwide exposure. The globalization of these movements can also be seen through the exponential increase in more institutionalized forms of social mobilization over the last fifteen years, such as Non-Governmental Organizations (NGOs). While the number of International NGOs (INGOs)—those organizations that operate in more than one country, such as Greenpeace or Oxfam—increased steadily during the post-war period, there was a clear acceleration during the 1990s: the number of INGOs is estimated to have quadrupled from 6,000 in 1990 to 26,000 in 1999 (Hoogvelt 2001: 267). Individual membership in these INGOs also appears to have accelerated. More importantly, a recent study shows that INGOs have become much more connected, both to each other and to international institutions such as the United Nations and the World Bank (Anheier et al. 2006: 4). As the authors of the study suggest, not only did the global range of INGO presence grow during the 1990s, but the networks linking these organizations became denser as well. The growth in the number of INGOs and in the density of their interactions has prompted some to argue that what we are witnessing is the emergence of a global civil society, a supranational sphere of social and political participation in which individuals engage in sustained forms of dialogue, debate, and negotiation among themselves, governments, and international organizations. While this form of international non-governmental interaction has historically existed, in transnational religions for example, the sheer scale and scope that international and supranational organizations have achieved in recent years is arguably something new (Anheier et al. 2006: 4). Whether this novelty amounts to the emergence of a global civil society is a debate that need not detain us here, but there is no question that social mobilization has truly reached global proportions. And one of the social movements that has gone global is the environmental movement.

Green Mobilization: from the Local to the Global

The mobilization of people interested in protecting the environment is not new. Numerous conservation associations and organizations were formed during the late nineteenth century, especially in Western Europe and North America, with the aim of protecting certain animal species in danger of extinction (Dalton 1993: 42-48; McCormick 1989: 10-20). However, the emergence of what we call the environmental movement in more contemporary times can be traced back to the 1960s, and quite directly, in fact, to the reaction spurred by the publication of *Silent Spring* in 1962. In that book, Rachel Carson detailed the adverse effects that the misuse of chemical pesticides and insecticides were having on humans and warned of an impending environmental crisis. The book attracted wide attention—it became a bestseller in the United States—and prompted the public to look more seriously at the environmental effects of human activity. The publication of the book was followed by a series of environmental disasters in North America and Europe[15] as well as by the release of other similarly alarmist works that predicted environmental crises (books by Paul Ehrlich and Barry Commoner), which together prompted people in the United States and Europe to organize and mobilize to demand better environmental protection. During the 1960s, environmental movements sprang up in many industrialized countries and, by 1970, what has been called the "environmental revolution" had begun (McCormick 1989: 47-68). Numerous environmental organizations were formed on both sides of the Atlantic as environmental problems became salient issues among the public and political leaders.

The environmental movement acquired an international dimension after 1970. While environmental mobilization had previously remained limited to local and national politics, during the 1970s and 1980s some NGOs adopted a more international approach,[16] and we witnessed an increase in the number of environmental INGOs (EINGOs), as Graph 1.1 shows.

The growth in the number of these organizations came about for several reasons. First, significant advances in scientific technology allowed scientists to study and analyze previously invisible forms of environmental damage and advance more convincing accounts of its origins and consequences (Held et al. 1999: 387). Political leaders and international organizations thus began to adopt a more scientific approach to environmental problems, thereby lending credibility to the demands advanced by environmental groups. The internationalization of scientific debates surrounding environmental issues was an important component of the formation and institutionalization of international environmental organizations (Haas 1990). Second, this new research pointed to the global consequences of environmental degradation. For example, when scientists first suggested in 1973 that CFCs break down in the stratosphere and release ozone-destroying catalysts, they raised the possibility of global environmental damage (Meyer et al. 1997: 630). As the global consequences of environmental degradation became evident, the realization that global solutions were needed provided environmental movements with an opportunity to take the global stage. Third, in addition to the weight that new scientific evidence added to their arguments, the adoption of environmental concerns by international institutions, particularly the United Nations, further afforded environmental groups international exposure and credibility. The 1972 United National Conference on the Human Environment, in Stockholm, and the subsequent establishment of the United Nations Environment Programme (UNEP) the following year, not only popularized environmental issues internationally, but they gave environmental groups a new international platform. Greatly helped by information technology and cheaper forms of transportation, the green movement had reached the international stage by the late 1980s, and transnational environmental mobilization expanded.

GRAPH 1.1
CUMULATIVE COUNT OF INTERNATIONAL ENVIRONMENTAL ORGANIZATIONS

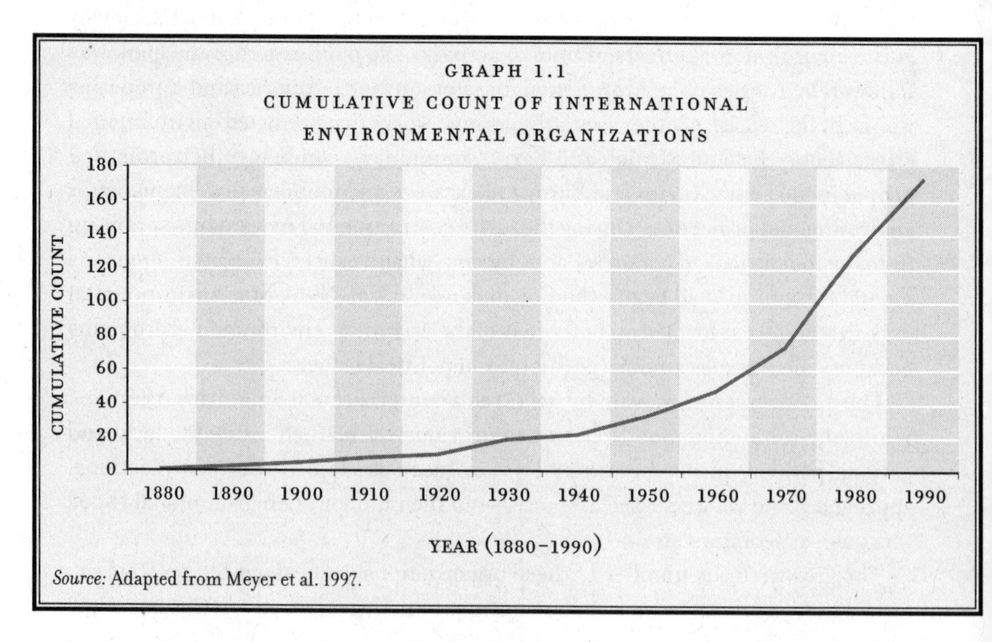

Source: Adapted from Meyer et al. 1997.

The globalization of environmental mobilization has also involved the Global South. Generally, environmental mobilization in developing countries has involved a struggle by marginalized people to protect their environmental means of livelihood, such as land, water, and forests. More recently, however, environmentalism in developing countries has expanded to cover a variety of issues, including deforestation, water quality, depletion of indigenous resources, human settlement, threats to public health, toxic contamination, and atmospheric pollution (Dwivedi 2001). Many of these environmental movements have traditionally been locally based. However, the process of globalization has contributed to the formation of transnational links between environmentalists from the North and the South. Importantly, the Stockholm conference contributed to the development of networks of IENGOs, and Northern-based ENGOs began to engage more systematically with development issues and ENGOs from the Global South. In many cases, this has occurred as a result of the expansion of activities of North-based NGOs; most of the largest ENGOs have established offices and chapters in numerous countries across the Global South. Greenpeace, Conservation International, and the World Wildlife Fund now operate in many developing countries.

In other instances, this interaction has been the result of partnerships created by Northern NGOs. ENGOs headquartered in the North that have worked in developing countries tended to work on conservation issues. This typically involved the operation of numerous programs of environmental conservation across the Global South. Many of these programs proved unsuccessful because they neglected the socio-economic conditions, specifically poverty, that led to the loss of biodiversity. The majority of the biologically richest areas are found in regions of the Global South in which the poorest people live, and they depend on their natural environments for their subsistence. Any attempts at conservation must therefore address the underlying issues of poverty and improve the quality of life of these peoples. After having supported numerous failed conservation projects, some ENGOs have changed the nature of their programs and have combined environmental conservation management programs with economic development initiatives (Wapner 2004: 131). These include IENGOs such as the World Wildlife Fund (WWF), the New Forests Projects, and the Association for Research and Environmental Aid (AREA). In many cases, these organizations form partnerships with locally based ENGOs and actors to develop programs of environmental protection, partnerships that result in the transfer of expertise, knowledge and, importantly, resources.

Increased cooperation among these actors has taken place against a backdrop of a considerable increase in funding to NGOs in developing countries. Such increase is partly the result of the decision taken by industrialized countries to channel a growing portion of their Official Development Assistance (ODA) through Global South NGOs as well as through transnational NGOs based in their countries. According to an OECD report, by the early 1980s all industrialized countries had adopted a system of co-financing projects implemented by their national NGOs and, by the late 1980s, it amounted to $2.2 billion, or more than 5 per cent of total ODA (OECD 1989, as cited in Wapner 2004: 133). These North-South transfers of expertise and resources have contributed to the strengthening of ENGOs in countries of the Global South, which become more capable of addressing environmental degradation and, ultimately, to take more active economic and environmental roles. These transfers, and the global alliances and the emergence of networks of international environmental organizations, are a result of growing global interconnectedness.

The extent to which these alliances and the empowerment of NGOs in developing countries has contributed to better environmental management and ultimately to redressing environmental degradation in the Global South is explored in the chapters that follow. But there is no doubt that the process of globalization has provided opportunities for non-state actors to address environmental issues nationally and internationally. At the international level, ENGOs have not only become active political players in their own right, but they have also contributed to raising environmental issues to the international political agenda and to the recognition that environmental issues are global in scale. Moreover, ENGOs have also been very influential in altering the manner in which people across the world view the environment; whereas previously environmental issues did not figure among top priorities in many countries, by the late 1980s citizens in numerous nations placed environmental concerns among the major issues facing their countries.

Global Environmental Governance

Since the 1972 Stockholm conference, the increased visibility of ENGOs at the international level, the "scientification" of environmental issues that occurred during the early 1970s, and growing awareness of environmental damage have resulted in the elevation of environmental concerns to the international political agenda, where they have remained ever since. But more importantly, these factors have also contributed to the establishment of a series of international institutions and the enactment of international treaties over the last 30 years that have, as their main objective, the improvement of natural-resource management and the reversal of environmental degradation. Since the 1972 Stockholm Conference and the institution of the UNEP in 1973, countries around the world have made collective efforts to address environmental problems. This is not to say that international institutions and laws did not exist before; in fact, an international convention on the regulation of whaling was signed as early as 1946, and so were other treaties regulating the transportation of toxic materials during the 1950s. But the number and density of international environmental treaties and agreements began to increase significantly since the Stockholm Conference. The Stockholm conference marked the first time that multilateral agencies, national governments, and NGOs had come together to analyze and discuss environmental problems and strategize about possible collective solutions. It therefore institutionalized the international importance of environmental concerns and set a precedent for international cooperation on environmental issues.

Starting with the enactment of the Convention on International Trade in Endangered Species (CITES) in 1973, the world has witnessed a rather impressive growth in international environmental agreements (see Table 1.1 for the most important ones). By 2000, there were over 130 multilateral environmental agreements and hundreds of bilateral ones.

Many of these agreements are considered "dead letters," and others can be considered rather weak or unenforceable. However, in many cases they have changed the behaviour of states and they have contributed to the alleviation of some environmental problems. The Montreal Protocol for the Protection of the Ozone Layer is a good case in point; through the establishment of agreements and reduction targets, global emissions of CFCs had been reduced from 1.1 million tonnes per year in 1986 to less than 150,000 tonnes by 1999 (Haas et al. 1993). Scientists now forecast that, should the reduction of emissions continue, the ozone layer will be fully recovered by 2050. In addition to these various treaties, a number of NGOs have established a range of initiatives to address environmental issues outside governments. These include initiatives involving ecolabelling and the certification of environmentally friendly products, supervised by the Forest Stewardship Council (FSC), which certifies products made from forests that are managed in a sustainable manner.

The various international agreements and treaties that have been adopted over the last 30 years, as well as the self-regulating systems that have been established outside governments, make up a global environmental regime within the larger framework of global environmental governance.[17] How have countries of the Global South been involved with global environmental governance? Developing countries have been at the centre of global environmental governance since Stockholm; many developing countries were active participants at the conference and signed the Stockholm declaration. They have also been signatories to numerous international agreements and treaties that have been adopted over the last three decades. However, up until the mid-1980s, deep divisions existed between industrialized and developing countries given that their priorities differed significantly. While industrialized countries and Northern environmentalists tended to emphasize the importance of issues relating to conservation, developing countries were mostly concerned with environmental issues that are inherently related to development and poverty, such as human health, water quality, and sanitation. By the mid-1980s, debate among environmentalists, development practitioners, and government representatives contributed to a change in perspective on the international environmental agenda as it became clear that environmental protection cannot be achieved without economic development. This change in perspective culminated in the publication in 1987 of the report *Our Common Future*, by the World Commission on Environment and Development (WCED), chaired by Norway's prime minister Gro Harlem Brundtland and established by the UNEP. In the report the

TABLE 1.1

MAJOR INTERNATIONAL ENVIRONMENTAL AGREEMENTS

Year	Agreement
1972	United Nations Conference on the Human Environment (UNCHE), Stockholm Declaration is adopted
1973	United Nations Environment Programme is created
1973	Convention on International Trade in Endangered Species (CITES) is established and enters into force in 1975
1979	Convention on Long Range Transboundary Air Pollution (CLRTAP) is established and enters into force in 1983
1982	United Nations Convention on the Law of the Sea (UNCLOS) established and enters into force in 1994
1985	Vienna Convention for the Protection of the Ozone Layer is established and enters into force in 1988
1987	Montreal Protocol on Substances that Deplete the Ozone Layer is established and enters into force in 1989
1989	The Basel Convention on the Control of Transboundary Movements of Hazardous Wastes and their Disposal is established and enters into force in 1992
1990	The Global Environmental Facility (GEF) is established
1992	United Nations Conference on Environment and Development (UNCED) is held and the Rio Declaration and *Agenda 21* are adopted
1992	United Nations Framework Convention on Climate Change (UNFCCC) is established and enters into force in 1993
1994	International Convention to Combat Desertification is established and enters into force in 1996

1997	Kyoto Protocol on Climate Change is adopted and enters into force in 2005
2000	Cartagena Protocol on Biodiversity is adopted and enters into force in 2003
2001	Stockholm Convention for the Elimination of the Persistent Organic Pollutants (POPS) is established and enters into force in 2004
2002	World Summit on Sustainable Development (WSSD) is held and the Johannesburg Declaration and the Johannesburg Plan of Implementation are adopted

Brundtland Commission, as it became known, coined the concept of "sustainable development," which was defined as "development that meets the needs of present generations without compromising the ability of future generations to meet their needs."

The report's main contribution was to frame environmental issues through the link that exists between development and the environment and to advocate that alleviating poverty and protecting the environment must go together. In the report, the WCED argued that poverty fuels environmental degradation because it leaves poor people little choice but to deplete natural resources to meet their basic needs, and, conversely, that poor people are the ones most vulnerable to environmental degradation because they are the least protected. For the WCED, the elimination of poverty had to accompany any effort to redress environmental degradation.

The Brundtland Report had a profound effect on the international environmental agenda for the next decade and on environmental management in developing countries. With the objective of devising how to implement sustainable development, the UN General Assembly decided to organize an international conference to study possible plans of action in Rio de Janeiro, Brazil, in 1992. The main purpose of the conference was to "elaborate strategies and measures to halt and reverse the effects of environmental degradation in the context of increased national and international efforts to promote sustainable and environmentally sound development in all countries."

The United Nations Conference on Environment and Development, the Earth Summit as it has come to be known, was held from June 5 to June 18 and was the largest conference ever held: it was attended by 45,000 people, 178 nations, 118 heads of state or government, more than 8,000 official delegates, 3,000 NGO representatives, and 9,000 journalists, among others. In addition to its unprecedented scale, the Earth Summit was very significant because it managed to secure a compromise among industrialized and development countries on "basic principles" to guide action on the environment and development through the adoption of *The Rio Declaration*. In that document, signatory countries agreed upon 27 guiding principles that should govern the future of environmental decision-making on issues such as national responsibilities and international cooperation on environmental cooperation, the need for development and the eradication of poverty, and the roles and rights of citizens, women, and indigenous peoples. Importantly for developing countries, Principle 7 affirms "the common but differentiated responsibilities" that industrialized and developing countries have in environmental protection. Whereas environmental issues had not figured on the political agenda 30 years earlier, through this document countries had collectively agreed on a number of guiding principles, a remarkable accomplishment indeed.

More importantly, the conference approved a truly path-breaking document called *Agenda 21*. The 400-page document contained 40 chapters that presented a blueprint across 115 program areas that were to be followed by all signatories in order to implement sustainable development. The document consisted of four broad areas: social and economic development, conservation and management of resources, strengthening the role of major groups, and means of implementation. To help implement this ambitious agenda, the Commission on Sustainable Development (CSD) was established by the UN. As part of the UN's Economic and Social Council (ECOSOC), the CSD was given the task of promoting and reviewing progress made on implementation and to help coordinate activities of UN agencies in this area. Because financial support was needed to help developing countries implement the agenda, the Earth Summit strongly supported the Global Environmental Facility (GEF), created in 1990, to provide funding to developing countries to support environmental objectives that countries would not otherwise be able to afford. Since its creation, the GEF has been operated jointly by the World Bank, the United Nations Development Programme (UNDP), and the UNEP and has pledged approximately $2 billion to developing countries over the last decade. Finally, the Earth Summit initiated several treaty processes, such as the UN Framework Convention on Climate Change and the Convention on Biological Diversity. Through these conventions, nations agreed to protecting climate change and to preventing dangerous anthropogenic interference with global climate change, and on the need for the conservation of biodiversity, the sustainable use of resources, access to genetic resources, and the transfer of technologies.

Even though the Earth Summit has generally been considered a success, it has been criticized because of some of its failures, such as its inability to produce agreements on the enactment of an Earth Charter or the development of a convention on forestry, and, ultimately, on its failure to make any of its recommendations binding. Nevertheless, the conference was of unprecedented importance. *Agenda*

21 set forth a policy framework for all countries to foster sustainable development practices, and it essentially set the agenda for years to come. For developing countries, *Agenda 21* proved important in that it linked poverty alleviation and ODA with the greater goals of environmental protection. Of significance was the agreement made by all countries to establish the required institutions and policies to foster the various programs, as well as the need to include non-state actors, such as NGOs, in decision-making processes regarding environmental management. NGOs were active participants during the summit. Their ability to influence Rio was in great part due to the coordination among them through extensive transnational alliances and networks of communication. NGOs were in fact instrumental in framing the language of many of the agreements that were negotiated. Importantly, they managed to commit signatories to opening up participatory spaces for NGOs in policy formulation at the national level. Once the Earth Summit was over, NGOs had not only brought environmental issues to more prominence internationally, but they had also had an impact on the formulation of international environmental treaties.

During the decade after the Earth Summit, numerous treaties were signed, such as the 1994 Convention to Combat Desertification, the 1991 Stockholm Convention on Persistent Organic Pollutants, the 2000 Cartagena Protocol on Biosafety, and perhaps most notably, the 1997 Kyoto Protocol. Nevertheless, a decade after the Earth Summit, momentum and international leadership to pursue the goals outlined at Rio subsided, and much of the ground was lost. The UN attempted to infuse new energy into the environmental agenda by holding a follow-up summit. In September 2002, the World Summit for Sustainable Development (WSSD) was thus held in Johannesburg, South Africa. The World Summit was held in part as a result of the general consensus that existed among world leaders and members of the NGO community that many of the Rio objectives had not been achieved. However, the summit proved much less successful than its predecessor since it produced vague statements, scant plans of action, and only very general agreements. In part, this was due to the United States' refusal to commit to timetables. As James G. Speth and Peter Hass have stated, the summit "failed to embrace new verifiable goals or to advance significantly efforts for the protection of the global environment" (2006: 78).[18]

The Johannesburg summit happened 30 years after the first world summit on the environment was held. Since the Stockholm conference of 1972, environmental issues have remained at the top of the international political agenda, and this has been largely the result of the active role NGOs have played. The road from Stockholm to Johannesburg is marked with a gradual but consistent rise in the importance of environmental issues, which has resulted in a remarkable establishment of environmental regimes within the broader context of globalization. In the following chapters, contributors examine the effects of this agenda on the Global South by presenting perspectives from 11 developing countries.

CONCLUSION

While the debate over the nature of globalization is likely to continue, the latest phase of this process appears to be distinct both in its scope and speed. Over the last three decades, global interconnectedness—political, economic, and social—appears to have occurred at a much faster pace and has reached more societies than ever before in a deeper way. We are truly living through an unprecedented global phenomenon. The debate over its effects will also likely continue; however, as this chapter has tried to show, there is no question that global economic integration has brought about numerous environmental challenges for the Global South; global economic expansion and integration have created higher levels of pollution and increased pressure on our natural resources, such as forests and fisheries, which are being depleted at rates higher than ever before. Economic integration has also facilitated the movement of toxic waste to developing countries, and the lowering of trade barriers has often resulted in companies seeking laxer environmental standards, thus increasing pressure on the environment in the Global South.

However, globalization is not limited to economic integration; the process has also facilitated increased interaction among individuals, organizations, and institutions across the globe, thereby bringing about opportunities to deal collectively with these challenges. This has certainly been the case of NGOs. The revolution in information technology has facilitated greater communication among people across the globe, allowing them in turn greater opportunities to organize and mobilize in defence of our natural environment. As a result, northern ENGOs have expanded their operations to the Global South and have increased their interaction with their southern counterparts, a phenomenon that has contributed to the transfer of resources and expertise on environmental issues to developing countries. Beyond contributing to more vibrant civil societies, these transfers have in turn opened up opportunities to strengthening environmental management in the Global South. The globalization of the environmental movement has also meant that NGOs have not only been influential in bringing environmental issues to the international political agenda, but they successfully contributed to building an international system of environmental governance. Let us now look at our 11 perspectives from the Global South.

Notes

1 The term Washington Consensus was coined by John Williamson in the late 1980s. See Williamson 1990.

2 For one of the best overviews of the debate, see Weiss 1999.

3 Globalism gained preponderance in the early 1990s as the end of the Cold War gave way to the triumphal view that free markets were the only viable economic system in the New World Order. Globalism became the hegemonic discourse of the decade as leaders around the world, including those from the Global South, embraced the new paradigm with gusto. By the turn of the twenty-first century, however,

globalism had come under attack on several fronts. At the theoretical level, various inconsistencies of the argument were revealed (Saul 2005; Steger 2002; Gray 1998). On a more concrete level, globalism lost strength from the withdrawal of support from large sectors of society around the world. In the North, this became manifest through the mobilization of opponents to this neo-liberal version of globalization in various cities in the developed countries where meetings of powerful international institutions were held. In the Global South, as disillusionment set in once the promised benefits of globalism failed to materialize, people began to exchange globalists for social democrats at the ballot box. Nowhere has such change of heart become more evident than in Latin America, a region in which most globalist leaders have been voted out of office in country after country over the last decade.

4 Trade intensity is a measure of the magnitude of global trading activity that can be estimated by the ratio of world trade to world output or by national trade to Gross Domestic Product (GDP) measures (Held et al. 1999: 151).

5 One of the challenges in comparing the contemporary period with the Gold Standard era is that figures of traded services are not available for the nineteenth century; they became available only from the 1980s onwards.

6 One of the most fervent supporters of globalization has been the reputable British weekly *The Economist*. In his farewell article, written in April 2006, the magazine's chief editor, Bill Emmott, advanced a very strong defence of globalization, arguing that its results have been mostly positive (2006).

7 This is not to say that all Marxist analyses take this rather simplistic approach. Based on Antonio Gramsci's work, some have for example argued that globalization cannot be reduced strictly to an ideological project as it is a multi-faceted, dynamic, and evolving process, the final trajectory of which cannot be predicted (see, for example, Dicken et al. 1997).

8 This is not to say that this is a completely new line of inquiry, but it seems to have attracted little scholarly attention before (see, for example, Baumol 1970; Baumol and Oates 1975; Bergsten 1973).

9 During the late sixteenth and early seventeenth centuries, many European countries, particularly England and the Benelux countries, underwent significant transformations with the emergence of agrarian economies based on capitalist economic production. Market incentives, technological innovations, and increased investment resulted in thriving agrarian economies that were dependent on the natural environment for food, clothing, and building materials, and for their transportation. As a result, entire ecosystems were devastated, forests cleared, and numerous species eradicated (Held et al. 1999: 383). Environmental degradation would again accelerate with the advent of the Industrial Revolution in the latter part of the nineteenth century. Because the early stages of this economic transformation was based on industrial sectors that were inherently "dirty," such as coal, oil, iron, and steel, the generation of pollutants reached unprecedented levels.

10 The case of Chile best exemplifies this situation. Chile is one of the countries of the Global South that implemented economic reform early on, in the mid 1970s, establishing a model based on the export of several core commodities, such as timber, fish, and fruit. The country experienced accelerated environmental degradation during the 1980s and 1990s (in terms of deforestation, the depletion of fisheries, the pollution of water systems, and soil degradation) which is largely attributable to an increased intensity of resource use due to a booming export economy (Altieri and Rojas 1999).

11 The evidence that human-induced greenhouses gases have increased and that they are effecting global climate change seems to be now indisputable. This conclusion has been reached by the most authoritative report on global warming to date, elaborated by the US National Academy of Sciences (http://www.nap.edu/html/climatechange/).

12 By 2004, the World Health Organization (WHO) estimated that an annual human toll of 150,000 lives had already been lost due to climate change (cited in Speth and Haas 2006: 25)

13 The term "collapsed fisheries" is usually defined as catches dropping below ten per cent of the recorded maximum (Worm et al. 2006: 788).

14 For an excellent review of the debate, see Brunnermeier and Levinson 2004.

15 These involve the millions of gallons of crude oil spilled in the English Channel by the running aground of the tanker *Torrey Canon* in 1967, a leak of toxins into the Rhine River, and the blowout of an oil-drilling platform off the Santa Barbara coast which destroyed parts of the California coastline, both in 1969.

16 NGOs such as Greenpeace and the Rainforest Action Network expanded their efforts to the international arena in an attempt to attract international attention. In the 1970s, for example, Greenpeace waged very visible anti-whaling campaigns that were captured on television and the images were seen by millions of people across the world. Their intention here was to change attitudes and raise environmental awareness.

17 International regimes are the formal and informal rules, norms, principles, and decision-making procedures that shape or limit the behaviour of nations in the international system, whereas governance refers to the sum of actions and processes, formal and informal, that many individuals and institutions take to decide what is their common interest and how they should act collectively.

18 For an excellent, succinct analysis of the summit, see Speth 2003.

References

Altieri, Miguel, and Alejandro Rojas. 1999. "Ecological Impacts of Chile's Neoliberal Policies." *Environment, Development and Sustainability* 1: 55-72.

Anderson, Kym. 1992. "The Standard Welfare Economics of Policies Affecting Trade and the Environment." In Anderson and Blackhurst (eds.), *The Greening of World Trade Issues*. Ann Arbor: University of Michigan Press.

Anderson, Thomas, and Daniel Leal. 1992. "The Free Market versus Political Environmentalism." *Harvard Journal of Law and Public Policy* 15 (Spring): 297-310.

Anheier, Helmut, Marlies Glasius, and Mary Kaldor. 2006. *Global Civil Society 2005/06*. London and Thousand Oaks, CA: Sage.

Arrow, Kenneth. 1995. "Economic Growth, Carrying Capacity and the Environment." *Science* 268(April 28): 520-21.

Bamyeh, M. 2000. *The Ends of Globalization*. Minneapolis: University of Minnesota Press.

Barber, B.R. 1996. *Jihad vs. McWorld*. New York: Ballantine.

Barlow, Maude, and T. Clarke. 2001. *Global Showdown: How the New Activists are Frightening Corporate Global Rule*. Toronto: Stoddart.

Barton, Jonathan, and Rhys Jenkins, eds. 2002. *Environmental Regulation in the New Global Economy: The Impact on Industry and Competitiveness*. Northampton: Edward Elgar.

Baumol, William. 1970. *Environmental Protection, Spillovers and Trade*. Englewood Cliffs, NJ: Prentice-Hall.

Baumol, William, and Wallace Oates. 1975. *The Theory of Environmental Policy*. Englewood Cliffs, NJ: Prentice-Hall.

Bellamy Foster, J. 1994. "Waste Away." *Dollars & Sense* 195 (Sept.-Oct.): 7.

Bergsten, Fred C. 1973. *The Future of the International Economic Order: An Agenda for Research*. Lexington, MA: Lexington Books.

Bhagwati, Jagdish. 1978. *Free Trade Regimes and Economic Development: Anatomy and Consequences of Exchange Control Regimes in Foreign Trade Regimes and Economic Development Series*, Vol. 11. Cambridge: Ballinger Publishing Co.

—. 1993. "The Case for Free Trade." *Scientific American* (Nov.):42-49.

—. 2004. *In Defense of Globalization*. New Delhi: Oxford University Press, 2004.

Brunnermeier, Smita, and Arik Levinson. 2004. "Examining the Evidence on Environmental Regulations and Industry Location." *Journal of Environment and Development* 13(1): 6-41.

Brzezinski, Z. 1993. *Out of Control: Global Turmoil on the Eve of the Twenty-First Century*. New York: Charles Scribner's Sons.

Burbach, R., and K. Danaher, eds. 2000. *Globalize This! The Battle Against the World Trade Organization and Corporate Rule*. Monroe, ME: Common Courage Press.

Cardoso, Fernando Henrique. 1996. "La globalizacón y el nuevo orden mundial." *Boletín editorial de El Colegio de México*, No. 68.

Charnovitz, Steve. 1993. "Free Trade, Fair Trade, Green Trade: Defogging the Debate." *Cornell International Law Journal* 27(4): 459.

—. 1994. "A Taxonomy of Environmental Trade Measures." *Geographic International Environmental Law Review* 6(1): 560-78.

Chichilinsky, Graciella. 1994. "North-South Trade and the Global Environment." *American Economic Review* 84(4): 851-74.

Clapp, Jennifer. 2001. "What the Pollution Havens Debate Overlooks." *Global Environmental Politics* 2(2): 11-19.

—. 2006. "Unplanned Exposures to Genetically Modified Organisms: Divergent Responses in the Global South." *The Journal of Environment and Development* 15(1): 3-21.

Cohen, Robin, and Shirin M. Rai. 2000. "Global Social Movements: Toward a Cosmopolitan Politics." In Robin Cohen and Shirin M. Rai (eds.), *Global Social Movements*. London and New Brunswick, NJ: Athlone. 1-17.

Dalton, Russell J. 1993. "The Environmental Movement in Western Europe." In Sheldon Kamieniecki (ed.), *Environmental Politics in the International Arena*. Albany: State of New York University Press. 41-68.

Daly, Herman. 1991. "Elements of Environmental Macroeconomics." In Robert Constanza (ed.), *Ecological Economics*. New York: Columbia University Press.

Dauvergne, Peter. 2005. "The Environmental Challenge to Loggers in the Asia-Pacific: Corporate Practices in Informal Regimes of Governance." In David Levy and Peter Newell (eds.), *The Business of Global Environmental Governance*. Cambridge, MA: MIT Press. 169-97.

Diamond, Larry J., and Marc F. Plattner. 1996. *The Global Resurgence of Democracy*. Baltimore: Johns Hopkins University Press.

Dicken, P., J. Peck, and A. Tickell. 1997. "Unpacking the Global." In R. Lee and J. Wills (eds.), *Geographies of Economies*. London: Arnold.

Dwivedi, Ranjit. 2001. "Environmental Movements in the Global South: Issues of Livelihoods and Beyond." *International Sociology* 16(1): 11-31.

Emmott, Bill. 2006. "A Long Goodbye." *The Economist* (April 6): 13-14.

Esty, Daniel C. 1993. "Rio Revisited: Turning the Giant's Head." *Ecodecision* (Sept.): 90-112.

Ferguson, Dieneke. 1995. *Curves, Tunnels and Trade: Does the Environment Improve With Economic Growth?* London: The New Economics Foundation for the World for Nature International.

Figueroa, Eugenio R. 2001. "Environmental Impacts of Globalization through Trade Liberalization in Agriculture." In Francesco di Castri, Robert Paarlberg, and Otto Solbirg (eds.), *Globalization and the Rural Environment*. Cambridge, MA: Cambridge University Press. 319-63.

Fukuyama, Francis. 1992. *The End of History and the Last Man*. New York: Free Press.

Giddens, Anthony. 1990. *The Consequences of Modernity*. London and Cambridge: Polity.

Gills, Barry, ed. 2000. *Globalization and the Politics of Resistance*. London: Palgrave.

Gowan, J. 1999. *Washington's Faustian Bid for World Dominance*. London: Verso.

Gray, John. 1998. *False Dawn: The Delusions of Global Capitalism*. London: Granta.

Grossman, Gene, and Alan B. Kruger. 1992. "Environmental Impacts of the North American Free Trade Agreement." In Peter M. Garber (ed.), *The Mexico-US Trade Agreement*. Cambridge, MA: MIT Press. 13-56.

Haas, Peter H. 1990. *Saving the Mediterranean: The Politics of Environmental Cooperation*. New York: Columbia University Press.

Haas, Peter J., Robert O. Keohane, and Marc A. Levy. 1993. *Institutions for the Earth: Sources of Effective International Environmental Cooperation*. Cambridge, MA.: The MIT Press.

Held, David, Anthony McGrew, David Goldbatt, and Jonathan Perraton. 1999. *Global Transformations: Politics, Economics and Culture*. Cambridge: Polity.

Hill, Kevin, and John E. Hughes. 1998. *Cyberpolitics: Citizen Activism in the Age of the Internet*. Lanham, MD: Rowman and Littlefield.

Hirst, P. and G. Thompson. 1999. *Globalization in Question: The International Economy and the Possibilities of Governance*. Cambridge: Polity.

Holm, Hans-Henrick, and Georg Sorensen. 1995. "International Relations Theory in a World of Variation." In Hans-Henrick Holm and Georg Sorensen *Whose World Order? Uneven Globalization and the End of the Cold War*. Boulder, CO: Westview. 187-206.

Hoogvelt, Ankie. 2001. *Globalization and the Postcolonial World: The New Political Economy of Development*. Baltimore, MD: The Johns Hopkins University Press.

Horsman, M., and A. Marshall. 1994. *After the Nation-State: Citizens, Tribalism and the New World Disorder*. London: HarperCollins.

Hufbauer, G. 1991. "World Economic Integration." *International Economic Insights* 2(3): 26-27.

ICAO (International Civil Aviation Organization). 1999. *Special Report on Aviation and the Global Atmosphere*. Cambridge: Cambridge University Press.

Kaplan, Robert. 2000. *The Coming Anarchy: Shattering the Dreams of the Post Cold War*. New York: Random House.

Korten, D.C. 1995. *When Corporations Rule the World*. West Hartford, CT: Kumarian Press.

Lash, William. 1994. "Environment and Global Trade." *Society* 31(4): 52-58.

Mani, M., and D. Wheeler. 1998. "In Search of Pollution Havens? Dirty Industry in the World Economy." *Journal of Environment and Development* 7(3): 215-47.

McCormick, John. 1989. *Reclaiming Paradise: The Global Environmental Movement*. Bloomington: Indiana University Press.

Meninato, Rolando. 2001. "The Impact of Biotechnology on South America." In Francesco di Castri, Robert Paarlberg, and Otto Solbirg (eds.), *Globalization and the Rural Environment*. Cambridge, MA: Cambridge University Press. 217-25.

Meyer, John W., David John Frank, Anna Hironaka, Evan Schofer, and Nancy Brandon Tuma. 1997. "The Structuring of a World Environmental Regime." *International Organization* 51(4): 623-51.

Mishra, R. 1999. *Globalization and the Welfare State*. Cheltenham: Elgar.

Muñoz, Héctor, and Robin L. Rosenberg. 1993. *Difficult Liaison: Trade and the Environment in the Americas*. Coral Gables, FL: North-South Institute, University of Miami.

O'Brien, M.B. 1993. "The General Agreement on Tariffs and Trade: An International Agreement's Effect on Local Environmental Law." *International Legal Perspectives* 5: 83-120.

OECD. 1989. *Development Cooperation in the 1990s: Efforts and Policies of Members of the Development Assistance Committee*. Paris: Organization of Economic Cooperation and Development.

Ohmae, K. 1991. *The Borderless World: Power and Strategy in the Interlinked Economy*. London: Fontana.

O'Rourke, K., and J.G. Williamson. 1999. *Globalization and History: The Evolution of the Nineteenth-Century Atlantic Economy*. Cambridge, MA: MIT Press.

Porter, Gareth. 1999. "Trade Competition and Pollution Standards: 'Race to the Bottom' or 'Stuck at the Bottom'?" *Journal of Environment and Development* 8(2): 133-51.

Roberts, Timmons, and Nikki Thanos. 2003. *Trouble in Paradise: Globalization and Environmental Crises in Latin America*. New York: Routledge Press.

Robertson, Roland. 1992. *Globalization: Social Theory and Global Culture*. London: Sage.

Sandbrook, Richard. 2003. *Civilizing Globalization: A Survival Guide*. Albany: State University of New York Press, 2003.

Sander, H. 1996. "Multilateralism, Regionalism and Globalization: The Challenges to the World Trading System." In H. Sander and A. Inotai (eds.), *World Trade after the Uruguay Round: Prospects and Policy Options for the Twenty-First Century*. London: Routledge. 17-36.

Sanderson, Samuel. 1992. *The Politics of Trade in Latin American Development*. Stanford, CA: Stanford University Press.

Saul, John Ralston. 2005. *The Collapse of Globalism and the Reinvention of the World*. Woodstock, NY: Overlook Press.

Schoenenbaum, Thomas J. 1997. "International Trade and Protection of the Environment: The Continuing Search for Reconciliation." *American Journal of International Law* 91(2): 268-313.

Scholte, Jan Aart. 2005. *Globalization: A Critical Introduction*. New York: Palgrave-Macmillan, 2005.

Speidel, Joseph J. 2003. "Environment and Health: Population, Consumption and Human Health." In David E. Lorey (ed.), *Global Environmental Challenges in the Twenty-First Century*. Washington: Scholarly Resources. 1-13.

Speth, James Gustav. 2003. "Perspectives on the Johannesburg Summit." *Environment* 45(1): 24-29.

Speth, James Gustav, and Peter M. Haas. 2006. *Global Environmental Governance*. Washington: Island Press.

Steger, Manfred B. 2002. *Globalism: The New Market Ideology*. Lanham, MD: Rowman and Littlefield.

Stern, Paula. 1996. Engineering Regional Trade Pacts to Keep Trade and American Prosperity on Fast Track." *The Washington Quarterly* 19(1): 211-22.

UNDP (United Nations Development Programme). 2005. *Human Development Report*. New York: Oxford University Press.

UNEP (United Nations Environment Programme). 2002. *Global Environment Outlook*. Nairobi: United Nations Environment Programme.

Wade, R.H. 2004. "Is Globalization Reducing Poverty and Inequality?" *World Development*, 32(4): 567-89.

Wapner, Paul. 2004. "Politics Beyond the State: Environmental Activism and World Civic Politics." In Ken Conca and Geoffrey D. Dabelko (eds.), *Green Planet Blues*. Boulder, CO: Westview. 122-39.

Weiss, Linda. 1999. *The Myth of the Powerless State*. Ithaca, NY: Cornell University Press.

Wilkinson, R.C. 1973. *Poverty and Progress: An Ecological Perspective on Economic Development*. New York: Praeger.

Williamson, John. 1990. *Latin American Adjustment: How Much Has Happened?* Washington, DC: Institute for International Economics.

Worm, Boris, Edward B. Barber, Nicola Beaumont, J. Emmett Duffy, Carl Fole, Benjamin S. Halpern, Jeremy B.C. Jackson, Heike K. Lotze, Forenza Micheli, Stephen R. Palumbi, Enric Sala, Kimberley A. Selkoe, John J. Stachowicz, and Reg Watson. 2006. "Impacts of Biodiversity Loss on Ocean Ecosystem Services." *Science* 314: 787-90.

Wright, Ronald. 2004. *A Short History of Progress*. Toronto: House of Anansi.

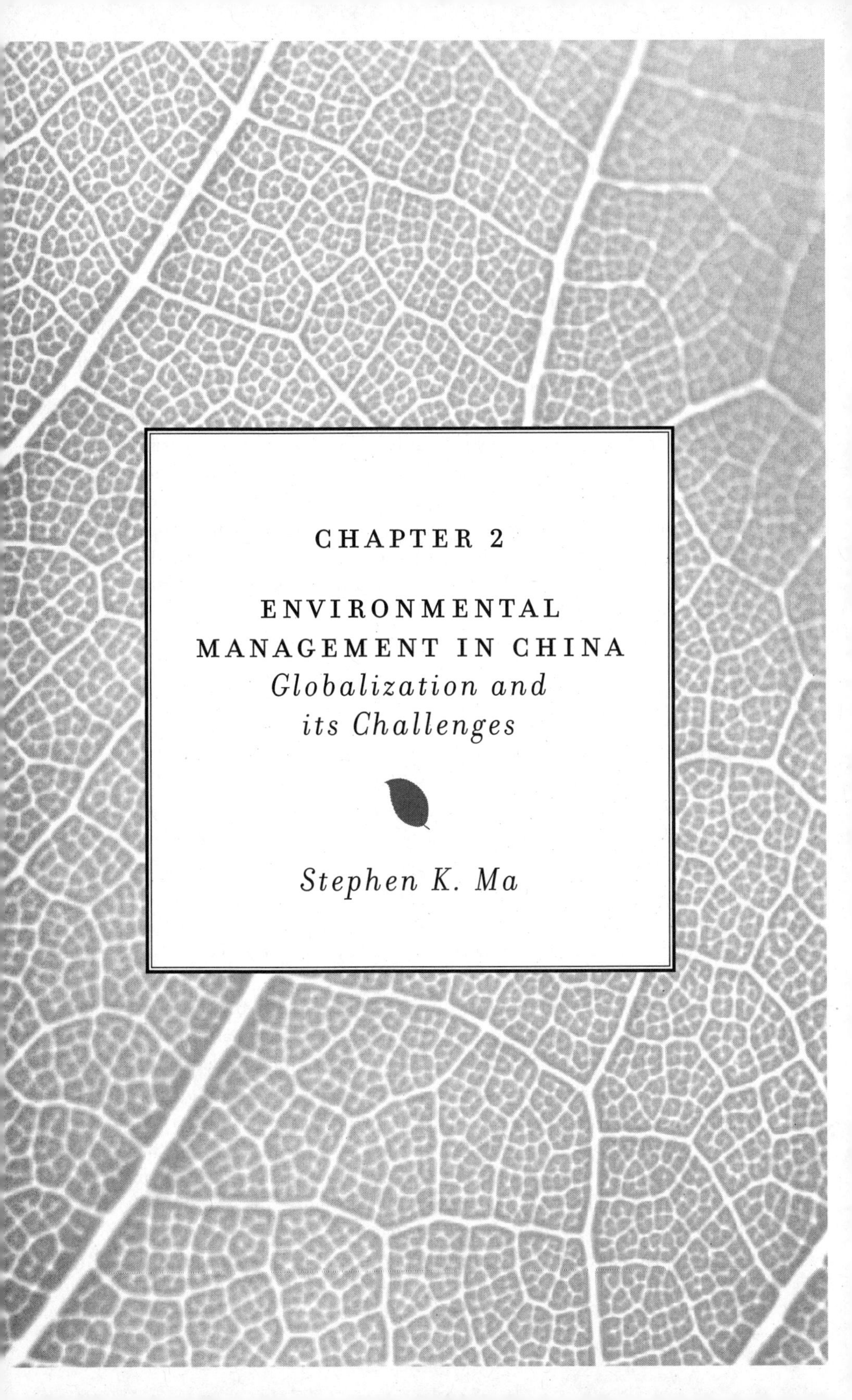

CHAPTER 2

ENVIRONMENTAL MANAGEMENT IN CHINA
Globalization and its Challenges

Stephen K. Ma

INTRODUCTION

In November 2005, the water supply to Harbin, the capital city of Heilongjiang Province in China's northeast, was shut down due to a huge benzene slick in the Songhua River. The Chinese government, which blamed China National Petroleum Corporation (CNPC) for the disaster, did not publicly confirm that the Songhua was poisoned with benzene until ten days after the spill was caused by a blast upstream at a chemical plant owned by a CNPC subsidiary. Later, Zeng Yukang, deputy general manager of CNPC, expressed "sympathy and deep apologies" to the people affected by the disaster. Although China has witnessed sizzling economic growth during the last twenty years, the nation ranks among countries with the smallest water supply per capita. The shutdown of the water supply to Harbin is not an isolated incident in China as hundreds of cities regularly suffer shortages of water for drinking or industry. Moreover, protesters in rural areas claim that pollution is ruining water supplies and damaging crops. Indeed, according to reports, "more than 70 percent of China's rivers and lakes were polluted" and "more than 90% of urban China already suffers from some degree of water pollution."[1] It was alarming to learn that "[e]nvironmental experts fear pollution from untreated agricultural and industrial waste could turn the Yangtze into a 'dead river' within five years."[2]

The impact of the pollution created by the spill of the Songhua River went beyond China's borders. As Harbin's water supply was cut off, panic ensued, people cleared supermarket shelves of bottled water, and the government was forced to truck in millions of bottles of drinking water and dig 100 new wells. Because the Songhua River runs into the Russian city of Khabarovsk, turning into the Heilong River, Russian authorities complained that they were not receiving enough information on the threat flowing toward them. Although Chinese officials claimed that Beijing had shared information and suggested that complaints were premature, the reality remains crystal clear: environmental challenge is no longer an isolated phenomenon or part of the internal affairs of a nation. As this example illustrates, environmental problems are international in nature.

In this chapter I seek to address three questions. First, what new environmental challenges has globalization brought to China? Second, has the international environmental agenda had an effect on the nation's policy-making? And third, how have civil society actors influenced China's environmental management? I argue that globalization has presented new environmental challenges to China and that being a member of the global village has had a strong impact on the nation's environmental management. Not only has globalization had an impact on China's environmental policy-making, but it has also prompted the involvement of non-state actors in China's environmental management. Indeed, these actors provide another avenue for motivating and mobilizing public participation in environmental protection.

BRIEF OVERVIEW OF ENVIRONMENTAL MANAGEMENT IN CHINA

Since the founding of the People's Republic of China (PRC) in 1949, the Chinese government has been dominated by the Chinese Communist Party (CCP). Its guiding ideology was Mao Zedong's thought. As one of the founding fathers of the CCP, Mao emphasized the importance of ideological indoctrination over professional knowledge, and self-reliance over foreign aid, among other things. Consequently, in the 1950s and 1960s China had only limited contact with the international community, and its international relations were mostly limited to other Communist or Socialist countries, headed by the former Soviet Union.

Apparently, it was exposure to the outside world that brought the issue of the environment to the attention of the Chinese government. According to Qu Geping, director general of the Environmental Protection Bureau in Beijing, environmental management began to be viewed as vital only when China took part in the Stockholm Conference on the Human Environment in 1972. One year later, the world's most populous nation held its first national conference on environmental protection, at which an initial program for protecting and improving the environment was charted. In order to facilitate environmental management, the State Council established the leading Group of Environmental Protection in 1974, and the National People's Congress promulgated the Law of Environmental Protection in 1979.

As early as the 1970s, and before China achieved conspicuous economic development triggered by its leadership's modernization programs, the nation was already facing a series of grave environmental problems. These included the atmosphere being polluted by sulphur dioxide; the waters of major rivers, lakes, and offshore seas being contaminated by toxic substances; severely polluted groundwater in many cities; a rapidly dropping water table; rising noise levels in urban residential districts; food crops, vegetables, and industrial crops being contaminated by industrial wastes; the spread of diseases related to environmental problems such as cancer, heart and lung diseases, fluorosis, and others resulting from toxicity of heavy metals; and damage to the natural environment and disturbance of the ecological balance.

To tackle these problems, the Chinese government had to find certain solutions whereby economic development would not be achieved at the expense of the environment, ecology or ecosystems. Environmental guidelines were formulated in 1972, urging reconciled relations among economic development and environmental protection, the multiple utilization of natural resources, and mobilization of the masses in environmental protection and improvement. More specifically, according to the guidelines, urban planning should encourage small and medium-sized cities while restraining the expansion of big cities. Moreover, industries should adopt new environmentally friendly technologies, minimizing the discharge of pollutants. Assessment of environmental impact for new, expanded, or renovated construction projects was required by law.

Of course, one of the most powerful instruments for environmental protection is legislation. In China, some environmental legislation dates back to the 1950s. For example, Sanitary Standards for Designing Industrial Enterprises were promulgated in 1956. Standards—including norms for the management of air, water, and soil—applied to industrial, agricultural, and residential areas. Three years later, Sanitary Regulations for Drinking Water were also developed by the government. However, despite these provisions, environmental issues were basically sitting on the back burner, receiving little attention from both state and society.

Reliance on legislation for environmental protection did not regain force until the 1970s, when the Regulations Concerning Environmental Protection and Improvement were formulated in 1973. These regulations constituted China's first legal provisions with comprehensive stipulations regarding the principles for environmental protection, and they played an important part in strengthening environmental management mechanisms. In the same year, two other legal provisions were enacted: the Regulations for Protection of Rare Wild Animals and the Standards for Discharge of Industrial Waste: Solid, Liquid and Gas; in addition, the Sanitary Standards for Designing Industrial Enterprises were revised. Two years later, the Sanitary Standards for Foodstuffs were formulated. Despite these legal provisions, little substantive progress toward better environmental management was made, for two important reasons. First, political infighting among China's elites continued ruthlessly as Mao's Cultural Revolution unfolded. Mao launched an "ideological crusade" aimed at preventing what he believed to be the gradual erosion of the revolutionary spirit fostered by the early guerrilla experience. He therefore waged a thorough "rectification campaign" in order to halt what he called the "further growth of bureaucratic tendencies" within the party and government. As a result, dissension among the top leaders led to a fierce power struggle between contending groups. The leadership was too preoccupied with ideological debates and political survival and could hardly attend to other issues such as environmental protection. Second, as illustrated by the then widely known slogan of "grasping revolution and promoting production," economic development took precedence over environmental protection during the 1970s.

The ten-year-long Cultural Revolution ended in 1976, closing an ugly chapter in Chinese history when the nation was largely locked in by its own radical leaders and isolated from the rest of the world. As China entered a new phase of its contemporary history, the Constitution of the People's Republic of China was revised and adopted in 1978. Its Article 11 stipulates that "[t]he state protects the environment and the natural resources and prevents and eliminates pollution and other hazards to the public." It was the first time in the history of the PRC that environmental protection was formally included in the nation's fundamental law.

At the end of 1978, the ruling Chinese Communist Party held its epoch-making Third Plenum of the Eleventh Central Committee. While paying tribute to Mao's "outstanding leadership," the meeting found that Mao had made mistakes, signalling the beginning of the drive to deemphasize Mao and to dispel the myth of his infallibility. Therefore, the CCP decided to forgo ideological debates and political purges and embark on the path of "four modernizations," with a focus on reform and opening the nation to the outside world. China's new leadership formed at the meeting concluded that the time had come for Chinese people to dedicate themselves to the study of new circumstances, things, and questions, and uphold the principle of seeking truth from facts.

As the Chinese leaders called on their people to heed the importance of science and technology, environmental legislation began to be formulated with renewed momentum. The Standing Committee of the Fifth National People's Congress approved the Environmental Protection Law of the People's Republic of China and put it forth for trial implementation in 1979. Other pieces of legislation enacted in the same year include the Forestry Law of the People's Republic of China, the Regulations for Protection of Aquatic Resources of the People's Republic of China, the Standards for Quality of Water for Farmland Irrigation, the Standards for Quality of Water for Fishery, and the Standards for Safety in Using Insecticides.

Opening China to the outside world led to a closer connection to the international community, which resulted in the adoption of several international treaties and agreements. For example, in 1980 China joined the Convention on International Trade in Endangered Species of Wild Fauna and Flora (CITES) (Qu and Lee 1984: 1-12). Such a move was an important step for the nation to take, given its previous international isolation; it finally joined forces with other countries to foster environmental protection. Nationally, in order to coordinate activities in environmental protection properly, the Ministry of Urban and Rural Construction and Environment Protection was finally created in 1982.[3]

New principles in environmental management were laid out during the 1980s as environmental protection became an integral part of national economic planning. Targets and norms for environmental protection were established in the long-term planning of the national economy. Supervision and inspection of enterprises were also required and the elimination of pollution and protection of the environment became part of performance assessment for enterprises. Moreover, multi-purpose utilization and recycling of resources were encouraged, fines for violating the standards of environmental protection were set, and the legal responsibilities of the offenders were investigated and pursued.

An important aspect of environmental management in China has been academic support. Though faced with a shortage of well-trained scientists in the field, China has still managed to pool forces to carry out studies in environmental protection from a number of sources. These include institutes under the Chinese

Academy of Sciences; institutes under the Chinese Academy of Social Sciences, research institutes of the departments under the State Council, universities and colleges, research institutes at the provincial level, research personnel in various industrial and mining enterprises, and research personnel in the environmental protection agencies. Even though they are scattered across the nation, these scientists and technicians could, should they join forces, improve the nation's environment management considerably (Qu and Lee 1984: 3–19).

In order to coordinate their activities, China established the Chinese Society for Environmental Sciences (CSES) in 1979. This society is a national association that consists of a volunteer staff providing scientific services in the areas of government decision-making, environmental management and environmental construction, and environmental sciences and technologies. The CSES has several working committees that deal with academic affairs, the dissemination of scientific knowledge, education in environmental science, the development of science, technology, and industry, international exchange, and consultation and assessment. In addition, it has nine branches that cover the areas of environmental management, environmental engineering, environment in national defence, assessment of environmental impact, atmospheric environment, natural protection, environmental technology, aquatic environment, and green packaging. Moreover, the thirteen professional committees under the CSES focus on areas such as environmental planning, standards and norms in environmental protection, environmental monitoring, environmental remote sensing and information, evaluation of environmental impact, protection of marine environment, enduring organic pollutants, groundwater environment, ecological agriculture, the science of environmental law, solid waste, nuclear safety and safety in radioactive environments, green packaging, clean production and recycling economics, and prevention of pollution by motor vehicles.[4] Clearly, by taking full advantage of the members' knowledge and expertise in environmental sciences, these committees and branches offer significant support to the nation's progress in environmental protection.

In China, the most important actor in environmental management is the State Environmental Protection Administration (SEPA). The institution has eleven bureaus. They are in charge of such areas as education and propaganda; planning and financial affairs; policies, laws, and regulations; administrative system and personnel; standards in science and technology; control of pollution; protection of natural ecology; management in nuclear safety; assessment and management of environmental impact; environmental inspection; and international cooperation.[5] The bureaucratic division of labour within SEPA seems to be well designed. Various aspects of environmental management—from domestic policy to international cooperation and from initial planning to final evaluation—seem to be covered properly. However, does this mean that the Chinese government is well prepared for the new challenges that the process of globalization has posed?

NEW ENVIRONMENTAL CHALLENGES POSED BY GLOBALIZATION

Globalization is, in Joseph Stiglitz's words, "a force that has brought so much good" (Stiglitz 2002: 4). By enabling countries to increase their exports, the process has driven their economic development. As a result, globalization has helped many countries grow far more quickly than they would otherwise have been able to do, and it has improved the living standards of many people around the world. Moreover, while reducing the sense of isolation felt in much of the developing world, it has given many people there access to the outside world and, more importantly, to knowledge and technology.

On the other hand, globalization has not necessarily brought about stability and an automatic reduction of poverty to all parts of the developing world. Given the growing economic and commercial interconnectedness, the collapse of an emerging market's currency means that other economies have become more susceptible and can potentially fall as well. Indeed, Russia and most of the other countries that have experienced a transition away from communism and toward market-driven economies have seen an increase in poverty instead of the promised prosperity. As Stiglitz has put it, "the West has driven the globalization agenda, ensuring that it garners a disproportionate share of the benefits, at the expense of the developing world" (2002: 6-7).

One of the prices the people in developing countries have to pay is in the area of environment. China is no exception. Integration with the global economy has "contributed to China's new status as a destination of choice for the world's most environmentally damaging industries" (Economy 2004: 63). These include petrochemical plants, semiconductor factories, and strip mining, among others. Integration with the global economy has also "provided an insatiable global market for China's resource-intensive goods such as paper and furniture" (Economy 2004: 63). Deng Xiaoping's reformist policies of opening the nation to the outside world, as described by Elizabeth Economy, have left "as large a footprint on the natural environment as did centuries of imperial, republican, and early Communist rule" (2004: 64).

Clearly, globalization can be a double-edged sword. Multi-national corporations (MNCs) often place the pursuit of their own economic interest over the care of the local environment. Although a number of MNCs have significantly elevated the level of environmental technology employed in Chinese enterprises, many others, driven by profits and with the complicity of corrupt officials, have taken advantage of China's weaker environmental laws and enforcement capacity to relocate their most polluting enterprises to mainland China. For example, according to Hilary French, "millions of used car batteries are sent from the US every year to smelters" in countries including China, "to be melted down for lead recovery" (2001: 21). French continues: "The smelting process exposes workers to dangerous lead levels, causing classic symptoms of lead poisoning, including headaches, dizziness, stomach cramps, and kidney pains" (2001: 21). Irresponsible environmental abuses by foreign businesses abound. In the late 1990s, the Chinese National Environmental Protection Agency "accused firms from Taiwan and South Korea of setting up shops in China in order to flee tougher environmental regulations at home" (French 2001: 22).

In addition, globalization has promoted trade and foreign investment. However, while trade and investment may facilitate economic development, their impact on a country's environmental situation can often be negative. First, the gain in global trade for many developing countries is achieved through the exploitation of local resources, which is often accompanied by environmental degradation. Second, trade seldom takes into account environmental costs, and it therefore undermines a country's ability to protect the environment. Third, competition for foreign investment may force developing countries to specialize in pollution-intensive industries, providing opportunities for developed countries to get rid of their most polluting industries and relocate them in others' backyards. Moreover, faced with globalization, governments in the developing world must try to protect certain national industries at any price. Very often, this will result in inefficiencies in domestic production and negligence in environmental protection (Economy 2004: 12).

Globalization has accelerated various kinds of economic activities, thereby dramatically increasing demand for resources such as water, land, and energy. For example, "[f]orest resources especially have been depleted, triggering a range of devastating secondary impacts such as desertification, flooding, and species loss" (Economy 2004: 62-63). In China, the sandstorms in Beijing, its capital city, have worsened rapidly in recent years, and many believe that the deteriorating environment in the neighbouring provinces is to blame. According to a report by the British Broadcasting Corporation (BBC) on March 20, 2003, when hit by the choking sandstorms, pedestrians donned masks and fled for shelter, and there were traffic delays as thick dust reduced visibility: "the Chinese capital is now less than 250 kilometres from the encroaching desert, prompting alarm from officials and from the public. Experts have warned that sandstorms could be a major problem for Beijing when it hosts the 2008 Olympics, despite costly efforts to halt the desert's advance."[6]

Globalization has also led to an increase in the concentration of population in cities along the coastal areas, which provide easier access to other members of the global village. According to Bin Lu, by the end of 2003 "the urbanization rate of China reached 40.5%, and the urban population has been over 500 million" (Lu 2004). By 2020, it has been estimated that "the urbanization rate will reach 57%, i.e. an annual increase of more than 18 million people" and that approximately "85-90% of it (about 16 million) will be migrating into the urban area from the countryside" (Lu 2004). As Lu continues, "adding more than 18 million urban populations per year" is another bottleneck that will further "the conflict between urban population growth and the capacity of the resource, energy and environment" (Lu 2004). Environmental problems such as environment pollution, traffic jams, habitation environment deterioration, social differentiation, lack of public sanitation indemnification, and destruction of historical culture are getting more and more serious (Lu 2004).

In addition, China's entry into the global village has meant that the world's most populous nation must join other members of the international community in dealing actively with some of the most pressing global environmental hazards. These include climate change, ozone depletion, and biodiversity loss. Climate change is increasingly threatening to the human life on our planet, especially in the Global South. As the World Bank points out, "sea level rise, warming temperatures, uncertain effects on forest and agricultural systems, and increased variability and volatility in weather patterns are expected to have a significant and disproportionate impact in the developing world, where the world's poor remain most susceptible to the potential damages and uncertainties inherent in a changing climate."[7] In fact, the future of the Kyoto Protocol on climate change is largely in the hands of the world's biggest contributors to greenhouse gas emissions. According to a July 2005 BBC report, China is the world's second-biggest emitter of greenhouse gases.[8] However, as a developing country it is not yet required to reduce its emissions. Since China accounts for a fifth of the world's population, increases in its emissions could dwarf any cuts made by the industrialized countries.

Nevertheless, per-capita consumption of energy in China is not currently too high. As the aforementioned report highlights, the average Chinese person consumes only from 10 to 15 per cent of the energy an average US citizen does. Of much concern is China's drastic economic development, which prompted many analysts to expect China's total emissions to overtake America's by mid-century. A major factor is the use of fossil fuels: China is the world's biggest coal producer, and its oil consumption has doubled in the last 20 years. The country faced power cuts in 2004 as soaring growth outstripped electricity generation. To be fair to the nation's leadership, they were quite frank about the gravity of the situation. They recognize that climate change could devastate their society. In addition, the Chinese government also ratified the Kyoto Protocol in 2002. In 2004, Beijing announced plans to generate 10 per cent of its power from renewable sources by 2010. However, whether the country will ever agree to internationally imposed emissions restrictions remains an unanswered question.

According to a news release by the United Nations Development Programme, "China is the world's largest producer and consumer of ozone-depleting substances (ODS)."[9] For example, as reported by the Commonwealth Scientific and Industrial Research Organization in Australia (CSIRO) in 1999, emissions of halon-1211, a fire retardant, were increasing at the rate of approximately 200 tonnes per year.[10] Its continued growth could be due to increased legal manufacture and release in China, which was responsible for about 90 per cent of the world's production of halon-1211.

According to a 1999 study on environmental degradation in Southwest China, with over 50 ethnic groups China is rich in cultural and natural diversity and is listed by biologists as a country with a megadiversity. In terms of biodiversity it comprises more than 30,000 species of higher plants and 2,100 terrestrial vertebrates. Several hundred species are endemic to China, the most famous being the giant panda, the lesser panda, and the Yangtze crocodile. Much of this variety is found in an area of Southwest China, where in a few locations the untouched ecosystems are among the most diverse living assemblies in Asia. However, since the 1950s, the forests have been indiscriminately felled, reducing forest cover from 30 to 13 per cent. This has threatened biodiversity, causing drastic declines in mammal and bird counts, recurrent flooding and erosion, and recurrent snow disasters. These not only threaten global climate, but undermine the livelihood of the local people.[11]

And so, as China's economy runs at full steam, it has become one of the world's largest contributors to all these problems. Their resolution definitely requires both its full commitment and sincere cooperation. More importantly, it also requires political wisdom on the part of the Chinese leadership.

Indeed, globalization is not merely a question of pure technology and markets; it is also political. It has to do with "a whole host of issues (especially in developing nations) that governments or their citizens otherwise might not discuss out of fear, ignorance, or political will. In the process, debates over environmental issues become surrogates for issues previously off the table. These issues most notably involve the state's role in determining (with economic interests) who has control over natural resources and the wealth they create, destroy, or redistribute" (Durant et al. 2004: 494). The role that the Chinese state will play in these battles over national development and natural resources can have an enormous impact on the process of globalization and the welfare of international community. Conversely, for the same reason, globalization will certainly produce effects on the policy-making process of the Chinese government.

THE EFFECTS OF GLOBALIZATION ON ENVIRONMENTAL POLICY-MAKING

As stated earlier, China took part in the Stockholm Conference on Human Environment in 1972, marking the beginning of a new chapter on environmental protection and ushering in a new era in the nation's environmental policy-making. Yet much more has taken place after two major political events in contemporary China: the end of the Cultural Revolution in 1976 and the official adoption of Deng's modernization program, which extricated China from its international isolation and restored relations with the rest of world.

It was against this background of profound political change that China held the Second National Environmental Protection Conference in December 1983. The meeting was considered "the critical point of change" for the nation's environmental protection, "as the leaders showed a new level of recognition of these problems" (Heggelund 2004: 139). Soon after the conference, and in order to coordinate all agencies involved in environmental protection, the Environmental Protection Commission was established in May 1984 under the State Council. To increase the authority of the Environmental Protection Bureau, the State Council renamed it the National Environmental Protection Bureau (NEPB) in December of that year. As a result, though still part of the Ministry of Urban and Rural Construction and Environmental Protection, the documents issued by the NEPB were given equal status to those issued by the Ministry.

Four years later, in 1988, the NEPB was granted further power; it became the National Environmental Protection Agency (NEPA), independent from the Ministry of Urban and Rural Construction and Environmental Protection. Taking advantage of its new status, NEPA convened the Third National Environmental Protection Conference, where five new environmental policy strategies were announced. These included the contract responsibility system with city government officials, the comprehensive clean-up system, the collective control system, the pollution permit system, and the dead-line clean-up system (Heggelund 2004: 140). These new strategies were obviously designed to tackle the increasingly serious environmental problems by calling for a comprehensive system of environmental protection and keeping local government officials accountable for their actions in the process.

Promoted to ministerial status in 1998, NEPA was officially renamed the State Environmental Protection Administration (SEPA). Meanwhile, the Environmental Protection Commission was abolished and its coordination functions were transferred to SEPA, which, as a full-fledged cabinet member in the central government, has emerged as the principal actor on China's environmental stage. The world's most populous nation has indeed gone through "a sea change in Chinese environmental protection attitudes and practices" since 1972 (Economy 2004: 217). At that time, there was no environmental protection apparatus, no environmental legal system, and only the smallest environmentally educated elite. Three decades later, what we have is a completely different context, where "numerous bureaucracies are engaged in protecting the environment, the legal infrastructure embraces virtually every aspect of the environment, and there is a vast, ongoing environmental education effort throughout the society" (Economy 2004: 218).

How did all these happen? The remarkable change has been largely attributed to globalization. As Economy has argued, "the international community has played an essential role in this transformation" (2004: 218). Indeed, the effects of globalization on China's environmental policy-making can be best observed in the following two areas. First, it has placed additional pressure on the nation's environmental protection. In June 2006, for example, the Chinese government's Information Office published a white paper entitled "Environmental Protection in China (1996-2005)." The document, composed of ten chapters, provides a systematic introduction to the unremitting efforts made by China in environmental protection over the past ten years. While the document states that "China stresses international cooperation in environmental protection," it also admits that "[as] the demand on resources from economic and social development is increasing, environmental protection is facing greater pressure than ever before."[12]

Zheng Bijian, Chair of the China Reform Forum—a non-governmental and non-profit academic organization that provides research on and analysis of domestic, international, and development issues related to China—wrote about the constraints China had to face in its modernization program in late 2005. As he acknowledged,

> The formidable development challenges still facing China stem from the constraints it faces in pulling its population out of poverty. The scarcity of natural resources available to support such a huge population—especially energy, raw materials, and water—is increasingly an obstacle, especially when the efficiency of use and the rate of recycling of those materials are low. China's per capita water resources are one-fourth of the amount of the world average, and its per capita area of cultivatable farmland is 40 percent of the world average. (Zheng 2005: 19)

Zheng's honest statement confirms once again the shortage of natural resources urgently needed for China's rapidly expanding economy. Although his admission refers to constraints in striving for national prosperity, it is obvious that globalization and the resultant exchange of information on environmental protection have been challenging China on "the efficiency of use and the rate of recycling" of natural resources.

In June 2006, *China Business News* (CBN) published an exclusive interview with Yang Chaofei, director of the Policies and Regulations Department of SEPA. Yang was straightforward in agreeing that "China has faced international pressure in relation to environmental issues for several years."[13] He further defined the five key issues in question. First, environmental security has become an important element of national security that can no longer be ignored. Second, the total volume of contamination is huge, as is its impact on the world environment. Third, environmental frictions with neighbouring countries are rising, and these must be dealt with properly. Fourth, a rise in demand for resources is affecting global supply. Finally, environmental problems have become a restriction on foreign trade. Meanwhile, Yang did not deny that "contamination that China is responsible for has an increasingly serious impact on the rest of the world, in particular neighboring countries." As he acknowledges, China's "rising demand for resources to fuel our rapid economic development is putting more pressure on the international resources market." Clearly, by identifying these forces and their impact on China, the Chinese government seems to demonstrate its determination to fight against environmental hazards resolutely.

Indeed, to be part of the global village means that there must be certain constraints on China's environmental policy-making. As pointed out by Vaclav Smil, "[n]o lasting change can come without recognizing the seriousness of environmental degradation and its critical linkages with economic progress and the quality of life" (1993: 194). Although China closed down some outdated, heavily polluting enterprises, installed modern controls on many new projects, and returned farmed slopeland to trees or grass, many of the remedies designed to improve the quality of the nation's environment "have been submerged by economic expansion and by individual quest for greater material well-being" (1993: 195).

In addition, globalization has also prompted China to double its efforts in searching for alternate means to protect the environment while at the same time contributing to economic development. In February 2005, the Chinese government passed a renewable energy law that both requires power operators to buy electricity from alternative energy providers and gives economic incentives to such producers. The new law, which came into effect on January 1, 2006, provides a host of practices to ensure that renewable energy can be produced, marketed, and used. The law not only "orders power grid operators to purchase 'in full amounts' resources from registered renewable energy producers within their domains. It also encourages oil distribution companies to sell biological liquid fuel on the sidelines." In addition, "power grid operators should buy renewable-source-generated power at directed prices calculated by the government. The extra costs incurred by this will be shared throughout the overall power network." Moreover, as stipulated in the law, the government should offer "financial incentives, such as a national fund to foster renewable energy development, and discounted lending and tax preferences for renewable energy projects."[14]

It is encouraging to realize that China is serious in its pursuit for renewable energy. As a recent article in US *News and World Report* states, "While China is most commonly known as a voracious consumer of energy with a spotty environmental record, the emerging industrial giant is quietly becoming a world leader in developing renewable energy sources and technology" (Fang 2006: 37). The new Renewable Energy Law "specifies tariffs that favor nonfossil energy sources such as wind, water, and solar power." In addition, building codes have been developed "mandating that all new construction dramatically improve energy efficiency" (Fang 2006: 38). The Chinese government made public an ambitious goal of "having 10 percent of the country's gross consumption be renewable by 2020—a huge increase from the current 1 percent" (Fang 2006: 37). If this developing country with severe energy shortages can manage to move steadily toward that goal, it will considerably alleviate the twofold worries in the international community about depleting energy sources and worsening pollution problems.

CIVIL SOCIETY INFLUENCE ON ENVIRONMENTAL MANAGEMENT

Opening China to the outside world has inevitably exposed its people to other cultures. One of the unintended consequences of Deng's reformist policies was the evolution of the nation's political culture, from "subject culture" to "citizen culture." Gabriel Almond and G. Bingham Powell, Jr., have identified three roles played by average people, namely as subjects, participants, and parochials (1988: 42). The development in China since the late 1970s and early 1980s has illustrated that more and more Chinese people no longer obey and follow their rulers as subjects. Instead, they began to participate in national affairs as citizens. As observed by Merle Goldman, the last decade of the twentieth century witnessed how "a small number of critical intellectuals in the establishment and those operating on its margins" were transformed ... "from 'comrades' into 'citizens'" (2005: 7). Among those who refused to sing the same tune as the government were Dai Qing, once a reporter at *Guangming Daily* and an adopted daughter of a famous marshal, and Li Rui, a former personal secretary of Mao Zedong. Both spoke out loudly against the construction of the Three Gorges Dam based on their environmental concerns (Economy 2004: 142-45). Yet the authorities turned a deaf ear to their cries. Not discouraged, many others followed in their footsteps. As mentioned earlier, the Songhua accident poisoned a Chinese river, and the incident had a multiplicity of effects. As reported in the *Los Angeles Times*, "[t]he release of millions of gallons of toxic liquid into a major city's water supply, China's biggest environmental accident in years, is shaping up as a wake-up call for a society that has made huge sacrifices for economic development."[15] Moreover, the incident "has laid bare many of China's fundamental problems, including corruption, official secrecy, wholesale destruction of the environment and a growing sense that many domestic problems can no longer be contained within China's borders." In addition, it also brought people's attention to the voices of actors outside the government, including non-governmental organizations (NGOs).

One of these voices came from Mei Jiayong, an official with Greenpeace China. As he commented, "[t]he pollution released is very toxic and can cause cancer.... Hopefully we'll be lucky this time. But this may be the tip of the iceberg for China's environment."[16] Greenpeace China was established in Hong Kong in 1997 and has since added offices in Beijing and Guangzhou. Like any Greenpeace office in the world, it focuses on the most crucial threats to the biodiversity and environment of our planet. Greenpeace China has been working hard on stopping climate change and developing renewable energy, promoting food safety and developing sustainable agriculture, eliminating toxic chemicals and promoting extended producer responsibility, and protecting ancient forests.[17] As environmental issues receive increasing attention from both the Chinese state and society, Greenpeace China is likely to have more influence on the nation's environmental management.

More importantly, Greeenpeace China is not a lone civil society actor in China's environmental protection. Since 1994 when the first environmental NGO, Friends of Nature, was launched, "[g]rassroots NGOs have sprung up in many regions of the country to address issues as varied as the fate of the Tibetan antelope, the deterioration of China's largest freshwater lakes, and mounting urban refuse" (Economy

2004: 21). As in many countries of the Global South, the marked proliferation in China of NGOs was in part due to the funding flowing from international sources. For instance, about half of Green River's revenues are supplied by the Hong Kong-based branch of Friends of the Earth (Economy 2004: 309). The NGO Green Volunteer League received a $10,000 grant from the Canadian International Development Agency (CIDA), and Action for Green was helped by the Global Greengrants Fund (Economy 2004: 158). There is little doubt that financial aid from the international community has played a key role in promoting civil society actors' involvement in China's environmental protection.

Pro-environmental voices have also come from the academic community. Wang Yukai, a Professor at Beijing's National School of Administration, has argued that "the [Songhua] accident has exposed many problems in China's crisis management system," and that "the government didn't give timely information, and its explanation changed several times. In their desire to prevent panic, they did just the opposite."[18] There are also consultants specialized in environmental protection in China, such as Ma Jun, an environmental consultant with the Beijing Bo Xin Consulting Co. Displeased with government response to the poisonous spill into Songhua, he condemned the bureaucrats for insisting on "denying a problem, a delay that only made things worse." His prediction was ominous: "as China's economy races ahead, the risks are rising at least as fast."[19]

Indeed, the impact of the environmental movement has been growing. As Economy has suggested, "[t]hey possess the full complement of skills necessary to organize effectively: technical expertise on the environment, strong backgrounds in journalism and media, and extensive ties to environmental activists both throughout China and abroad" (Economy 2004: 173). More importantly, the activities of these green communities help the government in several ways. They provide another avenue for motivating and mobilizing public participation in environmental protection. For example, the Chinese Association of Environmental News Journalists and Friends of the Earth in Hong Kong jointly established the Awards of the Earth in 1997 to encourage more participation in environmental protection. It is the first environmental award granted by NGOs in China.[20]

Moreover, through their own initiatives and often with little or no public funding, environmental NGOs (ENGOs) serve as inexpensive mechanisms for monitoring local pollution and offering environmental education. Notably, Song Kexin, a recipient of an Award of the Earth in 2006, founded the Green Future Association for Environmental Protection, the first NGO with a focus on environmental protection along the Yellow River Valley. Song's father died of stomach cancer in 1996. Believing that the death was caused by pollution from local factories, Song spent the next ten years investigating the source of contamination, which led to the closing of more than 100 heavily polluting local enterprises. Meanwhile, through the Green Future Association, Song organized a rally of over 10,000 people in 2002, urging local residents to safeguard their Mother River—Yellow River.[21]

ENGOs in China have also had a positive impact on the nation's environmental management by developing connections and cooperation with international communities. Greenpeace China is a prominent example. As a result of the connections Greenpeace has with numerous other organizations worldwide, Greenpeace China can work with NGOs in other countries to exchange information on environmental issues, to cooperate on environmental projects, to expose global environmental problems, and to force solutions for a green and peaceful future. In the process, Greenpeace China has formed alliances with many international actors, keeping China's environmental bureaucracy updated, and contributing to the understanding, harmony, and solidarity among members of the global village.

It is important to note how cooperation on environmental matters with the international community has helped China's globalization in economic terms. By introducing both the government officials and the citizenry to the more demanding environmental standards usually set in developed countries, the exchange of information and knowledge with the outside world has made it possible for China to learn a lot within a comparatively short period of time. As a result, China has had to "upgrade product quality and increase exports by rapidly adopting international manufacturing, measurement, and quality standards" (Ross 1988: 205).

Finally, activities by these ENGOs in China have made the nation's environmental management more responsive to local problems by serving as conduits of information flow between the government and the governed and drawing the attention of civil servants to the needs of local residents, which state officials have traditionally attempted to disregard. For instance, in March 2003, officials in Yunnan Province of Southwestern China began to plan the launch of the so-called Nu River project, a plan to build one of the biggest dams in the world. The project would produce more electricity than even the mighty Three Gorges Dam, which is a Chinese hydroelectric river dam which spans the Yangtze River and the largest hydroelectric river dam in the world, more than five times the size of the Hoover Dam. However, the Three Gorges Dam not only threatened a region considered an ecological treasure, but it also meant the relocation of numerous local villages. For decades, the ruling Communist Party had approved and executed such projects by fiat and with absolutely no input whatever from society. The Nu River proposal, already delayed for more than a year, has presented the Chinese government with a quandary of its own making: will it abide by its own laws? As *The New York Times* reported, a coalition led by Chinese ENGOs urged the central government to hold open hearings and make public a secret report on the Nu dams before making a final decision.[22] The revealing of this story was unprecedented. In a country where people cannot challenge decisions by their leaders, public participation is a fairly radical idea. But the groups contended cogently that new environmental laws grant exactly that right to information. Ma Jun, an environmental consultant in Beijing, was quoted as saying that it was a precedent-setting case and that "for the first time, there is a legal basis for public participation. If it happens, it would be a major step forward."[23]

China's leaders often claim to embrace the concept of rule of law: "For many people in China's fledgling civil society—environmentalists, journalists, lawyers, academics and others—the law has become a tool to promote environmental protection and to try to expand the rights of individuals in an authoritarian political system."[24] These rights include, among other things, the right to participate in public policy-making by pressing for more environmentally friendly state action. It has been argued that, in China, "NGOs were created because citizens felt that the state was not dealing effectively with a particular problem—such as environmental management—and they organized in order to use their numbers to put pressure on the state, or to respond themselves to the shortcomings of the state" (Vig and Axelrod 1999: 68). This was exactly how China's green community felt about the government in the Three Gorges Dam case and why they decided to stand up to the once omnipotent state bureaucracy.

The existence in China of ENGOs is in fact advantageous to the government. The positive impact that non-state actors have on the nation's environmental management has become more than apparent. As it has been suggested, "[i]n China, where the ruling Communist Party discourages or outright crushes any attempts at grassroots movements, environmental protection is one of the only areas of activism that is thriving" (Schafer and Ansfield 2005). That also explains why, for the first time in 2005, SEPA decided to issue awards for environmental protection to average citizens. Recipients of the award include Tian Guirong, a peasant who spent years collecting used batteries to prevent them from contaminating the environment; Wang Canfa, a lawyer offering legal assistance to victims of pollution; Zhao Yongxin, a journalist reporting the environmentally damaging anti-leaking project in Yuanming Garden; and Liang Congjie, a founder of Friends of Nature, China's first ENGO. Together with a government official, they were named People of the Year for Green China.[25] Though assisted by ENGOS, SEPA is unlikely to proceed with its tasks smoothly. It has been "engaged in a David-and-Goliath tussle against powerful interest groups," or "GDP lobby" as political insiders call it. The GDP lobby includes "key ministries and industry giants such as State Power, the Three Gorges Dam Group, CNPC and above all the State Development Planning Commission, a powerful super-ministry." To counterattack these powerful interest groups pursuing GDP growth at the expense of environment, the environmental community was pushing for adoption of a new measure called "green GDP," seeking to "gauge the health and sustainability of economic growth, not simply its speed" (Liu 2005). It is very interesting to see which side will prevail, the GDP lobby or the green GDP, and to explain why.

CONCLUSION

Globalization has presented new environmental challenges to China. Even though globalization has brought tangible economic benefit to many countries, economic development is not cost-free. One of the prices people in the Global

South, including China, have had to pay for economic growth has been in the area of the environment. Multinational corporations often place the pursuit of their own economic interest above the care of the local environment. Global trade for many developing countries is achieved through the exploitation of local resources, which often undermines a country's ability to protect the environment. In addition, competition for foreign investment has forced developing countries to specialize in pollution-intensive industries, providing opportunities for developed countries to get rid of their most polluting industries and relocate them in others' backyards. Moreover, efforts by governments in the developing world to protect certain national industries threatened by globalization often result in inefficiencies in domestic production and negligence in environmental protection.

Globalization has also led to a dramatically increased demand for natural resources and a concentration of population in cities along the coastal areas, which have easy access to other parts of the world. More importantly, China's entry into the global village has meant that the world's most populous nation must join other members of the international community in dealing actively with some of the most pressing global environmental hazards such as climate change, ozone depletion, and biodiversity loss.

As I have sought to show in this chapter, however, globalization has also had an important effect on China's environmental policy-making, in many cases a positive one: it has helped to bring about the birth of SEPA, a full-fledged cabinet member in the central government and the principal actor on China's environmental stage. Being a member of the global village has also meant that pressure on the nation's environmental protection has increased, as we have seen. Moreover, it has also prompted China to double its efforts in searching for alternate means to protect the environment while at the same time contributing to economic development.

In the process, non-state actors have influenced China's environmental management. They provide an alternate avenue for motivating and mobilizing public participation in environmental protection and they serve as inexpensive mechanisms for monitoring local pollution and offering environmental education. Moreover, and as this chapter has demonstrated, they have had a positive impact on the nation's environmental bureaucracy by keeping it updated through their connections and cooperation with international communities. They have also attempted to make the nation's environmental management more responsive to local problems. Clearly, green NGOs have become an indispensable force in the fight to protect the environment. It remains to be seen, however, whether, faced with strong opposition from the powerful "GDP lobby," the green movement's attempt at green GDP will materialize in years to come as China presses ahead with its fast-paced industrialization.

Notes

1 BBC News. World Wide Web: http://news.bbc.co.uk/go/pr/fr/-/2/hi/asia-pacific/4636371.stm. Accessed June 30, 2005.

2 BBC News. World Wide Web: http://news.bbc.co.uk/go/pr/fr/-/2/hi/asia-pacific/5029136.stm. Accessed May 30, 2006.

3 It is currently called the State Environmental Protection Administration.

4 General Information on Chinese Society for Environmental Sciences, http://www.chinacses.org/CN/xhgk.html. Accessed April 15, 2006.

5 The structure of the State Environmental Protection Administration (SEPA), http://www.zhb.gov.cn/dept/jgznjj/zns/. Accessed April 15, 2006.

6 BBC News, World Wide Web: http://news.bbc.co.uk/2/hi/asia-pacific/1883494.stm. Accessed July 10, 2006.

7 See: http://web.worldbank.org/WBSITE/EXTERNAL/TOPICS/ENVIRONMENT/EXTCC/0,,menuPK:407870~pagePK:149018~piPK:149093~theSitePK:407864,00.html. Accessed March 5, 2006.

8 BBC News. World Wide Web: http://news.bbc.co.uk/1/hi/sci/tech/3143798.stm. Accessed March 20, 2006.

9 See: http://www.undp.org.cn/modules.php?op=modload&name=News&file=article&catid=14&topic=23&sid=73&mode=thread&order=0&thold=0. Accessed June 3, 2005.

10 See: http://www.sciencedaily.com/releases/1999/03/990303163332.htm. Accessed February 18, 2006.

11 John Studley, "Progress, Biodiversity Loss and Environmental Degradation in SW China." See http://www.geocities.com/john_f_studley/China_re.htm. Accessed November 22, 2005.

12 See: http://english.people.com.cn/200606/05/eng20060605_271168.html. Accessed June 5, 2006.

13 See http://www.china.org.cn/english/2006/Jun/171410.htm. Accessed June 20, 2006.

14 *China Daily*, March 1, 2005.

15 *Los Angeles Times*, Nov. 25, 2005.

16 *Los Angeles Times*, Nov. 25, 2005.

17 See: http://www.greenpeace.org/china/en/about.

18 *Los Angeles Times*, Nov. 25, 2005.

19 *Los Angeles Times*, Nov. 25, 2005.

20 News release from the "Chinese Children's Hand in Hand, Building an Earth Village." World Wide Web: http://www.childrenandearth.org.cn/news/020101/earthaward.htm. Accessed on February 10, 2005.

21 *World Journal*, April 29, 2006.

22 *New York Times*, Dec. 26, 2005.

23 *New York Times*, Dec. 26, 2005.

24 *New York Times*, Dec. 26, 2005.

25 *World Journal*, Nov. 28, 2005.

References

Almond, Gabriel A., and G. Bingham Powell, Jr. 1988. *Comparative Politics Today: A World View*. 4th ed. Glenview, IL: Scott, Foresman and Co.

Durant, Robert F., Daniel J. Fiorino, and Rosemary O'Leary. 2004. *Environmental Governance Reconsidered: Challenges, Choices, and Opportunities.* Cambridge, MA: MIT Press.

Economy, Elizabeth C. 2004. *The River Runs Black: The Environmental Challenge to China's Future.* Ithaca, NY: Cornell University Press.

Fang, Bay. 2006 (June 12). "China's Renewal: Hungry for Fuel, It Emerges as a Leader in Alternative Energy." US *News & World Report.* 37-40.

French, Hilary. 2001. "Can Globalization Survive the Export of HAZARD?" USA *Today Magazine*. May 2001, Vol. 129, Issue 2672, 20-23.

Goldman, Merle. 2005. *From Comrade to Citizens: The Struggle for Political Rights in China.* Cambridge, MA: Harvard University Press.

Heggelund, Gorild. 2004. *Environment and Resettlement Politics in China.* Burlington, VT: Ashgate Publishing Co..

Liu, Melinda. 2005 (Dec. 6). "China's Katrina." *Newsweek.* Web-Exclusive Commentary. http://www.msnbc.msn.com/id/10354870/site/newsweek/. Accessed Feb. 10, 2006.

Lu, Bin. 2004. "Urban Problems to Be Overcome in Chinese Cities as Urbanization [Expedites]." http://csur.t.u-tokyo.ac.jp/ws2004/papers/A-Lu.pdf. Accessed Feb 25, 2006.

Qu, Geping, and Woyen Lee, eds. 1984. *Managing the Environment in China.* Dublin: Tycooly International Publishing.

Ross, Lester. 1988. *Environmental Policy in China.* Bloomington: Indiana University Press.

Schafer, Sarah, and Jonathan Ansfield. 2005 (Nov. 26). "Toxic Fallout." *Newsweek.* Web-Exclusive Commentary. http://www.msnbc.msn.com/id/10214757/site/newsweek/. Accessed March 5, 2006.

Smil, Vaclav. 1993. *China's Environmental Crisis: An Inquiry into the Limits of National Development.* Armonk, NY: M.E. Sharpe.

Stiglitz, Joseph E. 2002. *Globalization and Its Discontents*. New York: Norton.

Vig, Norman J., and Regina S. Axelrod, eds. 1999. *The Global Environment: Institutions, Law, and Policy.* Washington, DC: CQ Press.

Zheng, Bijian. 2005. "China's 'Peaceful Rise' to Great-Power Status." *Foreign Affairs* 84(5): 18-24.

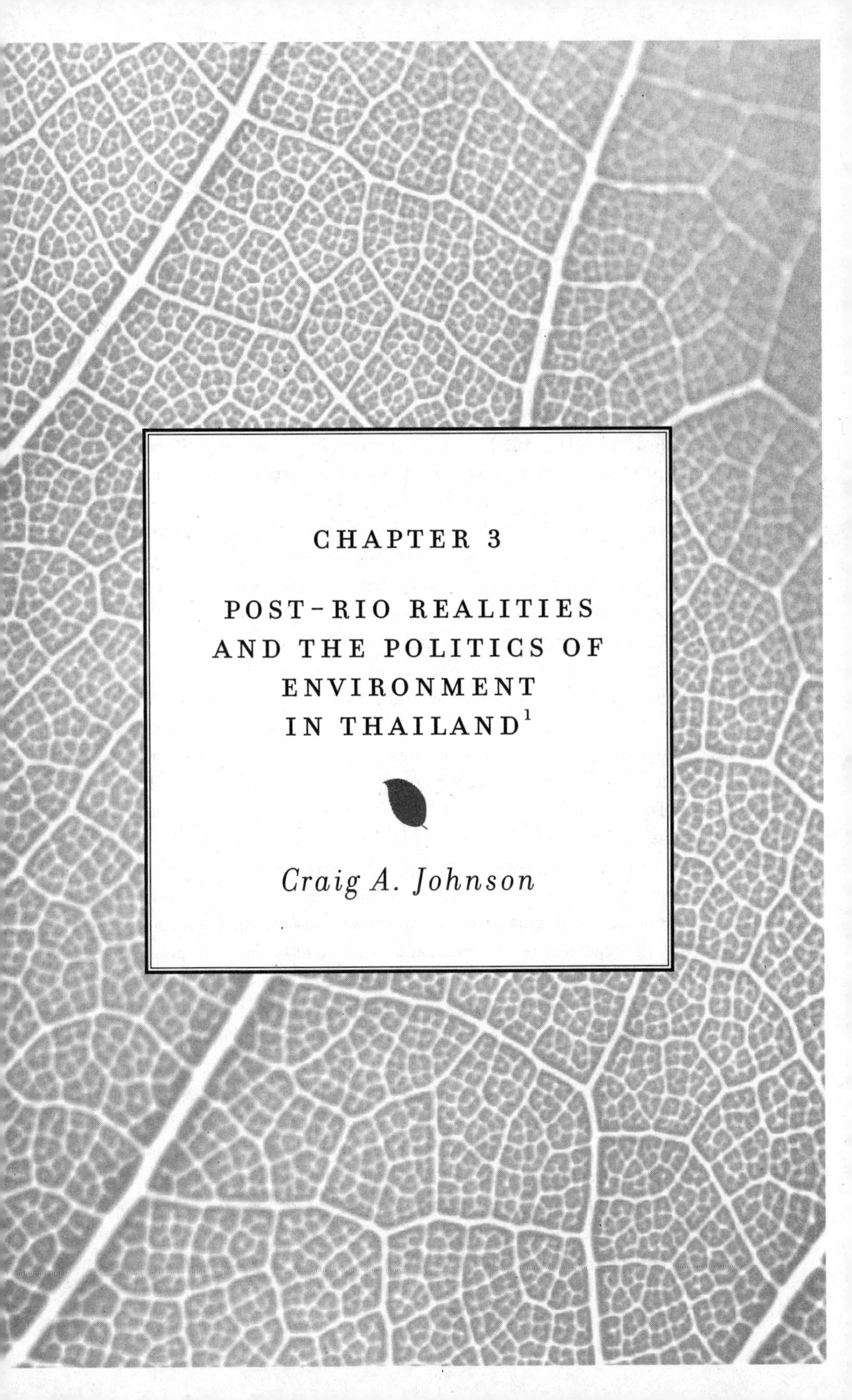

CHAPTER 3

POST-RIO REALITIES AND THE POLITICS OF ENVIRONMENT IN THAILAND[1]

Craig A. Johnson

INTRODUCTION

Toward the end of the twentieth century, Thailand earned the enviable distinction of having the world's fastest-growing economy (Unger 1998). Between 1984 and 1994, annual growth rates of 12 to 14 per cent were widely attributed to an investment regime that had become highly receptive to foreign investment and an economic strategy that promoted through currency devaluation, tax incentives and the establishment of Export Processing Zones (EPZS) for the production of export commodities (Krongkaew 1995; Phongpaichit and Baker 1995). In rural areas, the transformations of this period generated tremendous opportunity and wealth (Phongpaichit and Baker 1995; 1998). However, the rise of new markets (both within and outside Thailand) created strong incentives to "mine" the natural resource economy, producing a wide array of environmental costs and conflicts. Principal among these were the environmental costs of land and water degradation, the economic costs of land encroachment, and a wide array of social costs arising from the loss or degradation of neighbouring livelihood systems. By 1989, the extraction of forest resources had become so severe that the Thai government introduced a ban on all commercial logging (Phongpaichit and Baker 1995: 82).

Against this backdrop there appeared in the 1980s a growing number of highly visible political campaigns that were ostensibly aimed at addressing environmental issues, such as deforestation, urbanization, and the construction of hydroelectric dams. The Thai government's cancellation of an ambitious effort to construct the Nam Choan hydroelectric dam in 1988 was widely perceived as a "turning point" for Thailand's environmental movement (Hirsch and Lohmann 1989; Hirsch 1993; Rigg 1997). During and after this period, the Thai government put in place a number of ambitious environmental policies which on the surface appeared to challenge the power of neo-liberalism and free-market capitalism. Alongside the commercial logging ban, other environmental "victories" included the 1992 National Environmental Quality Enhancement and Protection Act, which gave government new powers of inspection, assessment, and (limited) enforcement; the Community Forest Bill, which aimed to resolve longstanding conflicts by allowing local communities to manage forest resources in protected areas; and the 16th Constitution, which in 1997 granted rural "communities" a constitutional right to manage and conserve natural resources.

As Philip Hirsch has pointed out, the ability to challenge and influence the authority of state and corporate power in these and other instances was highly dependent upon an important—and until this point unprecedented—coalition of peasants, Buddhist monks, students, academics, and urban-based environmental non-governmental organizations ENGOS (1993). Coalitions of this kind became increasingly common in the 1990s, culminating with the Assembly of the Poor, a coalition of non-governmental organizations (NGOS), farmers, peasants, agricultural labourers, village leaders, and rural politicians that descended on the streets of

Bangkok in 1995. The environmental victories of the 1990s were therefore an important watershed in Thai politics, in which NGOs, farmers' organizations, urban labour, and academics came together to resist the environmental and wider social costs of Thailand's neo-liberal boom.

However, the factors that led to the crystallization of the movement and its long-term implications on environmental politics and policy in Thailand are a matter of considerable debate. Neo-classical interpretations, for instance, attribute environmental protests of the 1980s and 1990s to a growing scarcity of natural resources in rural areas. Scholars working in this tradition argue that environmental conflicts were primarily the result of market failures in which governments failed to assign an appropriate value to scarce environmental resources through price mechanisms (Christensen and Boon-Long 1994; Christensen and Rabibhadana 1994). Framed in this way, the very existence of environmental movements reflects a more systematic failure on the part of the state to create policies that would establish entitlement over scarce resources, ideally through the enforcement of private property rights.

A second and decidedly more critical literature takes issue with market capitalism as an economic system and liberal democracy as a means of governing it. Scholars working in this tradition attribute the environmentalism of the 1980s and 1990s to the changing structure of Thailand's economy and to the contradictory forces that underlie its political economy. Central to this perspective is the idea that industrialization, urbanization, and the spread of commercial agriculture led to the displacement of marginal groups whose labour and livelihood were traditionally dependent upon the ability to access local resources. Pasuk Phongpaichit and Chris Baker, for instance, attribute the rural activism of the 1980s and 1990s to a more general history of surplus extraction and unequal exchange between a primarily urban-based class of traders, bureaucrats, and entrepreneurs and the upland frontier (1995). Along similar lines, Hirsch (1990) argues that the environmentalism of the 1980s and 1990s was reflective of a wider struggle over access to natural resources and the extraction of rural labour. Among certain authors (for instance, Hirsch and Lohmann 1989; Hirsch 1993), agrarian struggles over natural resources and the conservation efforts of environmentalists are assumed to be mutually supportive processes.

Finally, constructivist interpretations of this period challenge the idea that social movements can be reduced to "simple" social constructs, such as scarcity or class, and highlight instead the ethnic, regional, and religious cleavages that underlie Thailand's environmental movements (see, for instance, Forsyth 1996; Walker 2000; Vandergeest 2003).[2] Central to this perspective is the idea that environmental movements constitute an effort on the part of groups to forge claims to rights and resources by developing or adopting discursive terms that will strengthen their entitlement to state- and socially sanctioned forms of entitlement. The social construction of "community" is well-documented in the literature

(Li 1996; Johnson 2001; Johnson and Forsyth 2002), as is the use of racial and ethnic identity (Vandergeest 2003). Although they are broadly sympathetic to the needs of the communities and activists pursuing environmental agendas, scholars in this tradition also try to highlight the underlying motivations and contradictions of Thailand's environmental movements. One contradiction, which has received considerable scrutiny in the literature, is a conflict of interest that is perceived to exist between middle-class environmentalism, which is (at times) deeply concerned about issues of conservation, democratization, and social equity, and rural social movements, which are more centrally concerned with issues of land, livelihood, and entitlement.

In short, explanations for Thailand's environmental movement are varied and raise a number of questions regarding the factors that led to the mobilization of the 1980s and 1990s as well as the wider implications of the movement on issues of conservation and environmental policy. One question, which concerns the impact of social movements on environmental policy, is whether the mobilization of the 1980s and 1990s led to real and sustained improvements in the state's ability to regulate norms of environmental conservation and protection. A second and somewhat less tangible question concerns the impact that the movements had on the status of civil society organizations in Thailand and on the relationship between civil society and the state.

In this chapter I argue that Thailand's environmental movement provided an important vehicle through which groups previously excluded by market and state could advance a new political agenda in national and international settings. Facilitating this process was a transnational ideology (of community, self-sufficiency, and sustainability) that gave new legitimacy to social movements of different political and ideological stripes. But the environmentalism thus "constructed" was also the product of a particular historical struggle. Specifically, cycles of economic crisis and political conflict between dominant factions of commercial and military-bureaucratic power in Thailand created a new "opportunity structure" in which environmental activism could arise. The ability of the movement to challenge the authority of state and corporate power in these instances was limited in no small way by the strength and cohesion of the coalition and the extent to which the demands being made by the movement challenged fundamentally the underlying forces of neo-liberal development.

To advance this argument, I situate Thailand's environmental movement in a wider history of globalization, modernization, and agrarian change. The chapter proceeds as follows. In the first section, I trace the origins of Thailand's modern political economy and argue that processes of modernization, liberalization, and economic transformation led to new forms of environmental degradation in Thailand. The second section traces the origins and impact of Thailand's environmental movement. In the final section I argue that the oil shocks of the 1970s and the end of the Vietnam War led to a major period of social upheaval, creating new coalitions of peasants, students, and urban labour and facilitating a third phase of global environmentalism in which international NGOs and rural-urban coalitions used a new language of diversity, community, and sustainability to highlight the environmental costs of Thailand's neo-liberal boom.

COUNTING THE COSTS: MODERNIZATION, LIBERALIZATION, AND THE ENVIRONMENT

Although Siam's economic elite had certainly flirted with import substitution during the decades prior to and immediately following the Second World War, American economic assistance during the 1950s and 1960s helped to ensure that Thailand's political economy avoided the extreme forms of economic nationalism that had become so prevalent in large import-substituting economies, such as India and Brazil (Phongpaichit and Baker 1995: ch. 5; Tejapira 2006). At the end of the Second World War, Allied demands that Siam pay for its collaboration with the Japanese led to a US-brokered deal in which rice exports were provided in lieu of cash reparations. "Enforced exports" and the adoption of green-revolution technologies led to a cash-crop boom in the 1950s and 1960s. Originally confined to the fertile rice-growing plains of the Chao Phrya Valley, Thailand's green revolution very quickly spread to the northern river-valley systems, to the southern provinces, and to a far lesser extent to the semi-arid plains of the northeast. Between 1950 and 1980 the number of farm holdings expanded at a rate of 70,000 to 80,000 per year, and the rural population increased from 13.4 million to 31.8 million, most of this taking place in the north and northeast (Phongpaichit and Baker 1995: 55; Tejapira 2006).[3]

The political economy that emerged after the end of the Second World War was thus primarily an agricultural economy that thrived on exports, especially rice, but one that also established and retained a tariff regime that sheltered industrial manufacturing, providing a substantial source of government revenue.[4] Maintaining and very significantly exploiting this arrangement was Thailand's military-bureaucratic elite, which between 1957 and 1973 established near-total control over the bureaucratic state. In Bangkok and (to a lesser degree) in other urban centres, the political economy of authoritarianism entailed a long-standing—and relatively stable—arrangement whereby the military allowed commercial interests to flourish in exchange for a place on the boards of Thailand's growing commercial sector (see, especially, Akira 1996 [1989]: 42-48; Phongpaichit and Baker 1995: ch. 5).

Stable markets and favourable terms of trade generated considerable wealth for a small yet expanding class of commercial farmers. In rural areas, commercial agriculture became increasingly organized around the production, processing, and export of high-value commodities, such as rice, poultry, feedmeal, and prawns (Burch 1996; Goss et al. 2000). For those excluded or displaced by this process (i.e., those living in marginal areas or those whose lack of entitlement to land and other productive assets forced them into tenancy or agricultural labour), the political economy of liberalization and development was until the 1970s upheld and maintained by an authoritarian state that was consistently willing and able to use a combination of regulation, intimidation, and assassination to repress the organization of social movements in rural areas (Phongpaichit and Baker 1995; Tejapira 2006).[5]

By the 1960s, however, a growing population and increasing concentration of land led to conflict in the upland areas, reflecting the limitations of Thailand's land-titling system and the decline of the rural frontier. Industrialization, agricultural modernization, and an expanding population accelerated the depletion of natural resources in rural areas. Between 1961 and 1989, forest cover in Thailand was reduced from 53 to 25 per cent of all land area, leading to the commercial logging ban of 1989 (World Bank 2006). In coastal fishing, over-harvesting of marine fisheries reduced fishing yields by 90 per cent, and coastal areas were seriously degraded by the expansion of commercial fishing, shrimp aquaculture, industry, and tourism (World Bank 2006).

Thailand's post-war boom thus created an economy that thrived on a combination of agricultural exports and selective protection of import-substituting industries, especially cigarettes, liquor, and textiles. However, like many import-substituting economies, Thailand became increasingly dependent upon the ability to finance foreign imports by inflating the value of its currency and borrowing abroad. The second oil shock of 1979 revealed dramatically the internal contradictions of this "protectionist export" economy. In the words of Phongpaichit and Baker:

> The second oil crisis of 1979-80 invoked reality. The oil bill ballooned to 30 per cent of total imports in 1981 and 1982. The balance of payments lurched towards deficit. Debt service on the private and public sector foreign loans since 1975 soared ... External foreign debt as a percentage of GDP grew from virtually zero in 1973 to 21 per cent in 1980 and a peak of 38 per cent in 1986. (1995: 147)

Responding to what had become a major economic crisis, Thailand's central bank and finance ministry reined in public- and private-sector borrowing and crucially (for the liberalization that would follow) entered into a series of structural-adjustment loan agreements with the International Monetary Fund (IMF). Austerity measures were quickly followed by a decision in 1984 to devalue the Thai currency, the baht, by 14.7 per cent and to allow a further devaluation of 20 per cent over the next three years (Phongpaichit and Baker 1995: 150). Alongside policy reforms to reduce, *inter alia*, the level of import taxes, abolish a series of export taxes, and encourage foreign investment, the subsequent fall in world oil prices and a rising yen resulted in an export boom, which led to annual growth rates of 13–14 per cent (Phongpaichit and Baker 1995: 151; Tejapira 2006: 10–11).[6]

The liberalization of the export economy and the arrival of offshore (primarily Japanese) capital led to a number of structural transformations within the national economy, which have strong bearing on our understanding of the environmentalism that emerged in the 1980s and 1990s. First, the economy moved away from agriculture and into manufacturing and services. Prior to the 1984 devaluation, manufacturing in Thailand accounted for 36 per cent of national exports; by 1991, this figure had jumped to 67 per cent (Phongpaichit and Baker 1995: 152). During this same period, agriculture went from 49 to 20 per cent of all exports, echoing a wider transformation in which its share of GDP dropped from 27 to 14 per cent between 1970 and 1990 (Phongpaichit and Baker 1995: 152–53).

The impact on labour and employment was equally dramatic. Before the devaluation in 1984, a combination of economic policy, agribusiness, and private investment led to a boom in cash-crop agriculture. After 1984, the agricultural population continued to grow, but at a rate much lower than that of manufacturing. According to the government's Labour Force Survey, the annual expansion of the agricultural workforce went from 402,000 in 1980 to 125,000 in 1991; over the same period, annual additions for "non-agriculture," which includes manufacturing and services, went from 261,000 to 553,000 (cited in Phongpaichit and Baker 1995: 199). During the 1980s, the urban population expanded by roughly 4 million people (Phongpaichit and Baker 1995: 198).

The economic transformations that manifested themselves after 1984 had the combined effect of exacerbating existing rates of natural-resource depletion, and creating a new range of environmental problems arising as a result of the shift toward manufacturing and services and the (related) appearance of consumer-led demand for environmentally sensitive services, such as hydroelectricity, clean drinking water, water-treatment facilities, and access to land in urban and peri-urban areas (Arbhabhirama et al. 1988; Hirsch and Lohmann 1989; Lohmann 1991; Christensen and Boon-Long 1994; Phantumvanit and Panayotou 1990; Phongpaichit and Baker 1995; TDRI 1986; 1994).

By the end of the 1980s, concerns about the sustainability of Thailand's neoliberal boom were already being raised in a number of national and international fora. Universities and research organizations, such as the Thailand Development Research Institute and the Thailand Environment Institute, published a series of studies during this period (many of them in English), highlighting the environmental and economic costs of groundwater depletion, deforestation, fisheries depletion, and air and water pollution arising as a result of urbanization and industrial manufacturing (see, especially, Phantumvanit and Panayotou 1990; TDRI 1986; 1994). Similar concerns were raised by the national media (especially by the English-language press), by domestic NGOs, and by members of the Thai royal family.

Widespread in the literature of the time was the notion that liberalization, industrialization, and the commercialization of agriculture had combined to accelerate the rate of natural-resource depletion, a claim that has considerable merit. If we take the case of coastal fishing, for instance, we can see that the currency devaluation in 1984 had a profound impact on fisheries exports, which started to boom after 1985 (see Table 3.1).

In this instance, the combined effect of strong export markets, destructive technologies (especially trawlers and push nets), and a lack of effective environmental regulation exacerbated the depletion of Thailand's coastal fisheries. Government surveys, for instance, suggested that catch rates in the Gulf of Thailand and the Andaman Sea dropped from a high of 288 kilograms per hour in 1964 (Isvilalanda et al. 1990) to a range of roughly 55 kilograms per hour in the late 1980s.[7] Beyond overfishing, an equally serious problem involved the destruction of coastal ecosystems, which led to widespread conflict between small-scale inshore fisheries and larger offshore commercial fleets (Johnson 1997; TDRI 1998).[8]

By the mid-1990s, the globalization of Thailand's political economy had therefore led to the intensification of environmental degradation in resource-dependent industries, such as coastal fishing and commercial agriculture, and the appearance of new environmental problems, such as air and water pollution (Krongkaew 1995; Phongpaichit and Baker 1995). As Phongpaichit and Baker have argued, the economic transformations of this period led to at least two new forms of rural protest, both of which have a bearing on our understanding of environmental movements in Thailand (2001). One was a backlash from commercial farming interests, whose traditional security had been undermined by falling prices, rising debt, and structural adjustment policies, which reduced support for commercial agriculture. For large farmers, devaluation made their exports more competitive on world markets, but it also increased the price of imported products, especially oil- and petroleum-based fertilizers.

A second and very different form of protest came from the economic and ecological margins of the rural frontier. Included here were the hill tribes, small-scale farmers, and others, whose lack of entitlement to land and other productive assets forced them into tenancy, peasant agriculture, and/or agricultural labour (Phongpaichit and Baker 1995; Tejapira 2006). Political mobilization during this period was directed primarily toward the government's promotion of urban-industrial power generation (through the construction of hydroelectric dams), forced resettlement, and the resolution of resource conflicts involving villagers and commercial interests, especially eucalyptus tree plantations and prawn aquaculture (Hirsch and Lohmann 1989; Lohmann 1991; Hirsch 1993; Vandergeest et al. 1999).

During the 1980s and early 1990s, political responses to processes of environmental degradation and economic marginalization moved beyond primarily localized struggles over access and entitlement in rural areas to a new realm of national and international struggle over economic/environmental policy and constitutional rights of freedom, association, and access to natural resources in rural areas. Facilitating this process were three important factors: a growing level of interest among international environmental NGOs in Thailand, the Earth Summit of 1992, and a series of political and economic crises that created new opportunities for social mobilization in Thailand.

TABLE 3.1
FISHERIES EXPORTS AND IMPORTS (1981–1991)

Year	*Exports (metric tons)*	*Imports (metric tons)*
1981	320,325	47,174
1982	316,679	46,215
1983	344,899	58,942
1984	411,722	119,064
1985	466,219	152,707
1986	602,486	268,089
1987	663,650	227,327
1988	798,579	347,666
1989	875,293	455,755
1990	904,973	507,537
1991	902,575	663,971

Source: Directorate General of Fisheries and Department of Business Economics, cited in Infofish 1991: 28.

THE GREENING OF THAI POLITICS: ENVIRONMENTAL POLITICS AND POLICY IN THAILAND

Notwithstanding the "domestic" conflicts and transformations that were beginning to manifest themselves around this time, there appeared on the international agenda a number of transnational ideas and channels of funding that had new bearing on rural politics and urban-rural coalitions in Thailand. The Earth Summit of 1992, which led to framework agreements on biodiversity and climate change and (more importantly) new channels of international development financing, created a political climate in which new forms of rural organization and protest could arise (see below).[9]

During the 1980s, a number of international ENGOs, including Greenpeace and the World Wide Fund for Nature, became actively involved in Thailand, using a combination of rural development projects and international advocacy campaigns to highlight the environmental costs of Thailand's neo-liberal boom. Many of the projects that were supported by these and other ENGOs entailed development strategies that encouraged community-based management of local resources (Johnson 2002). Similar initiatives were supported by other international development agencies, including the World Bank, the Japanese International Cooperation Agency (JICA), the Food and Agriculture Organization (FAO), and the Manila-based International Center for Living Aquatic Resource Management (ICLARM).[10]

By the mid- to late 1980s, international NGOs and aid agencies had therefore facilitated the dissemination of a very wide range of images and policy ideals that embraced strong notions of community, sustainability, and environmental conservation. At the root of this thinking was a wider intellectual discourse, which embraced the idea that natural resources could be managed more effectively by local rural communities (see, especially, Agrawal and Gibson 1999). For domestic NGOs whose activists eschewed the hierarchy, bureaucracy, and authoritarian legacy of Thailand's central government, the emphasis on decentralization and local resource management resonated strongly with their own philosophy about rural self-sufficiency and local economic development, which embraced the notion that rural traditions and village organization could provide a strong element of resource husbandry and conservation. Moreover, the language of community and sustainability provided new channels of access to the international aid industry, whose institutional and financial support could advance the work of Thailand's NGOs.

It was during this time that Thailand's environmental movement achieved some of its most important victories, including the commercial logging ban (in 1989), the National Environmental Quality Enhancement and Protection Act (1992), the Community Forest Bill (which was originally drafted by the Royal Forestry Department in 1990), and the 16th Constitution, which in 1997 granted rural "communities" a constitutional right to manage and conserve natural resources. Particularly notable was Section 46 of the Constitution, which stipulated that

> Persons so assembling as to be a traditional community shall have the right to conserve or restore their customs, local knowledge, arts or good culture of their community and of the nation and participate in the management, maintenance, preservation and exploitation of natural resources and the environment in a balanced fashion and persistently as provided by law. (RTG 1997)

The inclusion of a clause that specifically recognized the right of local communities to manage natural resources was an enormous victory for Thailand's environmental movement (or at least for those who supported community-based management).[11] However, on the surface it also challenged the ability of commercial interests to exploit natural resources in rural areas.

To what extent, however, did the "victories" of this period influence the quality and long-term impact of environmental policy in Thailand?

In the period leading up to the Earth Summit, the Government of Thailand introduced and passed a series of legislative acts, which (in principle) expanded the ability of government to regulate environmental policy in Thailand. Principal among these (and summarized in Table 3.2) were the National Environmental Quality Enhancement and Protection Act (1992), the Energy Conservation Promotion Act, and the Conservation and Preservation of Wildlife Act. After the Earth Summit, the Thai government re-affirmed its commitments to the Rio Declaration and to Agenda 21, by signing and ratifying the UN Convention on Biological Diversity of 1992, the Cartagena Protocol on Biosafety (2000), the UN Framework Convention on Climate Change of 1992, and the Kyoto Protocol of 2002.

In short, Thailand's response to the Earth Summit and to the environmentalism of the 1980s and 1990s entailed the establishment of a number of agencies and institutions which in theory expanded the ability of the state to regulate and enforce norms of environmental protection and conservation.

However, critical assessments of environmental policy during this period suggest that inadequate resources and overlapping authority consistently undermined the ability of government to enforce these regulations (see, for instance, Christensen and Boon-Long 1994; Christensen and Rabibhadana 1994; Johnson 1997; Pichyakorn 2003). Indeed, if we consider some of the areas over which the Office of the National Environment Board has authority, we can see that its impact on actual policy has been modest at best. The World Bank, for instance, estimates that a third of Thailand's surface water is of "poor quality" (2006). The World Health Organization reports that levels of particulate matter and ground-level ozone are in excess of national environmental standards (2004). In the same report, it highlights the pollution of drinking water, especially in the Chao Phrya valley, and the spread of water-borne diseases, especially diarrhoea. Likewise, the Division for Sustainable Development in the United Nations Department of Economic and Social Affairs (UNDESA, CSD-14/15, 2005: 11) cites statistics suggesting that levels of dust and carbon monoxide are also well in excess of national environmental standards.

TABLE 3.2

MAJOR ENVIRONMENTAL LEGISLATION IN THAILAND

Institution/Act	*Year*	*Description*
National Environmental Quality Enhancement and Protection Act	1992	The Minister of Science, Technology and Environment with the approval of the Office of the National Environment Board has the power to regulate government agencies, public sector entities, and private companies. Under this jurisdiction, the Office of Environmental Policy and Planning and the Expert Review Committee have the power to request and review environmental impact assessments for consideration and approval. The legislation has a specific chapter for water pollution. All power plants were made subject to new emissions standards. Additionally all plants were required to implement emission control measures to comply with national ambient air quality standards (Thanh and Lefavre 2000: 138; http://www.unescap.org/DRPAD/VC/conference/bg_th_47_eia.htm)
The Energy Conservation Promotion Act	1992	Promotes energy conservation discipline and investment in energy-efficient technologies for industry (UNDESA, CSD-14/15 'Energy' 2005: 25) (WHO states the policy dates are 1997-2017. World Health Organization 2004: 5)
The Conservation and Protection of Wildlife Act	1992	Conserving waterways and welfare of wildlife (Pichyakorn 2003: 239)

Efforts to resolve problems and conflicts arising as a result of natural-resource shortages have to date suffered a similar fate, resulting in large part from overlapping authority and insufficient resources. As Pichyakorn has argued, Thailand has 32 laws relating to water, which are interpreted and enforced by 40 government agencies in nine different ministries (2003: 235). Along similar lines, the regulation of coastal aquaculture in Thailand is managed by multiple agencies, including the Department of Fisheries (which issues aquaculture licenses), the Land Department (which issues, defines, and recognizes land title), the Royal Forestry Department (which controls conservation and economic activity in mangrove forest areas), and the Ministry of Science, Technology and the Environment, which can in principle investigate and sanction the pollution of groundwater sources. Beyond legislation that imposed an outright ban on commercial activities, such as the commercial logging ban, the legislation that was passed during this period was defined with sufficient breadth and ambiguity that it allowed the continuation of business as usual in many sectors of the economy.

The battles being waged in Rio and on the streets of Bangkok thus embraced the language of environmentalism, and induced the state to introduce acts and institutions whose principal aim was to improve the protection and conservation of natural resources. However, the lack of coordination and insufficient investment in the enforcement of regulatory capacity suggests that the Thai state was only so willing to enact and enforce legislation that would challenge the interests of domestic and foreign capital, which by this time had become increasingly dependent upon the globalization of international trade and investment (López and Mitra 2000; Tejapira 2006).

In short, the environmental achievements of Thailand's environmental movement were limited at best. To what extent, however, did they influence the relationship between civil society and the state? And to what extent do they constitute a substantive change in Thai politics? In the following section, I argue that political struggles between dominant factions of commercial and military-bureaucratic power in Thailand created new spaces for political mobilization in Thailand, in which environmental and other forms of rural activism could arise. In so doing, I argue that the conditions that allowed the environmental movements of the 1980s and 1990s to happen in Thailand were directly dependent upon the forces and factors that led to the re-establishment of popular democracy in Thailand: economic crisis and neo-liberal reform.

THE NEO-LIBERAL ERA: ECONOMIC CRISIS AND POLITICAL REFORM

During the Cold War, US-led efforts to contain the spread of communism in Southeast Asia facilitated the conditions for authoritarian rule (Morell and Samudavanija 1981; Phongpaichit and Baker 1995; Tejapira 2006). Between 1957 and

1973, Thailand's military leadership suspended the National Assembly and outlawed political parties, farmers' organizations, and trade unions.[12] In rural areas, local administrative organizations (village councils, sub-district councils, district and provincial offices) provided an important means by which the state could monitor and control Thailand's rural population (Turton 1989; Hirsch 1990; Arghiros 2001; Tejapira 2006). Logistically, this was reinforced by the monthly meetings that village and sub-district headmen were required to attend, and by the Ministry of Interior bureaucrats who coordinated and controlled the provincial, district, and sub-district offices (Nelson 1998; Arghiros 2001).

The ability to maintain social order in rural areas was therefore predicated upon the power of Thailand's military and (by extension) that of the United States, the latter of which by the early 1970s had become increasingly uncertain. The Nixon Administration's decision to abolish the gold standard in 1971 and OPEC's decision to punish the United States, Western Europe, and Japan for supporting Israel during the Yom Kippur War in 1973 led to a major crisis in the world economy. For Thailand, the first oil shock increased the price of oil imports and reduced the demand on international markets for agricultural and other primary exports, cutting in half its terms of trade with the international economy (Phongpaichit and Baker 1995: 146). As was the case for many oil-dependent developing countries, geo-political instability in the early 1970s very quickly led to domestic upheaval and change.

During the 1960s and 1970s, peasant organizations became increasingly active in Thai politics, demanding, *inter alia*, formal recognition of land rights, improvements in rural tenancy relations and government support for paddy prices, which after the oil shocks of 1973 became increasingly unstable. In the early 1970s, it was estimated that the number of "violent incidents," involving clashes between government and insurgents, increased from one in 1965 to almost 200 per year in 1972 (Phongpaichit and Baker 1995: 294). According to official estimates, insurgents during this time were "active" in 41 of the kingdom's 71 provinces.

The government's response to rural insurgency entailed a combined application of development assistance and armed repression. Development projects were directed principally toward the rural institutions (village, district, and province) which were established during the colonial period. Counter-insurgency efforts, on the other hand, were directed principally toward the Communist Party of Thailand (CPT) and its military wing, the People's Liberation Army of Thailand. Substantially reinforcing the military's ability to undertake this effort was the US military, which by this time had become fully engaged in Southeast Asia: between 1951 and 1972, the Thai military received more than $1 billion from the US government; between 1964 and 1969, the number of US Army and Air Force personnel stationed in Thailand increased from 6,300 to 45,000 (Phongpaichit and Baker 1995: 276).

In 1973, a loose coalition of peasants, university students, intellectuals, labour, and business joined forces to topple the ruling Thanom-Phrapas regime, ending (temporarily) a military dictatorship that had controlled the state since 1957. The origins of the uprising were ostensibly a reaction to the decision of a Bangkok university to expel nine students for writing a "thinly veiled attack" on the ruling junta (Phongpaichit and Baker 1995: 302). However, as the conflict began to unfold, it quickly became clear that the views of Thailand's intelligentsia reflected wider feelings of anger and resentment within society, which included not only the traditional interests of peasants and labour but also those of an emergent class of commercial interests. Street protests ensued, leading to a series of violent confrontations (which claimed the lives of more than 100 university students) and a royal intervention that culminated in the exile of the military junta.

Between 1973 and 1976 the "revolutionary" government enacted legislation that favoured small-scale farming and urban labour, and introduced a series of rural development programs aimed at reducing debt and improving formal access to land. In 1974, peasant leaders in northern Thailand formed the Peasants' Federation of Thailand (PFT), an umbrella group of peasants and tenants that lobbied government to implement rent controls, land ceilings, and other forms of agrarian reform legislation. Crucially for the rural-urban coalitions that would later emerge, as we shall see below, the government also supported a series of "rural education programs" in which university students were granted funds to "educate peasants on their rights and duties in a democratic system" (Phongpaichit and Baker 1995: 303).

Insofar as it entailed the involvement of urban labour, intellectuals, and the rural poor, the 1973-76 period signified an important victory for the Left in Thailand. However, the conditions that facilitated this victory were necessarily dependent upon a coalition between groups that favoured a more radical redistribution of wealth and power in society and one that stood largely in defence of the status quo (Tejapira 2006). As Phongpaichit and Baker have argued, the ability of Thailand's intellectuals and university students to effect change during this period required the approval and support of the monarchy and of certain factions within the military (1995: 302).

Shortly after the student-led victory of 1973, internal struggles within the communist movement and further geo-political instability led to a right-wing backlash, which exposed the contradictions of this temporary alliance. In 1975, the Ford Administration fulfilled Richard Nixon's promise to pull US troops out of Vietnam, reducing substantially the level of military spending in Thailand. In a context of mounting insurgency and industrial action, the loss of military funding gave right-wing forces (which included the "traditional" military-business alliance) an opportunity to re-consolidate political power. Between April and August 1975, right-wing forces orchestrated a campaign of intimidation and violence, which led to the assassination of 18 leaders in the PFT (Phongpaichit and Baker 1995: 307-08). By October 1976, clashes between police and demonstrators had claimed the lives of more than 1,300 civilians, leading to a new period of social repression and military rule.

The resurgence of right-wing and military power in Thailand exposed dramatically the limitations of the left-wing coalitions that had temporarily controlled the state in 1973. At the heart of the right-wing crackdown was an effort on the part of the traditional military-business alliance to re-establish the conditions under which it had thrived since the 1950s (Tejapira 2006). The anti-communist agitation that led to the end of civilian rule was orchestrated by (certain) military and business leaders. But it also entailed the involvement of farmers, urban professionals, and even students whose participation had been so instrumental in the civilian victory of 1973. The shifting loyalties of the urban professional class suggest that the urban and rural class coalitions that formed during the 1970s were at most a marriage of convenience, whose ability to affect change was limited by class loyalties and appeals (usually on the part of the right) to nationalist sentiment (Reynolds 2002 [1991]; Tejapira 2006).

However, the geo-political and economic changes that had led to the upheaval of 1973 had set in motion a series of economic and political transformations that would end forever the military-business alliances of the post-war era. The oil shocks of the 1970s and the neo-liberal reforms of the 1980s had a profound impact on Thai politics. First, the structural shift away from agriculture and into industry led to a more subtle political transformation, in which commercial interests began to assert themselves in national politics and in direct competition with Thailand's military-bureaucratic elite. Central to this transformation was the newfound ability of business to obtain access to foreign capital, which became increasingly available after the Plaza Accords of 1985 (which forced Japan to inflate the value of the yen by selling its reserves of US exchange) and after a series of policy decisions in the 1990s that deregulated Thailand's financial sector, allowing domestic banks and financial institutions to borrow on international markets. The emergence of a new and independent class of commercial interests soon led to conflict over economic policy and political power.

Second, the neo-liberal era facilitated the rise of a new class of provincial politicians and entrepreneurs. Until the early 1980s, commercial power was largely concentrated in Bangkok. After liberalization, the establishment of export-processing zones in outlying areas led to a decentralization of economic power in Thailand. Scholars of this period (e.g., Phongphaichit and Baker 1995; Robertson 1996) document the rise of the *jao pho* ("local godfather"), whose newfound wealth reshaped the nature of politics and elections in rural areas. Prior to this period, elections offered important opportunities to extend networks and exert political influence by trading votes for rural "services," such as roads or irrigation canals (Robertson 1996; Nicro 1993). After the 1980s, such practices continued, but they were increasingly supported and sustained by the private capital of local godfathers, in the form of either discretionary funds or ministerial expenditure (Robertson 1996; Nicro 1993; Tejapira 2006).[13]

Finally, economic globalization led to new forms of political liberalization and political action. Benefiting from a relaxation of authoritarian rule and an international aid regime that had become increasingly receptive to the idea of channelling aid through NGOs, Thailand's NGO sector ballooned in the 1980s (Phongpaichit and Baker 1995: 384-89; Hirsch 1993). Many of these NGOs were staffed by former activists and members of the CPT, whose outlook on life was decidedly hostile to the forces of global capitalism and state-led development.

Throughout the latter half of the 1980s the business-dominated Cabinet purged military officials from provincial governorships and key ministerial posts (Laothamatas 1992; Phongpaichit and Baker 1995). Of particular importance in this respect was the struggle over the power of appointment within the National Assembly. Before 1988, the head of the Senate reserved all rights of appointment and nomination within the House of Representatives. In effect, this gave the military the power to decide all ministerial posts within Cabinet, including the prime minister. After 1988, Members of Parliament enacted new legislation that made the Speaker of the National Assembly the President of Parliament, shifting the power of nomination from the Senate (which was dominated by the military) to the House of Representatives (which was controlled by elected Members of Parliament, most of whom were businessmen) (Phongpaichit and Baker 1995: 349-51; Laothamatas 1992).

Struggles to control powers of appointment led to another military coup, in which a faction of military officers (calling itself the "National Peace Keeping Council") (NPKC) used allegations of corruption and "excessive wealth" to oust Prime Minister Chatichai Choonhaven in 1991, provoking yet another series of public rallies and demonstrations. Echoing the experience of 1973, street protests turned violent, and in May 1992 a military crackdown claimed the lives of more than 100 pro-democracy activists. Once again, the King was forced to intervene, negotiating the removal of the NPKC and a return to civilian rule. In the wake of the crisis, the King asked another "business-friendly" politician—Anand Panyarachun—to form a "caretaker" government that would facilitate the transition to civilian rule.

In rural areas, peasant leaders used the instability of the interim period to advance a new political agenda based on principles of non-violence and democratic change. In June 1992, a group of villagers from the impoverished northeast region of Thailand escalated their challenge to a government resettlement program by marching from the city of Khon Kaen to Bangkok (Baker 2000: 18). The march attracted widespread attention and led to a new period of rural mobilization. Using demonstrations, blockades, and an increasingly extensive international campaign, rural-urban coalitions used images of community and "alternative development" to push their (varied) interests onto the policy agenda and into the national press (Baker 2000; Hirsch 1993; Rigg and Nattapoolwat 2001). By 1994, there were over 1,000 recorded protests in Thailand, culminating in the Assembly of the Poor, a coalition of NGOs, farmers, peasants, agricultural labourers, village leaders, and rural politicians that descended on the streets of Bangkok in 1995.

As Chris Baker has argued, the Assembly of the Poor marked a transitional moment in Thai politics, whereby "traditional" struggles over land, labour, and access to natural resources were reformulated in a way that could embrace the changing nature of Thailand's rural political economy (2000). Capturing images of community, sustainability, and self-sufficiency, the environmentalism of the Assembly provided a "common ground" that could appeal to the concerns of middle-class activists and at the same time pursue "traditional" rural concerns about land, labour, and livelihood (Baker 2000: 6, citing Praphat Pintoptaeng). Of the 125 problems that were presented to the government in 1997, 93 concerned questions of access to land and forests, 16 were about dams, 8 were about slums, and one each was devoted to alternative agriculture, health and safety in the workplace, and small-scale fishing (Baker 2000: 16).[14] Underlying the environmentalism that emerged during this period was an ideological orientation that was decidedly hostile to the forces of economic liberalization and global trade.

The pro-democracy movements of the early 1990s therefore created a political climate in which multiple forms of protest and political agitation could arise. After the events of 1992, the Thai government was under immense pressure to show that it was both willing and able to reform the political system and to tackle corruption. Similar to the rural-urban coalitions of the 1970s, the environmental movements of the 1980s and 1990s became highly dependent on a series of strategic alliances between rural organizations and urban activists. However, unlike the early 1970s, political pressure to implement change came not only from "traditional" constituencies of farmers, generals, intellectuals, and business, but also from a new and increasingly powerful lobby of international investors, bond-rating agencies, and multinational corporations.

As Kasian Tejapira has argued, the forces driving Thailand's political reform process were rooted in two different and "at times conflicting" strands of social mobilization (2006: 20-22). One was an effort on the part of Thailand's commercial business interests to assert themselves politically and to consolidate the liberalization of Thailand's economy. A second—which Tejapira calls "the democratic-reform strand" (2006: 21)—involved the mobilization of farmers, NGOs, and social activists, whose principal efforts were aimed at promoting "people's politics" and addressing government corruption and localized struggles over land and resources.

Until 1997, democratic reform efforts were directed principally toward the establishment of liberal democratic principles in national and sub-national political institutions. In 1994, for instance, the government passed the Tambon Restructuring Act, which delegated modest powers to a new class of elected sub-district councillors (Arghiros 2001). In the latter half of the 1990s the government also started drafting a new constitution (the Kingdom's sixteenth since 1932), and a series of reforms aimed at fighting corruption and enhancing public participation in national and local government. Inviting input from academics, politicians, and civil society organizations, the constitution aimed to address the problems of vote buying and corruption that had become so prevalent during the neo-liberal era (Tejapira 2006).

As Tejapira points out, the forces driving Thailand's liberal democratic reform process were limited in terms of both their collective strength and their political agenda (2006: 23). The urban-rural coalitions that emerged during this period were large, but they were also unwieldy. Embracing a discourse that "tended to de-emphasize class and class conflict," Tejapira argues (2006: 22), they lacked an ideology and agenda that could challenge the forces of neo-liberalism and effect substantive change. Far more important, he argues, was the financial crisis of 1997.

By the end of 1997, the value of Thailand's currency had dropped to an all-time low of 59 to the US dollar, precipitating a major crisis for the banks and finance companies whose foreign loan commitments had doubled in the space of three months. The resulting credit squeeze led to the suspension of 58 financial institutions, 7,000 documented bankruptcies, and over 55,000 bankruptcy claims (Phongpaichit and Baker 2001). Toward the end of 1997, Thailand was forced to negotiate a US $17-billion bail-out loan from the IMF, which entailed new restrictions on government spending and constitutional reform (Tejapira 2006; Hewison 2005).

As Tejapira has argued, the neo-liberal reforms of the late 1990s engendered strong feelings of anger and resentment toward the IMF, the WTO, and the "neo-liberal globalization project" (Tejapira 2006: 23). Between 1997 and 1998, Thailand's GDP shrank by 10.5 per cent, and an estimated one million workers were put out of work (Tejapira 2006: 23). By the end of 1998, the combination of high interest rates, ongoing business failures, and the deregulation of the banking sector led to a "fire sale" of domestic assets to primarily foreign capital (Hewison 2005).

In the wake of the crisis, social activists started using images of community, self-sufficiency, and economic nationalism to justify a more disengaged relationship with the world economy. Chattip Nartsupha, one of Thailand's most outspoken social critics, argued that the 1997 economic crisis justified a need to

> ... look again at the local peasant community as the basis of the economic life and culture of the people, and to support a new direction in national development, beginning with strengthening from the bottom up, from the level of the family, the local community, and the agrarian economy. (quoted in Rigg and Nattapodwat 2001: 956)

It was in this context of political and economic crisis that Parliament met to consider the newly drafted constitution. Facing the wrath of NGOs, academics, activists, big business, international investors, and the IMF, whose US $17-billion bail-out package was now on the books, Thailand's parliamentarians were under immense pressure to pass the constitution and to demonstrate that Thailand was willing and able to reform its political and financial sectors. In the fall of 1997, the National Assembly passed into law a constitution that would enhance the power and independence of the judiciary, improve transparency and accountability during elections, and introduce new rounds of devolution.

In one respect, the constitutional changes of the late 1990s were deeply empowering for poor and politically marginal groups in rural areas. In my own field research, which was conducted during this period (Johnson 2001), villagers in Southern Thailand were (impressively) well apprised about laws relating to forests and fisheries, and about new changes in the 1997 constitution. Moreover, the strategic use of this information and knowledge led to new forms of social networking and political action that had direct bearing on national lobbies, such as the Assembly of the Poor (Johnson 2001; Baker 2000; Johnson and Forsyth 2002).

That having been said, we should not lose sight of the ways in which the Thai state was able to use the discourse of community and environmental conservation to appease *and defuse* more radical calls for self-determination and land redistribution in rural areas (Phongpaichit and Baker 1995). Indeed, what makes the reform process so interesting is the Thai government's apparent lack of commitment to any substantive political change (Tejapira 2006). Designed to democratize and empower sub-district (*tambon*) councils, for instance, the 1994 Tambon Restructuring Act and the 1997 Constitution failed to remove any substantial power from the non-elected, centrally appointed Ministry of Interior bureaucracy (Arghiros 2001). Likewise, efforts to encourage customary fishing rights and community forestry failed to overcome the political and commercial interests that thrive on the status quo (Johnson and Forsyth 2002). This became apparent in 1998 when commercial vessels in Songkhla province used the 1997 Constitution to defend their "community's" right to manage *and exploit* anchovy stocks in the Gulf of Thailand. The Thai state thus embraced the discourse of decentralization, community, and sustainability to legitimate the very same instruments of authority and control that it used to maintain order and stability in rural areas (cf. Hirsch 1990; Vandergeest 1991; 2003).

CONCLUSION

Theories of ecological modernization propose strong and multi-directional links between popular democracy and environmental conservation. Ulrich Beck for instance, has argued that environmentalism is "radically" democratic in the sense that it challenges the ability of science, bureaucracy, and the scientific establishment to allocate risk within society, opting instead for popular and contested determinations of what constitutes "acceptable risk" (1992). Along similar lines, Dryzek et al. (2002) have argued that environmental movements may influence the political orientation of the state but that states structure the political conditions under which these social movements may arise.

In the preceding analysis I portray a state that (at the best of times) used metaphors of community and participation and (at the worst) more violent means to establish and maintain a political system of appropriation and control in rural areas. I also portray a civil society that promoted and, at times, exploited these metaphors to widen, or at least maintain, a basic level of independence from the social intrusions that have defined the more authoritarian periods of Thailand's political history. The environmentalism that emerged during the 1980s and 1990s provided an important vehicle through which groups previously excluded by market and state could advance a new political agenda in national and international settings. Facilitating this process was a transnational ideology that gave new legitimacy to environmental movements (of different political and ideological stripes) and a democratic transition that began to manifest itself in the 1970s. Prior to this period, political stability in rural areas was largely the result of a macro-economic policy that had been highly favourable to large commercial farmers and a political strategy that was consistently willing to repress various forms of agrarian unrest.

What makes the environmentalism of the 1980s and 1990s so striking is that the aims of the movement (land rights, access to natural resources, and economic security) were not substantially different from those of earlier agrarian struggles in the 1970s (Baker 2000). Why then was the Thai state so willing to tolerate rural activism on "environmental issues" when it had so violently opposed very similar calls for land redistribution and agrarian reform in the past?

In this concluding section I offer two possible explanations. First, the liberalization of Thailand's political economy led to new forms of power within civil society and a new type of relationship between civil society and the state. Violent struggles over basic freedoms of association and speech redefined the moral authority of government, which created a political climate in which "new" rural social movements could thrive. Second, transnational environmental discourses, particularly ones concerning community, self-sufficiency, and sustainability, re-shaped the aims and ideologies of rural social movements, connecting them with a wider circle of allies and providing access to a new and substantial source of international development financing. The argument here is not that international NGOs like WWF and Greenpeace were all that powerful, but that the environmentalism thus created produced a discourse and an agenda that was less threatening to commercial and military power in Thailand.

This raises a second question. If the agenda was less threatening, does this not suggest that the movement failed to challenge the forces that drive environmental degradation and conflict in Thailand? The achievements of this period (e.g., the commercial logging ban, the National Environmental Protection Act, the cancellation of Nam Choan, and, especially, the 1997 Constitution) were substantial, and they should not be easily dismissed. However, it is difficult to disregard the fact that the Constitution, the Community Forest Bill, and even the commercial logging ban contained sufficient breadth and ambiguity to allow the continuation of many environmentally (and socially) harmful practices. Commercial logging interests accepted the logging ban because they could exploit other (more substantial) forests in Laos and Myanmar. The Constitution's recognition of community rights to manage local resources was a victory for some of Thailand's NGOs, but it also allowed commercial trawlers to plunder Thailand's coastal areas.[15]

A final—and related—question therefore concerns the nature and long-term viability of the class coalitions that led to the environmentalism of the 1980s and 1990s. Interpretations of this period often attribute environmental "victories" to the ability of farmers, social activists, intellectuals, and other "civil society leaders" to mobilize large numbers of people during highly visible political campaigns. Tracing the historical mobilization of environmental politics in Thailand, we can certainly associate the environmental victories of this period with large rallies involving thousands of people. However, to suggest that large demonstrations and images of community would provide a common ground on which civil society organizations (of very different interests and persuasions) could form a lasting coalition understates a fundamental contradiction that exists between an urban social movement whose economic needs and cultural experiences are rooted in the material consumption of industrial society, and a primarily rural constituency whose marginalization is directly the result of these patterns of production, consumption, and exchange.

Unlike the events of 1973, which were primarily the result of student unrest and agrarian distress resulting from an acute economic crisis, the violence of 1992 was more directly the result of a structural conflict between the military and a newly emergent business class of politicians (Laothamatas 1992; Phongpaichit and Baker 1995; Tejapira 2006). The pro-democracy movements and the political reform process that emerged in the wake of political and economic crisis created the conditions under which urban-rural coalitions could exert new pressure on government. However, the coalitions that thrived in this context also entailed a "marriage of convenience," in which post-industrial struggles against materialism and late industrialization could co-exist with traditional agrarian struggles over land, labour, and livelihood. Coexistence in this instance entailed an agenda that could appeal to large numbers of urban and international environmental activists. Out of political expediency, the agenda that was thus produced avoided more "radical" and class-specific aims, such as redistributive land reform and the enforcement of health and safety standards (Tejapira 2006).

Whether these processes and transformations constitute a "victory" or a "failure" for Thailand's democracy is of course a matter of interpretation and debate. As Tejapira (2006) has pointed out, the political reform process of the 1990s was as remarkable for its inability to undermine or address the power and inertia of the bureaucratic "electocracy" as it was for the (re-)establishment of constitutional law. That having been said, it is difficult to imagine the battles and agendas of the 1980s and 1990s being waged on the streets of Yangon or Phnom Penh or, for that matter, Bangkok before 1980. In this respect, and notwithstanding the anti-democratic tenor of the current regime, the political transformations of the late twentieth century (of which Thailand's environmentalism was part) must surely constitute a substantive opening of social and political life.

Notes

1 Note that this chapter was written before the military coup in September 2006. The author would like to thank Jordi Díez and two anonymous reviewers for their comments on earlier drafts of this chapter and Trevor Bennett for providing sources and statistics concerning environmental degradation and legislation in Thailand. Any errors of interpretation or analysis are of course my own.

2 Constructivist interpretations in the social sciences question the assumption that social institutions, such as societies, communities or nations, exist independent of the ways in which people interpret, define, and "construct" these realities through discursive practices. A classic treatment of these issues in sociology is *The Social Construction of Reality* by Peter Berger and Thomas Luckmann (1966), which stands alongside the work of post-modern theorists, such as Jacques Derrida and Michel Foucault, whose work is reviewed and subjected to critical scrutiny in *the End of Development*, by Trevor Parfitt (2002).

3 Although some of this growth can be attributed to natural rates of increase, the vast majority were drawn into the upland areas from other parts of the kingdom (Phongpaichit and Baker 1995: 55).

4 It is estimated that import tariffs accounted for 30 per cent of the government's total tax revenue in 1971 (Phongpaichit and Baker 1995: 145).

5 In this respect, the regulation of land was particularly important. Toward the end of the nineteenth century, the rise of an international rice economy and an increasingly profitable market in land encouraged the creation of a formal land-rights system, which could protect the investments of land-holding families and an incipient bourgeoisie (Feeny 1982; Ingram 1955; Phongpaichit and Baker 1995). However, because land was abundant and state resources scarce, peasants had little incentive to pay the requisite land taxes (Vandergeest and Peluso 1995: 404). Despite repeated attempts to register non-titled areas, the Thai state failed to keep pace with migration and an expanding population, exacerbating conflicts over land and natural resources in rural areas (Hirsch and Lohmann 1989; Hirsch 1993; Christensen and Rabibhadana 1994: 646; Vandergeest and Peluso 1995: 402-07, 411-13; Sato 2000).

6 Note that Japan's decision to raise the value of the yen in 1985 was a direct result of American efforts to redress its trade imbalance with Japan.

7 Catch rates can be calculated in a number of ways. One very basic way is to fish for a set period of time and compare the catch with landings captured (using the same technology and same period of time) in the past. A more sophisticated method is to calculate all of the costs of a fishing boat (e.g. fuel, nets, etc.) and compare them with the total value of the catch. The value of using this measurement is it illustrates change in economic value over time. The findings from this survey were generated using the former methodology.

8 Similar conflicts were widely documented in the 1980s and early 1990s involving a series of disputes over water between upland-shifting cultivators and lowland agriculturalists, the construction of hydroelectric dams and large-scale diversions of water to serve the needs of commercial agriculture and urban populations in the Central Plains and the northern river valleys.

9 Of course the ideas that gave rise to sustainable development long pre-dated the Earth Summit of 1992. Although the Club of Rome report highlighted problems of sustainability and carrying capacity, it was not until the publication of the Brundtland Commission report, *Our Common Future*, in 1986 that sustainable development emerged as an international source of political (and financial) mobilization.

10 Throughout the 1990s, the World Bank supported a series of studies and initiatives aimed at encouraging community-based management in Thailand (Midas Agronomics 1995). Under the auspices of the Bay of Bengal Programme (BOBP), the FAO supported a series of projects, consultations, and studies to explore

the viability of encouraging community-based resource management in Southern Thailand (Chong et al. 1998). In 1998, JICA initiated plans to study the viability of supporting a fishing rights initiative in Southern Thailand. At a regional level, the International Center for Living Aquatic Resource Management (ICLARM) conducted workshops and published a quarterly journal, promoting a more decentralized style of coastal resource management in the Southeast Asian region (see ICLARM 1999).

11 To suggest, however, that all of Thailand's NGOs supported community-based management of natural resources would be misleading. For an analysis of the various differences and conflicts that emerged among NGOs over the drafting of Thailand's Community Forest Bill, see Johnson and Forsyth (2002).

12 The ban on political parties was lifted in 1968.

13 Kasian Tejapira has described the political system of the 1980s and early 1990s as an "electocracy perched on top of a centralized bureaucratic state," involving principally the national electorate, local canvassers, party factions, and the Cabinet (2006: 14).

14 Although subsequent governments would later renege on the agreements achieved during this process, the Assembly was able to force their demands through to a cabinet ratification in 1997. Centrally involved in this process were NGOs, academics, and other civil society representatives who "served as interpreters, sometimes translating between local dialect and central Thai, more often translating between the style of officialdom and the style of the village" (Baker 2000: 22).

15 Here it is vital to stress that commercial logging and fishing in Thailand were producing primarily for the export market, suggesting that (irrespective of the transnational coalitions that formed during this period) the globalization of Thailand's political economy made it increasingly difficult to regulate and protect the environment.

References

Agrawal, Arun, and Clark Gibson. 1999. "Enchantment and Disenchantment: The Role of Community in Natural Resource Conservation." *World Development* 27(6): 629-49.

Akira, Suehiro. 1996 [1989]. *Capital Accumulation in Thailand: 1855-1985*. Bangkok: Silkworm Books.

Arbhabhirama, Anat, Dhira Phantumvanit, John Elkington, and P. Ingkasuwan. 1988. *Thailand: Natural Resources Profile*. Oxford: Oxford University Press.

Arghiros, Daniel. 2001. *Democracy, Development and Decentralization in Provincial Thailand*. Richmond, UK: Curzon Press.

Baker, Chris. 2000. "Assembly of the Poor: The New Drama of Village, City and State." *South East Asia Research* 8(1): 1-30.

Beck, Ulrich. 1992. *Risk Society: Towards a New Modernity*. London: SAGE Publications.

Berger, Peter, and Thomas Luckmann. 1966. *The Social Construction of Reality: A Treatise in the Sociology of Knowledge*. London: Penguin Books.

Burch, David. 1996. "Globalized Agriculture and Agri-Food Restructuring in Southeast Asia: The Thai Experience." In D. Burch et al. (eds.), *Globalization and Agri-Food Restructuring: Perspectives from the Australasia Region*. Aldershot: Avebury. 323-44.

Christensen, Scott, and Aruya Boon-Long. 1994. *Institutional Problems in Thai Water Management*. Bangkok: Thailand Development Research Institute.

Christensen, Scott, and Akin Rabibhadana. 1994. "Exit, Voice and the Depletion of Open Access Resources: The Political Bases of Property Rights in Thailand." *Law and Society Review* 28(3): 639-55.

Chong, Kee-Chai, Somsak Chullasorn, Jate Pimoljinda, and Suchat Sanchang. 1998 (March). "Stimulating Community Bonding in Phangnga Bay, Thailand." *Bay of Bengal News*.

Dryzek, John, Christian Hunold, and David Schlosberg. 2002. "Environmental Transformation of the State: the USA, Norway, Germany and the UK." *Political Studies* 50: 659-82.

Feeny, David. 1982. *The Political Economy of Productivity: Thai Agricultural Development, 1880-1975*. Vancouver: University of British Columbia Press.

Forsyth, Timothy. 1996. "Science, Myth and Knowledge: Testing Himalayan Environmental Degradation in Thailand." *Geoforum* 27(3): 375-92.

Goss, Jasper, David Burch, and Roy Rickson. 2000. "Agri-Food Restructuring and Third World Transnationals: Thailand, the CP Group and the Global Shrimp Industry." *World Development* 28(3): 513-30.

Hewison, Kevin. 2005. "Neo-Liberalism and Domestic Capital: The Political Outcomes of the Economic Crisis in Thailand." *The Journal of Development Studies* 41(2): 310-30.

Hirsch, Philip. 1990. *Development Dilemmas in Rural Thailand*. Oxford: Oxford University Press.

—. 1993. *Political Economy of Environment in Thailand*. Manila: Journal of Contemporary Asian Publishers.

Hirsch, Philip, and Larry Lohmann. 1989. "Contemporary Politics of Environment in Thailand." *Asian Survey* 29(4): 439-51.

ICLARM (International Center for Living Aquatic Resource Management). 1999 (Aug.). "ICLARM Hosted the International Workshop on Collective Action, Property Rights and the Devolution of Natural Resources Management." *Fisheries Co-Management News* 8.

Infofish. 1991. *Fishery Export Industry Profile: Thailand*. Kuala Lumpur: Asian Development Bank.

Ingram, James. 1955. *Economic Change in Thailand since 1850*. Stanford, CA: Stanford University Press.

Isvilalanda, Samporn, Thanwa Jitsangnam, Ruangrai Tokrisna, and Sukhoom Rowchai. 1990. *Management Policy of the Capture Fisheries in Thailand: Its Development and Impacts*. Bangkok: Kasetsart University, Department of Agriculture and Resource Economics.

Johnson, Craig. 1997. "Cooperation and Conflict in an Open-Access Resource: An Analysis of Thailand's Coastal Fisheries." *Background report to the Thai National Marine Rehabilitation Plan*. Bangkok: Thailand Development Research Institute.

—. 2001. "Community Formation and Fisheries Conservation in Southern Thailand. *Development and Change* 32(5): 951-74.

—. 2002. "State and Community in Rural Thailand: Village Society in Historical Perspective." *The Asia Pacific Journal of Anthropology* 2(2): 114-34.

Johnson, Craig, and Tim Forsyth. 2002. "In the Eyes of the State: Negotiating a 'Rights-Based Approach' to Forest Conservation in Thailand." *World Development* 30(9): 1591-1605.

Krongkaew, Medhi. 1995. "Contributions of Agriculture to Industrialization." In Medhi Krongkaew (ed.), *Thailand's Industrialization and its Consequences*. London: Macmillan. 33-65.

Laothamatas, Anek. 1992. *Business Associations and the New Political Economy of Thailand*. Oxford: Westview Press, Institute of Southeast Asian Studies.

Li, Tania Murray. 1996. "Images of Community: Discourse and Strategy in Property Relations." *Development and Change* 27(3): 501-27.

Lohmann, Larry. 1991. "Peasants, Plantations and Pulp: The Politics of Eucalyptus in Thailand." *Bulletin of Concerned Asian Scholars* 23(4): 3-17.

López, Ramon, and Siddhartha Mitra. 2000. "Corruption, Pollution, and the Kuznets Environment Curve." *Journal of Environmental Economics and Management* 40: 137-50.

Midas Agronomics. 1995. *Pre-Investment Study for a Coastal Resources Management Program in Thailand.* Background report prepared for the World Bank, Washington, DC.

Morell, David, and Chai-anan Samudavanija. 1981. *Political Conflict in Thailand: Reform, Reaction, Revolution.* Cambridge: Oelgeschlager, Gunn and Hain.

Nelson, Michael. 1998. *Central Authority and Local Democratization in Thailand.* Bangkok: White Lotus.

Nicro, Somrudee. 1993. "Thailand's NIC Democracy: Studying from General Elections." *Pacific Affairs* 66(2): 167-82.

Parfitt, Trevor. 2002. *The End of Development: Modernity, Post-Modernity and Development.* London: Pluto Press.

Phantumvanit, Dhira, and Theodore Panayotou. 1990. *Natural Resources for a Sustainable Future: Spreading the Benefits.* Bangkok: Thailand Development Research Institute.

Phongpaichit, Pasuk, and Chris Baker. 1995. *Thailand: Economy and Politics.* Oxford: Oxford University Press.

—. 1998. *Thailand's Boom and Bust.* Bangkok: Silkworm Books.

—. 2001. "Democracy, Capitalism and Crisis: Examining Recent Political Transitions in Thailand." Unpublished paper presented at Princeton University. http://pioneer.netserv.chula.ac.th/~pPhongpaichit/papers.htm. Accessed Aug. 2, 2006.

Pichyakorn, Bantita. 2003. "Involvement of Non-State Actors in the Development of Water Law in Thailand: A Role that is Ignored?" *Non-State Actors and International Law* 3: 231-50.

Reynolds, Craig. 2002 [1991]. "Introduction: National Identity and its Defenders." In Craig Reynolds (ed.), *National Identity and its Defenders: Thailand Today.* Bangkok: Silkworm Books.

Rigg, Jonathan. 1997. *Southeast Asia: The Human Landscape of Modernization and Development.* London: Routledge.

Rigg, Jonathan, and Sakunee Nattapoolwat. 2001. "Embracing the Global in Thailand: Activism and Pragmatism in an Era of De-agrarianisation." *World Development* 29(6): 945-60.

Robertson, Philip S., Jr. 1996. "The Rise of the Rural Network Politician: Will Thailand's New Elite Endure?" *Asian Survey* 36(9): 924-41.

RTG (Royal Thai Government). 1997. *The Constitution of the Kingdom of Thailand.* English Translation by the Office of the Council of State. Published in the *Government Gazette* Vol. 114, Part 55a. October 11, 1997.

Sato, Jin. 2000. "People in Between: Conversion and Conservation of Forest Lands in Thailand." *Development and Change* 31: 155-77.

TDRI (Thailand Development Research Institute). 1986. *The Status of Coastal and Marine Resources of Thailand.* Bangkok: TDRI.

—. 1994. "Thailand's Drought Crisis." TDRI *Quarterly* 9(1): 28-29.

—. 1998. *Background Report for the Thai Marine Rehabilitation Plan 1997-2001.* Bangkok: Thailand Development Research Institute.

Tejapira, Kasian. 2006. "Toppling Thaksin." *New Left Review* 39: 5-37.

Thanh, Bui Duy, and Thierry Lefavre. 2000. "Assessing Health Impacts of Air Pollution from Electricity Generation: The Case of Thailand." *Environmental Impact Assessment Review* 20: 137-58.

Turton, Andrew. 1989. "Thailand: Agrarian Bases of State Power." In Gillian Hart, Andrew Turton, and Ben White (eds.), *Agrarian Transformations: Local Processes and the State in Southeast Asia.* Berkeley: University of California Press. 53-69.

UNDESA (United Nations Department of Economic and Social Affairs). 2005. *National Information—Thailand CSD—14/15 Thermatic Profiles Atmosphere/Air Pollution.* National Reporting Guidelines. New York: Division for Sustainable Development, UNDESA, United Nations.

Unger, Danny. 1998. *Building Social Capital in Thailand: Fibers, Finance and Infrastructure.* Cambridge: Cambridge University Press.

Vandergeest, Peter. 1991. "Gifts and Rights: Cautionary Notes on Community Self-Help in Thailand." *Development and Change* 22(3): 421-43.

—. 2003. "Racialization and Citizenship in Thai Forest Politics." *Society and Natural Resources* 16: 19-37.

Vandergeest, Peter, Mark Flaherty, and Paul Miller. 1999. "A Political Ecology of Shrimp Aquaculture in Thailand." *Rural Sociology* 64: 573-96.

Vandergeest, Peter, and Nancy Lee Peluso. 1995. "Territorialization and State Power in Thailand." *Theory and Society* 24: 385-426.

Walker, Andrew. 2000. "The 'Karen Consensus': Ethnic Politics and Resource-Use Legitimacy in Northern Thailand." *Asian Ethnicity* 2(2): 145-62.

World Bank. 2006. *Thailand Environment* http://web.worldbank.org/WBSITE/EXTERNAL/COUNTRIES/EASTASIAPACIFICEXT/EXTEAPREGTOPENVIRONMENT/0,,contentMDK:20266329~menuPK:537827~pagePK:34004173~piPK:34003707~theSitePK:502886,00.html. Accessed July 5, 2006.

World Health Organization. 2004. *Thailand—Environmental Health Country Profile* http://www.who.int/en/. Accessed July 5, 2006.

CHAPTER 4

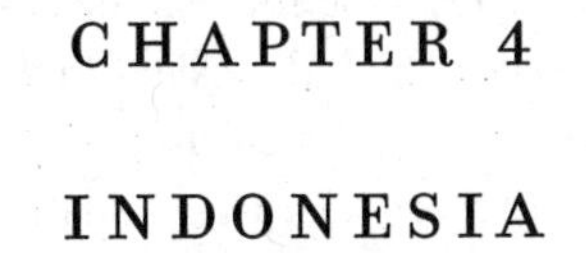

INDONESIA
Addressing Development and Environmental Management in the Context of Globalization[1]

Bruce Mitchell

INTRODUCTION

In the introduction to the volume, Jordi Díez and O.P. Dwivedi observe that growing global interaction, or "globalization"—which involves accelerated social, economic and cultural interaction among societies—has had multiple effects on the way in which environmental problems are addressed, both nationally and internationally. They argue that as globalization processes unfold, and as new environmental governance regimes emerge, we need to learn how nations and civil societies have, or have not, been able to address environmental problems. As a result, they conclude that it is important to examine the complex relationships among development, environmental management, and globalization.

In this chapter, experience with development and environmental management in Indonesia, as influenced by both domestic and internationalization processes, is examined. The chapter presents two case studies: the development of a "sustainable development" strategy prepared in the Indonesian province of Bali during the 1990s, and the enhancement of environmental management tools in the province of Sulawesi in the early 2000s. The analysis of these two cases yields several key lessons. They both show that, in Indonesia, development decisions and their subsequent impact on the natural environment are significantly influenced by external forces. In the cases presented here, it becomes clear that decisions made by tourists regarding where they will holiday, as well as by consumers regarding what seafood they want to eat, have an impact on the environment in general and environmental management practices in particular. Second, the case studies presented here show that culture plays an important role in developing and implementing management strategies, indicating in turn that culture should and can be equally prominent with environment and economy when thinking about "sustainable development." In this context, non-governmental organizations (NGOs) can play key roles in helping to create approaches that are sensitive to local culture. Furthermore, NGOs often can take action in addressing environmental problems much more quickly than government agencies. Third, the cases also demonstrate that governance and institutional arrangements can be created to facilitate more open, transparent, and participatory processes of decision making.

The chapter is divided into three sections. The first section provides an overview of the context for development and environmental management in Indonesia as a background to understanding our case studies. The two subsequent sections outline experiences regarding preparation of a sustainable development strategy for Bali, and enhancement of environmental management in Sulawesi.

DEVELOPMENT AND ENVIRONMENTAL MANAGEMENT IN INDONESIA: THE CONTEXT

Indonesia declared independence in 1945, after more than 300 years of being a Dutch colony, and being occupied by Japan during World War II, starting in March 1942. A few days after the Japanese surrender in August 1945, Sukarno, one of the earliest nationalist leaders, along with two other key figures, Mohammad Hatta and Sutan Sjahrir, declared Indonesia to be independent. However, the Dutch had assumed that they would return to govern their colonial territory after the end of the war and, for the next four years, the Indonesians fought for their independence. The archipelago's independence was finally achieved under The Hague Agreement of November 2, 1949, when Dutch control of all Indonesian territories was ceded, with the exception of the western half of New Guinea. Sukarno was the first president of the new nation.

The leadership of this new nation, which had been focused on achieving political independence for four years, then had to turn to practical matters of economic development and government. This was no small challenge for a nation extending more than 5,000 kilometres from east to west and including more than 17,000 islands whose land mass totalled just less than two million square kilometres. Fewer than half of the islands were, and still are, occupied by humans, but there was a multitude of regional languages in use across the archipelago. Today, Indonesia, with over 220 million people, is the fourth largest country in the world in terms of population—after China, India, and the US—with over 60 per cent living on the island of Java. The challenges of uniting a country spread over such a large area and involving many language groups were recognized very early. For example, in October 1928, a group of Dutch-educated nationalists, headed by Sukarno, founded the Indonesian Nationalist Party (PNI). One of its key platforms was the idea of "one nation, one people, one language," and the PNI became an advocate of Bahasa Indonesia, a derivative of Malay, to be the country's official language. "Malay" became "Indonesian" and the official national language at the time of the declaration of independence in August 1945, although as already noted above, real independence was not achieved until 1949.

During late 1965 and early 1966, Indonesia underwent violent turmoil as the regime of President Sukarno was challenged by a group of military officers, led by Soeharto, then a relatively obscure army major general. Over a period of three months, hundreds of thousands of people in Java, Sumatra, and Bali were killed, justified as necessary to end a threat of domination by Communists. Soeharto became the effective head of state in March 1966 after this violent period, and was subsequently officially declared president two years later. He called his regime the "New Order," a regime that lasted for 32 years, until 1998, and emphasized economic development and the building of strong political institutions. However, as Adam Schwartz has noted, the most distinguishing feature of Soeharto's New Order was "an authoritarian, military-dominated government" (1999: 2).

The context for development and environmental management in Indonesia has changed significantly since the 1970s (Schwarz 1999; Vatikiotis 1999; Babcock et al. 2003; Setiawan 2005). From the early 1970s until the late 1990s, annual Gross Domestic Product (GDP) growth averaged 7.1 per cent, which placed it well above the 4.5-per-cent average for developing countries as a group over the same period. High growth rates were especially noticeable from approximately 1985 until the mid 1990s. Foreign investment was encouraged and received, export-oriented industries flourished, and the banking sector was liberalized. Public investments were made to support universal primary education and basic health care throughout the nation, and, overall, optimism prevailed.

Difficulties were also present, however. Although political and social stability generally prevailed, unrest continued in Aceh and Irian Jaya provinces, and the independence movement in East Timor, which had gained strength, was ruthlessly repressed by the regime. The Soeharto government, having come to power in 1966, argued that stability was a necessary prerequisite for prosperity and used this rationale to justify significant investment in the armed forces along with a highly centralized government structure. The international financial community responded positively, with donor support, primarily through loans, increasing year over year.

Regarding the environment, optimism existed, serious problems notwithstanding. A Ministry of Environment had been established after the Stockholm Conference in 1972. A series of environmental laws and regulations were also introduced. For example, in March 1982, legislation concerning the Basic Provisions for the Management of the Living Environment was promulgated. It took a holistic perspective, defining the living environment to include the organic and inorganic natural environment, the human-made environment, and the social environment. Moreover, in June 1987, the country initiated its Environmental Impact Assessment (EIA) process through the introduction of a regulation specifying that both environmental and social impacts were to be addressed. In June 1990, responsibility for EIA was transferred to the newly created Environmental Impact Management Agency, or BAPEDAL, which was given the responsibility to "control environmental impacts using ecological principles in the utilization of natural resources such that the negative impacts of development do not alter environmental functions" (Doberstein 1995: 356). Foreign donors also provided increasing funding for institutional capacity-building and enhancement related to the environment. Canada was one of the first foreign donors to provide support to enhance environmental management capacity.

Environmental NGOs (ENGOs) also emerged in the 1980s and took on active roles. One of the most notable Indonesian ENGOs is WALHI, or the Indonesian Forum for the Environment, which is the largest environmental umbrella organization comprising NGO and community-based organizations in the country. As of 2004, WALHI had almost 450 member organizations active in 25 provinces, with a focus on social transformation, sustainability of life and livelihoods, and people's sovereignty. A core goal for WALHI is to protect the natural environment and local communities from injustice incurred or imposed through initiatives driven in the interest of economic development.

What became known as the "Asian Financial Crisis" in 1997, however, led to upheaval and large-scale social, economic, and political changes in Indonesia.

An economic crisis befell Indonesia during August 1997 when the national currency, the rupiah, dropped sharply. As Bobi Setiawan later observed, the impact was devastating, since "in 1998, inflation reached an annual rate of 78 per cent, but in reality prices of basic foods soared by 100 per cent to 400 per cent. Meanwhile, real wages of the average worker in Indonesia dropped by 40 per cent" (2005: 45). The situation was further exacerbated by the worst drought in 50 years. The financial problems continued, and the rupiah experienced a series of devaluations. Setiawan comments that "the devaluation had major negative consequences for Indonesian industrial sectors with a high dependency on imported materials, such as textiles and electronics, and other sectors with growing short-term external debts, such as real estate and construction" (2005: 45) . An important outcome of this situation was massive layoffs in the corporate and industrial sectors. By 1999, once the worst of the crisis had passed, Indonesia's GDP stabilized, but it had fallen in real terms by nine per cent below its 1996 level (*The Economist* 2000: 6).

On the political front, the crisis culminated in the overthrow of the Soeharto government in May 1998, an event that was primarily the result of a massive mobilization led by students and other groups frustrated with the president's autocratic style of governance. A series of relatively weak, but democratically elected, governments has followed since then. Following the change of regime, several initiatives were undertaken to create and implement "decentralized and deconcentrated" forms of governance.[2] These primarily entailed a transfer of decision- and policy-making authority and responsibility, including for environmental matters, to the local levels of government. However, despite these important changes, it soon became apparent that the prosperity of the Soeharto era had been superficial. As the old regime collapsed, several traits from the Soeharto area were revealed, such as an uncontrolled debt, widespread corruption, and the concentration of benefits in the hands of a very small elite. It also became apparent that rapid and destructive harvesting of natural resources had been allowed and that the pollution of aquatic systems had been common. Further, the lack of capacity for monitoring and enforcement that was a characteristic of the previous government meant that environmental laws, policies, and regulations had been largely ineffective.

An important component of the evolution toward a more democratic and decentralized system of government has been the emergence of significant regional autonomy, which has included environmental issues. As Setiawan has explained, the "utilization of natural resources now depends totally on the policies of local governments" (2005: 46). Setiawan illustrates this with reference to forestry, noting that local governments have received responsibility to manage forest concession rights, while the central government's role is limited to developing standards and guidelines for its management. Similar arrangements exist for mining as well as for oceans, as local governments have been granted more authority to manage natural resources within a 12-mile limit of their areas. These changes are consistent with the "subsidiarity principle" which holds that decision-making authority should be held by the level of government closest to the consequences of decisions. However, such changes are not without problems. Local governments often worry only about benefits and costs within their own jurisdiction, and do not concern themselves with issues associated with ecological systems (catchments or river basins, or bioregions) extending across various political or administrative boundaries.

It therefore becomes clear that fundamental change has been occurring in Indonesia in terms of environmental management, and economic development more generally. In theory, the reforms that have accompanied the country's democratization and decentralization processes should allow local governments to manage their own environments as well as to collaborate with other local governments, given that one local government often will be unable to manage all aspects on its own (Setiawan 2005: 47). Nevertheless, these changes pose a real challenge because they require local governments to manage their environments and regions in a cooperative manner. In Setiawan's words, "the decentralization process offers opportunities to move toward more local participation in resource allocation decisions, greater accountability by regional governments, a refocusing of central agencies on policy and oversight, and ultimately, more efficient and more sustainable use of natural resources" (2005: 48). Setiawan goes on to say that

> at the same time, democratization can also facilitate the participation of individuals and civil society organizations in decision-making processes related to environmental issues. The trend toward democratization should enable people to directly control the management of the environment and natural resources by the state. In other words, democratization can serve as an effective mechanism to control the tendency of government, including local government, to over exploit natural resources for the sake of economic development. (2005: 48)

Furthermore, there is no doubt that Indonesia faces serious environmental problems that were not addressed adequately during previous decades. For example, while Indonesia has the most extensive rainforest cover in all of Asia, it lost more than 28 million hectares of forest between 1990 and 2005; the country lost almost 24 per cent of its total forest cover during this time, placing it second only to Brazil in terms of loss of biologically rich primary forest. To provide context, in the 1960s approximately 82 per cent of the country was covered by rainforest. That proportion had dropped to 68 per cent in 1982, 53 per cent in 1995, and 49 per cent in 2006.[3] The primary causes of deforestation are illegal logging, mining, the expansion of large-scale agricultural activity, and settlements in forested areas. Many of these activities have been due to efforts to harvest more timber, extract more mineral products, and increase food production to meet export opportunities and earn foreign currency. In regard to water quality, many of Indonesia's rivers are seriously degraded. One of the major reasons is lack of, or insufficient, sewage systems in urban centres.

Indonesia is often cited as one of the worst countries in Asia in terms of sewage and sanitation facilities. Loss of forest lands and degraded water are but two examples that illustrate that Indonesia faces serious environmental challenges.

Against this backdrop of the political and economic change that has taken place in the country over the last decade, we can now turn to two experiences of environmental management in Indonesia. The next section reports on an experience in the Indonesian province of Bali in the early 1990s. The subsequent section covers an experience in one province in Sulawesi in the early 2000s. These cases are instructive for several reasons. First, they cover periods of time before and after the shift that started in 1998 toward democratization, deconcentration, and decentralization of governance throughout the country. Second, they represent different landscapes and cultures, reminding us to be careful in generalizing. Notwithstanding these significant differences, both Bali and Central Sulawesi represent areas with significant population pressures, evolving economies, influences from outside Indonesia, and serious environmental degradation.

BALI AND SUSTAINABLE DEVELOPMENT

Emil Salim, the first minister of the Ministry of Population and Environment for Indonesia, was one of the 21 members of the World Commission on Environment and Development (WCED), which published *Our Common Future* in April 1987. Salim was a high profile member of the WCED, having organized various events in Indonesia when the commission was in the process of working on the report. As Díez and Dwivedi explain in the Introduction, *Our Common Future* coined the term "sustainable development," a concept that has largely guided the international environmental agenda since then.

A five-person team from the University of Waterloo, Canada, visited Bali during April 1987, with colleagues from Gadjah Mada University in Yogjakarta in central Java, to explore opportunities to collaborate in an "action-oriented development study," whose main purpose would be to integrate environmental and economic considerations into overall regional development planning for the province. The team met with Rendha, head of the Regional Development Planning Agency for Bali Province. Rendha explained that Bali used five-year development plans, that he and the Governor of Bali were very interested in examining how the new concept of sustainable development discussed in the recently published Brundtland Report could be incorporated into the next five-year development plan for the province, and that they would be pleased to cooperate with the Waterloo-Gadjah Mada team toward that end. As a result, what became known as the Bali Sustainable Development Project (BSDS) was launched in 1989 (Martopo and Mitchell 1995).

In early discussions, Rendha made several key observations. First, he commented that the priority in the Province of Bali was the development of the economy, with an emphasis on agriculture and tourism. He made it clear that the intention was to achieve this in a way that treated culture as a regional asset, taking advantage of the small industry in the province, particularly the handicraft industry. The interest in expanding tourism was supported by the central government, which wished to diversify the economy of Indonesia beyond reliance on natural resources, such as oil, natural gas, forestry, and minerals. The central government viewed tourism in Bali as a key component for bolstering the national economy. The Balinese provincial government supported this view, but worried that emphasizing tourism could make Bali vulnerable to the natural fluctuations of tourism travel. Such reality was dramatically revealed during 1991, when Iraq invaded Kuwait and in turn was attacked by US armed forces, leading to a dramatic downturn that year in travel by people from Western to Muslim countries. Rendha also noted with concern the increasing rate of urbanization, especially adjacent to the major city of Denpasar.

Second, and perhaps most important in terms of sustainable development, Rendha highlighted the significance of Balinese culture as the foundation for development planning in the province. He and his colleagues argued that the starting point for sustainable development had to be culture, rather than the environment or the economy, because any policies, strategies or practices not consistent with Balinese culture would not succeed. He used an analogy, likening Balinese culture to a kite. In his view, both the structure of a kite and Balinese culture remained constant, even though either could be buffeted by winds or forces from various directions. The implication was that, regardless of how strong external forces might be, the basic integrity of Balinese culture was resilient and enduring, but also had to be continuously monitored and supported. The BSDS team agreed with Rendha. This is exemplified by the following comment: "most sustainable development literature has had remarkably little to say regarding the significance of culture. For Bali, the culture is so strong and its influence so pervasive that culture was incorporated into the BSDS definition of sustainable development, and the culture itself became an aspect of Bali to be sustained" (Bater et al. 1999: 115).

Subsequently, the Governor of Bali underlined that the Indonesian government used a "trilogy" to guide economic development efforts. This, he said, required attention to growth, stability, and equity. Furthermore, he explained that the central government had been preparing an eco-development policy with the objective of exploiting natural resources in a sustainable manner, so that they would be available to future generations. Such an approach was consistent with the one adopted by the Brundtland Commission, which, as we have seen, advances the notion of intergenerational equity in sustainable development. Finally, the Governor indicated that ideas for a sustainable development strategy would become important inputs into the second long-term development plan (1994 to 2019) for the province.

Despite his stated intentions and objectives, the Governor recognized that many important challenges existed in Bali. For instance, he explained that the amount of forest cover in the province was decreasing rapidly, that serious beach erosion was taking place, that industry was over-relying on natural resources from outside the province, and that commercial, industrial, and tourism development was over-concentrated in one district (Bandung). The Governor also noted that human resource skills were low, and that opportunities for women needed improvement. I now turn to the case study itself.

The province of Bali is located immediately east of Java and at eight degrees south latitude, and it extends over 563,286 hectares (about the size of Prince Edward Island). When the Bali Sustainable Development Strategy process began, Bali had a population of approximately three million people (Vickers 1989; Bater 1995). It also had an average of 1.5 million tourists per year. More than 90 per cent of the Balinese population is Hindu, with just over five per cent identified as Muslim. There are also small numbers of Christians and Buddhists. The overwhelming majority of Balinese Hindus makes the province unusual in a predominantly Muslim nation. The Balinese actively and publicly profess their faith and live by their culture. The imprint on the landscape is highly visible, ranging from large temples in villages to small temples on family property, offerings placed daily in visible locations, and many rituals practiced daily, weekly or on other regular cycles. The Balinese adhere to the concept of *Tri Hita Karana*, which stipulates that balance should be sought at three levels: between humans, between humans and the environment, and between humans and the gods. Another key concept, *Desa Kala Patra*, specifies the importance of achieving harmony among space, time, and situation. These central concepts influence very strongly their outlook and behaviour and, needless to say, they are much more amenable to the achievement of sustainable development than many core ideas of Western cultures.

These concepts led to support for "cultural tourism" by the team given the task of elaborating the new development plan, as it adopted the idea that "Bali is not for tourists, but tourists are for Bali." In other words, while expansion of tourism was viewed by both the central and provincial governments as a way to enhance economic development and diversify the economy, a parallel intent was that growing numbers of tourists should not harm Balinese culture. In that regard, Manuaba (1995)—a faculty member at Udayana University, the principal provincial university in Bali—explained the principles underlying culturally sustainable tourism:

1. Tourism development should be guided by a planning process involving broad community participation, to ensure balance among economic, social, cultural and environmental objectives;

2. The relationship among tourism, the natural environment, and the cultural environment must be managed so that the environment is sustainable in the long term;

3. As tourism is a significant activity, it must not be allowed to damage the natural resource base, create prejudice for future employment, or bring an unacceptable impact such as xenophobia to the community;

4. Development of tourism should have a moderate rate of growth and should be small to medium in scale;

5. In any location, harmony must be sought between the needs of the visitor, the place, and the local community;

6. The success of any proposed action depends on the interrelationship of government, the host community, and the tourism industry;

7. Education which leads to socio-cultural awareness at all levels in the community, especially regarding tourism activities and including the teaching of appropriate behaviour to tourists, must be seriously organized;

8. Legislation to ensure the protection of culture must be introduced and enforced, while facilitating the optimum development of suitable forms of tourism-based activity;

9. Investors and tourists must be educated about and respect local customs, norms, and values. Matters which possibly create negative cultural impacts should be avoided. In contrast, positive cultural impacts should be encouraged.

Manuaba offered the above principles to show how culture should be a foundation for sustainable development and tourism in Bali. Adopting this approach, the team concluded that sustainable development had to include two main components. First, it had not only to guarantee the continuity of natural resources (basic life supports) but also to preserve the continuity of cultural resources (from values and legends, to ceremonies and structures). Second, it should guarantee not only the continuity of economic production but also the continuity of culture itself. However, "continuity of culture" does not preclude the possibility that aspects of culture may change over time, since culture is dynamic.

Taking these two considerations into account, the team concluded that sustainable development for Bali "must be concerned not only with the balance between the demand for and supply of resources for the good life—but with the culturally-based balances within the heritage of Bali" (Mitchell 1995: 538). Specifically for Bali, sustainable development was interpreted by the team to include three core features: the continuity of natural resources and production, the continuity of culture and the balance within culture, and development as the process that enhances the quality of life. Taking these features as a guide, all development options were in turn assessed against seven specific criteria: ecological integrity, efficiency, equity, cultural integrity, community, integration/balance/harmony, and development as the realization of potential. The outcome of this process was the "Bali Sustainable Development Strategy" (BSDS). The plan was custom-designed for the province of Bali and intended to be incorporated into the subsequent five-year development plan for the province (Martopo and Mitchell 1995; Bater et al. 1999).

While the BSDS emerged primarily through a partnership between an Indonesian-Canadian university team and provincial government agencies, ENGOS also were involved, as we have seen, especially since some of the Indonesian university colleagues and staff were active members of these organizations. Of particular interest is the fact that individuals worked with their local NGOS to begin implementing ideas from the BSDS, in some instances even before some government agencies began to take action. Such initiatives by individuals and NGOS highlight the fact that often government does not need to be the main initiator of sustainable development activities.

In early 2006, a colleague from Gadjah Mada University who had participated in the BSDS commented in a personal message that the most important contribution of the project may have been to spread consideration of environmental issues and sustainable development widely throughout Bali. In his view, this represented a fundamental and long-lasting contribution.

CENTRAL SULAWESI: A CASE OF MOLIBU[4]

The BSDS occurred prior to the Asian financial crisis of 1997. In this section, attention turns to another experience that took place in the early 2000s once political change in Indonesia had been unleashed. This case relates to the use of a Future Search Conference in the province of Central Sulawesi and highlights the growing role of local governments and communities in environmental management under a different political context. I also use this example because I had first-hand experience as a participant in the project in which this initiative was taken.

The coastal zone of Banawa Selatan, located in the district or *kabupaten* of Dongalla, near Palu (the provincial capital of Central Sulawesi), has experienced significant changes over the last two decades. As Derek Armitage and Achmad Rizal have stated, the primary force for change has been the conversion of mangrove forest ecosystems into aquaculture ponds, or *tambak* (2005: 227). According to Armitage and Rizal, outcomes have included loss of mangrove forest habitat and related biodiversity, coastal zone erosion, decline in coastal fisheries, loss of local resource control, and increased levels of socio-cultural complexity, tension, and conflict between local communities and external interests. They argue that the process of transformation in the coastal zone of Banawa Selatan is nested in a complex set of interactions created by: (1) desires for foreign exchange by the Indonesian government; (2) requirements for domestic protein to support a growing population; (3) flawed bureaucracies; (4) unclear legal frameworks; (5) ethnic division of resource use and control with economic externalities; and (6) state-sanctioned worldviews and ideology supportive of intensive aquaculture production and privatization of resource use over traditional resource approaches. (2005: 226)

The first and sixth points emphasize the pervasive influence of globalization on local approaches to economic development and environmental management. Armitage and Rizal conclude that, collectively, these drivers have undermined local property rights, traditional resource management regimes, livelihoods, and opportunities to nurture resource use and management in the coastal zone consistent with ideals of ecological, socio-economic, and cultural sustainability.

Banawa Selatan is ethnically diverse, including Kaili, Buginese, Mandarese, Sundanese, and Javanese peoples. Such diversity has been the outcome of decades of inward migration and settlement programs. Small-scale resource harvesting is the base of the economy in this area, including subsistence and commercial fishing, gardening, coconut farming (for copra), and cocoa production. They are all accompanied by small-scale trading. Family household incomes range from US $35 to $1275 per month, and the coastal zone resources, including fisheries and mangrove systems, have traditionally been used and managed as common property resources. The mangrove systems have been especially important, as both men and women can harvest and use the products from them, including firewood, building materials, dye for fishing nets, and material for fishing boats, as well as fish and other wildlife.

As noted earlier, Indonesia has moved since 1998 toward a more democratic and decentralized form of government. The application of the Future Search Conference, which develops a vision of a desired future as well as an action plan to achieve the vision (Schwass 2004: 406), was introduced to Banawa Selatan by Agus Rahmat, Rodger Schwass, Tim Babcock, and Jack Craig, Indonesian and Canadian participants in the development project of which this was one component. The Future Search Conference used in Banawa Selatan "was intended to demonstrate a more effective, and more democratic, form of decision-making concerning the use of local natural resources than had hitherto been practiced in Indonesia" (Rahmat et al. 2005: 323). In the view of Rahmat and his colleagues, while Indonesia had been exposed for over 30 years under the Soeharto regime to the discourse of participatory development planning and management, the reality had been that "bottom-up planning" had normally not been productive, mainly due to the very tight control exercised by a rigid bureaucracy connected to the dominant political party that was unwilling to give up or share its power. The concept of empowerment, or *pemberdayaan*, had been frequently invoked during the 1990s, but it usually became diluted to mean top-down government assistance intended for the "good of the people."

In the following discussion, I present the notion of a Future Search Conference. I follow this with an explanation of how it progresses in practice, and what the pre-conditions are for an effective Future Search Conference. I then describe what actually happened when it was implemented in Banawa Selatan.

What is a Future Search Conference? Marvin Wiesbord (1992) provides an excellent review. As noted above, Schwass (2004: 405-06) explains that it is a tool to help develop a vision for a desirable future, and an action plan to achieve that vision. According to Schwass (2004: 407-08), a Future Search Conference usually consists of five stages:

1. Scan external issues facing an organization, region or community, and the wider environment in which it exists, with particular attention to trends and issues in the recent past before projecting those into the future for 15 years. Effort also is given to predicting likely new developments in the future that help to estimate the nature of the probable future environment;

2. Analyze the internal situation within the organization, region or community, including the recent past, present situation, and the probable internal future for the group or area itself, if nothing significant changes or is done. The purpose is to understand the shortcomings of the group or region in relation to the external environment;

3. Develop a desirable future or vision for the next 10 to 15 years, focusing on the ideal that can be achieved;

4. Identify constraints which could hinder or impede achieving the desirable future, and determine ways in which they can be overcome and new opportunities created;

5. Formulate strategies and action steps as the basis for a long-term action plan, including how to align people, agencies and resources with action steps.

Schwass also argues that eight general conditions should exist before using the Future Search Conference process (2004: 406-07):

1. Organizational support is present "at the top," along with willingness and resources to implement recommendations which emerge.

2. All group members should have access to the same information, including global and societal trends that could affect the future of the group, community or region.

3. Group members must be confident that their involvement and input could not negatively affect their future, and that all input will be treated equally.

4. All participants should be prepared and able to suspend judgement until all facts have been provided, and to find common ground.

5. Participants need to be convinced that their ideas will receive serious consideration, emphasize possibilities rather than limitations, and accept a sense of individual and shared responsibility for the future.

6. Participants should believe that, during the process, time is being used and managed effectively.

7. Participants should be encouraged to participate in all sessions, especially senior managers who will ultimately carry responsibility for implementation.

8. When several organizations are involved, creation of a steering group of senior people helps to show commitment, as well as to create opportunities for follow up.

Taking these considerations into account, a Canadian and Indonesian university-based team developed a Future Search Conference designed for the Banawa Selatan region, with the objective of achieving sustainable development practices. The team was aware of the conditions and knew key people and organizations well (Armitage and Rizal 2005). Members of the team were also aware that local people understood the issues and problems and were willing to make changes to address them.

The joint Indonesian-Canadian university team took the lead role in initiating the Future Search Conference, but quickly identified local partners in the provincial and local governments. In addition, the value of NGOs was recognized from the outset of the process. As a result, a project office was established in the headquarters of a local NGO in downtown Palu because many local people on the conference steering committee lived in that area. A conscious decision was also taken to involve two local NGOs: *Ibnu Chaldun* and *Yayasan* BEST. Their selection was based on the fact that they possessed extensive contacts with local people in Banawa Selatan and had an understanding of local issues. Moreover, the team considered that NGOs would make a significant contribution because several of their members were originally from Banawa Selatan and were therefore fluent in the local languages. The language competency was extremely valuable because some of the people living in the highland area in Banawa Selatan do not speak the national language, Bahasa Indonesia.

The team's work in developing a Future Search Conference for the region involved a sequence of steps: (1) establishing that the preconditions for a successful process were in place; (2) creating a team to plan and carry out the process; (3) identifying and involving stakeholders at all stages of the process; (4) preparing background papers to identify key issues, options, and actions by drawing on both local and scientific knowledge; (5) preparing and distributing pre-conference packages to all participants; (6) publicizing the Future Search Conference; (7) conducting the Future Search Conference; (8) preparing an action plan; and (9) implementing the action plan. I focus on some of these stages in the following discussion.

In terms of identifying and involving stakeholders, the team organized a group of individuals from the Environmental Study Centre at Tadulako University, in Palu, as well as from local NGOs. This group of people introduced the Future Search Conference concept to the eight local communities. The team made an analogy to the local concept of *molibu*. *Molibu* refers to an indigenous process of decision-making that is characterized by its consultative nature. Armitage and Rizal explain that *molibu* is a verb and means "to meet" (2005: 257). They also use *nolibu* as a noun, which means "a meeting." Both words are used by the indigenous Kaili people in Banawa Selatan who speak the Unde dialect. The team members took care to ensure that village leaders understood that not just the elite should participate, but that a truly representative cross-section of community interests was represented—and this was achieved. In addition, politicians (the State Minister of Environment) and high-ranking civil-service bureaucrats (e.g., the Director General of Marine and Spatial Planning, and the Secretary of the Ministry of Forestry) in Jakarta were invited and all participated. Their presence gave profile and credibility to the process.

Heads of villages and heads of two government departments prepared position papers about the current economic, social, health, educational, and environmental situations of the villages and reviewed relevant government policies. The position papers, as well as invitations, were hand-delivered to participants at least one week before the Search Conference itself. The Conference was publicized through announcements on local radio stations, in local and national newspapers, and by banners that were posted in the provincial (Palu) and regional (Donggala) capital cities.

The Future Search Conference lasted three days, from April 3 to 5, 2001. The first two days were used to reach agreement on issues and the third day to develop an action plan. Key issues were divided into eight categories (Table 4.1). Based on the insight gained from agreeing on the issues that were most relevant, the participants then reached a consensus on a vision for Banawa Selatan by declaring that "by the year 2015, Banawa Selatan will have developed into a harmonious, prosperous and well educated community with skilled people at all levels who respect the equal roles of men and women in managing natural resources in a sustainable manner" (Rahmat et al. 2005: 337). Based on this vision, participants identified key components of an action plan, including clarifying forest boundaries in Banawa Selatan, involving communities in conserving their forests through replanting and conservation land management, introducing or enhancing distance education at the elementary level, and developing high-school education facilities.

Of particular significance is the fact that, one week after the Future Search Conference was held, the head of the Donggala District invited the organizing committee, which included NGO members, to meet to discuss an implementation strategy for the action plan. The meeting was attended by heads of all government departments in the District. An agreement was reached by those at this meeting to accomplish the following: (1) stop all mangrove cutting in Banawa Selatan; (2) instruct all fish-pond owners to build a green belt around their ponds; (3) move part of the current mangrove planting project from Parigi sub-district, in the eastern part of Donggala District, to Banawa Selatan; (4) provide 100 sacks of cement to build a breakwater along the coastal area associated with two villages; and (5) ask all heads of agencies in Donggala District to give attention to environmental issues in all decisions and programs.

What have been the policy outcomes from the implementation of the ideas generated during the Future Search Conference? While the eventual impact will not be apparent for some time, at the community level mangrove replanting and the construction of a small breakwater were initiated within one year of the Conference. Moreover, by 2005 one of the key Indonesian facilitators was able to report that the mangrove replanting had progressed well, especially in the villages of Tolongano and Bambarimi, the two main villages in which the Project began its activities. Perhaps more significantly, a solid relationship between communities in Banawa Selatan and members of the Environmental Study Centre at Tadulako University was created. The communities requested the Environmental Study Centre (ESC) to continue the relationship, realizing that the ESC members provide expertise and bring contacts not available in their communities.

Time is still needed to assess the success of the program. However, the Future Search Conference process demonstrated to the local people that their knowledge and experience were valued and that they could be incorporated into decisions related to resource and environmental management. With the advent of a more decentralized and deconcentrated political system, the legitimacy for local knowledge through the process provides one building block for a more locally based, participatory approach to management. As the designers of the Future Search Process commented, "The fact that agreements were reached, with all sides appearing to appreciate or at least better understand each other's positions, and with follow-up

TABLE 4.1

KEY ISSUES IN BANAWA SELATAN IDENTIFIED DURING THE FUTURE SEARCH PROCESS

A. ENVIRONMENT

1. Coastal abrasion
2. Erosion and flooding
3. Damage to mangroves
4. Degradation of upland forests
5. River pollution (agricultural waste from paddy cultivation)

B. HEALTH

1. Insufficient fresh water
2. Inadequate health facilities
3. River pollution

C. AGRICULTURE AND PLANTATIONS

1. Plantation problems (insects and fungi attacks)
2. Damage to irrigation canals
3. Decline in plantation production (cacao and clove)
4. Unsuitable shifting agricultural practices
5. Inability of farmers to pay credit
6. Unutilized land
7. Low farmer prices for clove and cacao, determined by the buyers

D. FISHERY AND MARINE ISSUES

1. Environmentally unfriendly fishing techniques (using bombs and poison)
2. Traditional fishing equipment
3. Poverty among fishing families
4. Conflict between traditional and modern fisher folk

E. FORESTRY

1. Illegal logging
2. Extensive area of forest concessions
3. Disuse of traditional systems of forest management
4. Forest exploitation in new areas still permitted by government
5. Lack of forest access by local people living nearby

F. EDUCATION AND TRAINING

1. Inadequate educational facilities
2. Inability of many local people to pay costs of education
3. Lack of teachers (most are outsiders and remain only briefly in the area)
4. Relatively low levels of education

G. TRANSPORTATION

1. Lack of a bridge in Bambarimi to link it to four other villages
2. Isolation of most hamlets in Banawa

H. SOCIOCULTURAL AND LEGAL ISSUES

1. Widespread availability of pornographic videos
2. Conflict over farming areas (often between outsiders and local people)
3. Weak law enforcement
4. Disuse of customary law
5. Relatively powerless local economic institutions at the village level

(Source: Rahmat et al. 2005, 336-37)

actions actually taking place, was truly remarkable" (Rahmat et al. 2005: 341). A further observation by the team also deserves attention:

> A part of the success of the conference in attracting and maintaining the involvement of such a large cast of disparate actors was certainly due to the strategy of ... clothing the affair in the symbolism and idioms of "traditional culture." An elaborate "traditional" opening ceremony, heavily featuring local costumes, customs and music, was designed and carried out,.... Early on, ... the team decided to name the event "molibu" after a local, but mostly dormant, mode of local consensus-building and decision-making at the village level. This clearly resonated well with local sensibilities at several levels, so much so that the Bupati [Head] of Donggala subsequently proclaimed that the process, or at least the term "molibu," be used across the district for similar purposes. (Rahmat et al. 2005: 342)

As a result, as with the Bali Sustainable Development Strategy reviewed above, a key lesson derived from the experience in central Sulawesi is that culture can be a central foundation for sustainable development.

CONCLUSION

At the outset of this chapter, we saw that "global interaction," or "globalization," has had multiple effects on the way in which environmental problems are addressed. The experience in Indonesia confirms the validity of this statement. Decisions taken outside of Indonesia, such as which tourist destinations will be chosen, or what food will be eaten, directly affect people and places in the country. Bali, seeking to diversify its economy and to contribute to the diversification of the national economy, has been promoting tourism. In so doing, it has also made its economic well-being vulnerable to decisions made by tourists, who can well decide to visit other island destinations. External influences also affect the country in other ways. For example, the demand for local products places pressure on Indonesia's natural environment. Consumers in other countries wanting sea products such as shrimp have provided a stimulus for local villagers in Central Sulawesi to cut down mangrove systems to create ponds in which intensive aquaculture can be pursued. Once mangrove systems are removed, coastal areas become vulnerable to erosion and degradation, and the waste from the aquaculture ponds can jeopardize water-supply sources for villagers. Further, as former common property resources are re-allocated to become owned or controlled by individuals or firms, traditional access to resources and the benefits from their products and services can be lost. There is therefore no doubt that international forces, which encourage larger shrimp-growing operations requiring more capital than individuals normally have, do have an effect on the country's economy and environment.

Indonesia has embarked on a journey intended to change quite significantly its system of government, moving from a top-down, directed system to one that is more democratic, decentralized and participatory. As Bobi Setiawan noted earlier in this chapter, the new political reality of Indonesia requires that local governments work together collaboratively and cooperatively to address environmental issues and problems that transcend local administrative boundaries (2005). This is never easy. Furthermore, as we have seen in this chapter, a clear lesson from experience in Indonesia is that if sustainable development is to be achieved, attention must be given to culture (Manuaba 1995; Knight et al. 1997; Boyle 1998). Sustainable development is often interpreted as requiring integration of both the environment and the economy. However, as the experiences in Bali and Central Sulawesi presented in this chapter suggest, culture also plays a fundamental role in the elaboration of sustainable development strategies. These two experiences also underline the fact that governments and institutions can effectively facilitate participatory processes of decision-making, and that these are more likely to flourish in an open rather than in an autocratic system.

If a participatory approach is a core part of sustainable development, then initiatives to ensure bottom-up involvement by local stakeholders are essential. The experience in Banawa Selatan highlights that, when decision-making processes are linked to traditional ways of consensus-building, the chances are good that they will resonate with local leaders and communities. The Future Search Conference process was explicitly linked to traditional cultural practices in the Banawa Selatan region, and that approach made local people comfortable with and accepting of the process.

Balancing needs, interests and aspirations at local, regional, and national levels is never an easy task. The task is further complicated when globalization factors influence behaviour and decisions. Moreover, it must be noted that external demands will continue to apply pressure on Indonesia's natural environment. To illustrate, Indonesia exports more tropical timber than any other country in the world, an economic activity that earns the country approximately US $5 billion a year. It is therefore not surprising to read reports from Indonesia suggesting that up to 75 per cent of the logging in the country is illegal, given that the motivation to earn foreign exchange is great. A second example refers to the "fallout" from the 1997 financial crisis. In 2002, then-president Megawati Soekarnoputri dissolved the agency responsible for environmental impact assessment (BAPEDAL), without indicating clearly which other agency would take on that function. Logically, the Ministry for the Environment assumed such responsibilities, but he was not assigned necessary law-enforcement powers. These complications notwithstanding, the experiences in Indonesia discussed here provide reason to be optimistic that sustainable development strategies, custom designed with reference to local contexts, can be conceived, designed, and implemented.

Globalization will continue to create challenges for Indonesia, and it is not apparent that the decentralized and deconcentrated approach identified above will always be effective in dealing with some global issues. For example, when the Kyoto Accord was signed in December 1997, it required 138 developed nations plus the European Community countries to commit to reducing greenhouse gas emissions by 5 per cent relative to 1990 levels by no later than 2012. Developing countries were not included in the Accord, because of the view that their per-capita greenhouse emissions were relatively low, and they had major needs to strengthen their economies to ameliorate poverty and poor livelihood conditions. However, Indonesia, along with China, India, and Brazil, have sizeable populations, and as their economies grow so do their greenhouse gas emissions. For example, Indonesia is now the third-largest producer of carbon emissions, after the United States and China, and this has been particularly accentuated due to forest fires in Kalimantan.

International pressure can be expected to grow steadily for Indonesia to reduce its greenhouse gas emissions. For action to occur, major leadership will have to be provided from the national level, as it is unlikely that local government officials will become leaders on this front. As a result, while there are many attractive aspects of Indonesia's shift to give more authority and responsibility to provincial and local levels, the central government will have to continue to examine and experiment with alternative governance models to ensure it is a credible participant in global issues such as climate change.

Notes

1 I appreciate the contributions of Tim Babcock, who shared his significant insight and experience from decades of involvement in development projects and research in Indonesia.

2 In Indonesia, these two terms are used together to highlight the fact that responsibility and authority are being moved from the centre, or national government, to the provincial and local governments.

3 See http://rainforests.mongabay.com/20indonesia.html.

4 The concept of Molibu will be expanded upon below.

References

Note: the following reference list contains a number of sources not directly cited in the chapter. They are included to provide ideas for further reading and research.

Armitage, Derek R. 2002. "Socio-institutional Dynamics and the Political Ecology of Mangrove Forest Conservation in Central Sulawesi, Indonesia." *Global Environmental Change* 12(3): 203-17.

—. 2003. "Traditional Agroecological Knowledge, Adaptive Management and the Socio-politics of Conservation and Development in Central Sulawesi, Indonesia." *Environmental Conservation* 30(1): 79-90.

Armitage, Derek R., and Achmad Rizal. 2005. "Collaborative Management in the Coastal Zone of Banawa Selatan, Central Sulawesi: Adat Practices, Institutions and the Regional Autonomy Process." In Susan Wismer, Tim Babcock, and Baharuddin Nurkin (eds.), *From Sky to Sea: Environment and Development in Sulawesi*. Waterloo, ON: University of Waterloo Department of Geography Publication Series No. 61: 227-64.

Babcock, Tim, Bill Found, Bruce Mitchell, and Susan Wismer. 2003. "Human and Institutional Enhancement for Environmental Management in Indonesia: The Experience of the University Consortium on the Environment, 1987-2002." *Canadian Journal of Development Studies* 24(1): 73-88.

Babcock, Tim, Bruce Mitchell, Baharuddin Nurkin, and Susan Wismer. 2005. "Sulawesi: The Environment, the People and the UCE Project." In Susan Wismer, Tim Babcock, and Baharuddin Nurkin (eds.), *From Sky to Sea: Environment and Development in Sulawesi*. Waterloo, ON: University of Waterloo Department of Geography Publication Series No. 61: 3-38.

Bater, James H. 1995. "Bali: Place and People." In Sugeng Martopo and Bruce Mitchell (eds.), *Bali: Balancing Environment, Economy, and Culture*. Waterloo, ON: University of Waterloo Department of Geography Publication Series No. 44: 3-18.

Bater, James H., Leonard Gertler, Haryadi, Drew Knight, Sugeng Martopo, Bruce Mitchell, and Geoffrey Wall. 1999. "Capacity Building for Environmental Management in Indonesia: Lessons from the Bali Sustainable Development Project." *Journal of Transnational Management Development* 4(3/4): 107-34.

Boyle, John. 1998. "Cultural Influences on Implementing Environmental Impact Assessment: Insights from Thailand, Indonesia, and Malaysia." *Environmental Impact Assessment Review* 18(2): 95-116.

Doberstein, Brent. 1995. "Improving EIA: A Regional Perspective." In Sugeng Martopo and Bruce Mitchell (eds.), *Bali: Balancing Environment, Economy, and Culture*. Waterloo, ON: University of Waterloo Department of Geography Publication Series No. 44: 351-72.

The Economist. (2000, July 8). "A Survey of Indonesia." 1-16.

Gertler, Leonard. 1993. "One Country, Two Concepts: Variations on Sustainability." *Canadian Journal of Development Studies Special Issue*: 103-22.

Knight, Drew, Bruce Mitchell, and Geoffrey Wall. 1997. "Bali: Sustainable Development, Tourism and Coastal Management." *Ambio* 26(2): 90-96.

Li, Tania. 2002. "Local Histories, Global Markets: Cocoa and Class in Upland Sulawesi." *Development and Change* 33(3): 415-37.

Li, Tania, ed. 1999. *Transforming the Indonesian Uplands: Marginality, Power and Production*. Amsterdam: Harwood Academic Publishers.

Manuaba, A. 1995. "Bali: Enhancing the Image through more Effective Planning." In Sugeng Martopo and Bruce Mitchell (eds.), *Bali: Balancing Environment, Economy, and Culture*. Waterloo, ON: University of Waterloo Department of Geography Publication Series No. 44: 29-42.

Martopo, Sugeng, and Bruce Mitchell, eds. 1995. *Bali: Balancing Environment, Economy, and Culture*. Waterloo, ON: University of Waterloo Department of Geography Publication Series No. 44.

Mitchell, Bruce. 1995. "Sustainable Development Strategy for Bali." In Sugeng Martopo and Bruce Mitchell (eds.), *Bali: Balancing Environment, Economy, and Culture*. Waterloo, ON: University of Waterloo Department of Geography Publication Series No. 44: 537-66.

Rahmat, M. Agus, Rodger Schwass, Tim Babcock, and Jack G. Craig. 2005. "Applying Search Conference Methodology to Community-based Management in the Banawa Selatan Coastal Zone." In Susan Wismer, Tim Babcock, and Baharuddin Nurkin (eds.), *From Sky to Sea: Environment and Development in Sulawesi*. Waterloo, ON: University of Waterloo Department of Geography Publication Series No. 61: 323-51.

Rizal, Achmad. 2002. "A Visioning for the Future of the People of Banawa Selatan, Central Sulawesi, Indonesia." Unpublished doctoral dissertation. University of Waterloo, Department of Geography.

Schwartz, Adam. 1999. *A Nation in Waiting: Indonesia in the 1990s*. 2nd ed. Boulder, CO: Westview Press.

Schwass, Rodger. 2004. "Developing a Vision." In Bruce Mitchell (ed.), *Resource and Environmental Management in Canada: Addressing Conflict and Uncertainty*. Don Mills, ON: Oxford University Press. 403-19.

Setiawan, Bobi. 2005. "A View of the UCE Project in its Indonesian Context." In Susan Wismer, Tim Babcock, and Baharrudin Nurkin (eds.), *From Sky to Sea: Environment and Development in Sulawesi*. Waterloo, ON: University of Waterloo Department of Geography Publication Series No. 61: 39-58.

Sloan, Norman A., and A. Sugandhy. 1994. "An Overview of Indonesian Coastal Environmental Management." *Coastal Management* 22(3): 215-33.

Vatikiotis, Michael R.J. 1999. *Indonesian Politics under Suharto*. 3rd ed. London: Routledge.

Vickers, Adrian. 1989. *Bali: A Paradise Created*. Berkeley-Singapore: Periplus Editions.

Wiesbord, Marvin, ed. 1992. *Discovering Common Ground: How Future Search Conferences Bring People Together to Achieve Breakthrough Innovation, Empowerment, Shared Vision and Collaborative Action*. San Francisco: Berrett-Koehler Communications Inc.

World Commission on Environment and Development. 1987. *Our Common Future*. Oxford and New York: Oxford University Press.

CHAPTER 5

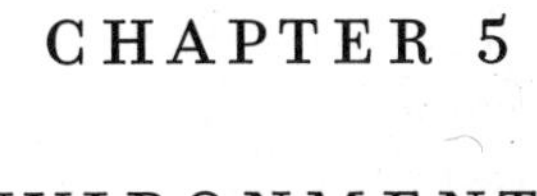

ENVIRONMENTAL CHALLENGES FACING INDIA

O.P. Dwivedi

INTRODUCTION

India's first environmental decade began in 1972, when its Prime Minister, Indira Gandhi, took the world stage in Stockholm, Sweden, during the United Nations Conference on Human Environment and declared, "We do not wish to impoverish the environment any further and yet we cannot for a moment forget the grim poverty of large numbers of people. Are poverty and need not the greatest polluters?" (Gandhi 1972). How to strike a balance between eradicating poverty and securing basic needs—food, health, drinking water, and shelter—in addition to attaining sustainable development, including upgrading the quality of the environment, have been major challenges facing India since that statement was made more than three decades ago. During this period, the country has faced a host of environmental problems: a continuous degradation of productive land in varying degrees (due to increased salinity and alkalinity, desertification, water logging, deforestation, etc.); shortages of fuel-wood and fodder for rural needs, which adds to pressure on forests; the depletion of forest cover, threatening the survival of indigenous bio-diversity and affecting habitat for wildlife; excess and unwise use of pesticides and fertilizers; and ill-advised agricultural practices, including monoculture, which place further stress on the fragile environment. In addition to all these alarming issues, India also faces challenges from the side-effects of industrial development and human re-settlements, including rapid urbanization and environmental refugees.

Understanding the interconnectedness of these issues through the country's legal and institutional arrangements is the main focus of this chapter. In addition to describing briefly the aforementioned environmental issues facing India, I first examine the major pollution-control laws created to protect the environment and then discuss the institutions established to implement and enforce such laws, outline the reasons for their ineffectiveness and the impediments that underlie certain institutional weaknesses, investigate the role of non-governmental organizations (NGOs) and other voluntary associations (including religion-based movements) that compel government agencies to act, trace the emergence of environmental justice and highlight selected court cases that have also forced governmental agencies to act, describe the country's involvement in the international environmental protection movement, and, finally, make some observations about India's efforts to secure an appropriate quality of life and well-being for its citizens. Essentially, I argue that, despite India's extensive environmental-planning and sustainable-development strategy, along with its industrial- and agricultural-development policies, a substantial change in the greening of India is not noticeable mainly because of the ineffective compliance and enforcement mechanism that is unable to prosecute offending industries, and also because of the corrupt atmosphere.

ENVIRONMENTAL CHALLENGES FACING INDIA

India faces numerous environmental challenges. I summarize each in turn in the following paragraphs:

(1) In 2001, India's population reached 1.027 billion (see Table 5.1 below), and is increasing at a rate of approximately 2 per cent annually, which means about 20 million people are added to India's population each year. In addition, India has probably the largest cattle population on earth: about 500 million domesticated animals that have only 13 million hectares for grazing (Dwivedi 1997: 7). The multiplying population of both human beings and animals is putting tremendous pressure on the environment in India. In the race for survival, both animals and human beings suffer. The worst consequence is the state of malnutrition (or undernutrition) of India's children: of approximately 166 million children found to be undernourished and underweight worldwide, India has by far the highest level—57 million, or 34 per cent, of the world total (UNICEF 2006: 3). If one adds those women and men who are equally malnourished, the number will reach a staggering proportion.

(2) The most brutal environmental assault that the country witnessed during the twentieth century concerns the nation's limited commons, such as grazing land, forests around villages that provided fodder and fuel-wood, land reclaimed from community ponds and rivers, private gardens consisting of fruit trees, and empty spaces around rural and urban communities. The relentless devouring of these commons by the poor and deprived has left the country bankrupt in publicly used commons and greenbelts.

(3) The forest cover is dwindling due to over-grazing and harvesting of trees for commercial use and domestic-fuel purposes, as well as to illegal encroachments. According to India's Planning Commission, the actual forest cover in the country in 1999 was 637,293 sq. km., approximately 19.39 per cent of the total geographic area of the nation compared to 640,819 sq. km. (19.49 per cent) coverage in 1987 representing a slight reduction of 0.10 per cent (India 2002). Though slight, such a rate of decrease has not only resulted in the extinction of some rare plant, animal, and microbial species, but it has also contributed to soil erosion, floods, and groundwater depletion. The continuing reduction in forest cover has also contributed to a proportional reduction in carbon sink and carbon sequestration (India 2002: 1055).

(4) India has hundreds of large industries, but only a few of them have any adequately installed water- and air-pollution abatement plants; and where such plants exist, many of these treatment plants do not operate throughout the year. In addition, there are many thousands of medium-size industrial plants that do not have any pollution-abating units. Furthermore, in many cities, industrial activities are located in city cores and are too close to the residential areas; emissions from these industries are causing many respiratory diseases. The situation gets further aggravated because of emissions from three-wheelers, cars, and trucks.

(5) India is the largest producer and consumer of pesticides in South Asia, and the most productive agricultural regions of the nation show a severe pesticide pollution problem, including the use of insecticides, herbicides, and fungicides. For example, more than 665 people died in 1989–90 because of pesticide poisoning, and thousands of cases of crippling resulting from pesticides have been recorded in the country (Venkataramani 1992: 129). In addition, the large-scale use of synthetic inorganic nutrients, such as fertilizers, has caused the widespread growth of algae, which has led to oxygen depletion and related problems due to the leaching effect of such nutrients, especially during the monsoon season.[1]

(6) Urbanization and migration from rural to urban areas have been steadily growing during the past fifty years, as Table 5.1 indicates. However, it is estimated that the rate of urbanization is going to be higher between 2001 and 2026, attaining an equal proportion with the rural population by 2026. The implication of this

TABLE 5.1

INDIA'S URBANIZATION, 1951–2051

Census Year	*Population (millions)*	*Urban Population (millions)*	*% of Urban Population*	*Rural Population (millions)*
1951	361.0	62.4	17.3	298.6
1961	439.1	78.9	18.0	360.2
1971	548.2	109.1	19.9	439.1
1981	683.3	159.5	23.3	523.8
1991	846.3	217.6	25.7	628.7
2001	1027.0	285.3	27.8	741.7
2026 (PROJECTED)	1419.3	505.8	35.6	913.5
2051 (PROJECTED)	1579.0	700.0	44.3	880.0

Source: For information between 1951 to 2001 census data, see India 2001a; for census estimates for 2026 and 2051, see Dyson et al. 2004: 103, 116, 120.

change will be greater at the municipal level for the quality of life, and will place numerous additional stresses on the environment.

(7) The Wildlife Tragedy. India has an impressive array of 88 national parks and 490 sanctuaries covering an area of 1,560,000 sq. km. (India 2002: 1065). In addition to tigers, whose estimated population in these parks and sanctuaries is around 3,500, there should be over 27,000 wild elephants, 300 Asian lions, and approximately 1,700 rhinoceros. But media reports show that the country's wildlife is disappearing fast. For example, the population of tigers from the Sariska Tiger Sanctuary, an area that is fenced off and controlled by the Forest Department, has nearly vanished (Thapar 2005). This is happening because poachers (who generally operate with the possible connivance of local forest department staff) are not controlled, and even if they are caught they rarely receive a speedy trial. It is obvious that those who should be guarding these wild animals are either ineffective or unaccountable. There is also the problem of a lack of coordination between the national and state government agencies with respect to the proper management of the country's wildlife (Thapar 2005).

ENVIRONMENTAL MANAGEMENT: THE LEGAL AND INSTITUTIONAL FRAMEWORK

India has enacted more than two dozen laws to protect its environment and prevent the consequences of environmental degradation on human health and property. Despite these laws, by 1970 India was faced with a major environmental crisis, and immediate and drastic actions were needed to cope with its magnitude. An analysis of these laws is significant for three reasons. First, it helps us to understand the evolution of India's environmental legal process: each major enactment points to a different set of political calculations, policy perspectives, and a different interpretation of costs and benefits. Second, these laws form the nexus of all environmental activities in the nation; despite being specific in nature, together they have built a common context for all environmental policy actions. Finally, these laws and rules constitute an emerging style of environmental regime in India (Dwivedi 1997). Let us look at India's main environmental legislation.

The Water Pollution Control Act

Even though water management is under state government jurisdiction, the federal government has the authority to intercede and legislate on a state's internal matters, provided that two-thirds of all state legislatures agree. For example, when the federal government realized that water pollution in the country was increasing at an alarming rate and had acquired nationwide significance while the state governments were inactive, it took the initiative to introduce water pollution control legislation, called the *Water (Prevention and Control of Pollution) Act*, in 1974. Section 2(e) of this Act defines water pollution as "such contamination of water ... or such discharge of any sewage or trade effluent or of any other liquid, gaseous or solid substance into water (whether directly or indirectly) as may, or is likely to create a nuisance or render such water harmful or injurious to public health or safety, or to domestic, commercial, industrial, agricultural or other legitimate uses, or to the life and health of the animals or plants or of aquatic organisms" (India 1974). It is a most comprehensive definition covering nearly all situations of water pollution.

Under this legislation, there is a Central Board of Water Pollution Prevention and counterparts at the provincial level. The board is responsible for coordinating the activities of, and resolving disputes among, state-level boards; providing technical assistance to the Ministry of Environment and Forests; carrying out, or sponsoring, research and investigation on problems regarding water pollution; setting water pollution control standards for the discharge of sewage and industrial effluents; and conducting inspection of sewage or industrial effluents, including municipal plants for the treatment of sewage or trade effluents (Sections 16 and 17). In addition, the state boards are granted the same rights within their territorial jurisdictions. Under the 1988 amendments to the Water Act, all industries—new or old—are required to obtain environmental clearance. In addition, penalties were doubled and the powers of the boards were more specifically defined.

A major flaw of this legislation is that it has no provision for prosecution of a government department or agency for an offence. This is because officials can use an escape clause by claiming that such an offence was committed without their personal knowledge or that they exercised due diligence to prevent the commission of the offence (India 1974: section 48). Other shortcomings of the Water Act are that its language is not all-inclusive, penalties are minimal, time limits are not set, public input has not been sought, and third-party rights are non-existent. Moreover, the legislation leaves out municipalities which are primarily responsible for treating residential waste, it does not include a provision for the affected public to directly launch prosecution against polluters, and the penalties that it established are low as polluting industries find the cost of defiance cheaper than the cost of compliance (Khator 1991). On the whole, the effectiveness of the Water Act remains dismal. This was in fact acknowledged by the Planning Commission of India, which stated that "three-fourths of surface water resources are polluted and 80 per cent of the pollution is due to sewage alone" (India 2002: 651). Furthermore, out of the 300 largest cities in the country, only 70 have partial sewage treatment facilities, and while these cities generate 15,800 million litres per day (MLD) of waste water, their treatment capacity is hardly 3,750 MLD. The Commission also stated that "most of the cities have only primary treatment facilities. Thus, the untreated and partially treated municipal wastewater finds its way into water sources such as rivers, lakes and ground water" (India 2002: 646).

The Air Pollution Control Act

In 1981, the federal government enacted the *Air (Prevention and Control of Pollution) Act*. The legislation defined air pollution as the presence of "any solid, liquid or gaseous substance (including noise) present in the atmosphere in such concentration as may be or tend to be injurious to human beings or other living creatures or plants or property or environment" (India 1981: section 2d). Similar to the Water Act, the Air Act also specifies penalties and prosecution guidelines. The Act also stipulates that no one can, without the previous consent of the State Board, operate any of the following new industrial plants in an air pollution control area: iron, steel, or nonferrous metal foundries; mining and ore processing; coke ovens and coal processing; petroleum, petrochemical, power and boiler, chemical and allied industries; ceramic, cement, and textile industries; processing plants for animal and agricultural products; and plants for recovery from and disposal of wastes, including incinerators. However, there is a provision (unlike the water pollution legislation) in the Air Act for prosecution of a government department or agency for an air-pollution-related offence, although the management of such an offending party can, as with the Water Act, use an escape clause by claiming that such an offence was committed by subordinates without their personal knowledge, or that they exercised due diligence to prevent the commission of the offence. The Air Act also suffers from the same shortcomings that the Water Act does in the areas of language, penalties, time limits, public input, and third-party rights. In addition, the administrative support for the implementation is also questionable.

The Environmental Protection Act

A gas leak from Union Carbide's Plant in Bhopal caused a major environmental disaster in December 1984. Forty tons of highly toxic gas—Methyl Isocynate (MIC), which was manufactured and stored in the Union Carbide Corporation factory in Bhopal—leaked, killing an estimated 3,000 people and sickening approximately 200,000 others (Baxi and Paul 1986). The catastrophe generated extraordinary interest both in India and abroad. The realization that an absence of adequate laws to deal with industrial disasters and to cover other aspects of environmental pollution (not mentioned under the Water Pollution or Air Pollution Acts) would bring other national catastrophes impelled the federal government to pass the *Environmental Protection Act* (EPA) in 1986. The Act has three major objectives: to be a supplementary rather than a new law; to provide a clear focus of authority; and to "plug loopholes" in the existing laws. The Act provided the first official definition of the term "environment" as including "water, air and land and the inter-relationship which exists among and between water, air and land, and human beings and other living creatures, plants, micro-organism and property" (India 1986: section 21). Furthermore, the term "environmental pollutant" was defined as "any solid, liquid or gaseous substance present in such concentration as may be, or tend to be, injurious to environment" (India 1986: section 26).

One special feature added to this legislation is the one allowing the federal government to delegate authority and to use existing resources to implement its provisions as well as to have, unlike the Water Act or the Air Act, countrywide jurisdiction. Other features include doubling the penalties, and the third-party right to take the offending party to court after giving sixty days notice to the central government (India 1986: section 196). Finally, the legislation authorized government to issue directions to any person, officer, or authority, including the power to direct closure, prohibition, or regulation of any industry, operation, or process, or stoppage or regulation of any supply of electricity, water, or any other service; and to delegate powers to any officer or state government or authority (India 1986: section 3:2). Another important feature was that no distinction was made between offending government departments and private companies. However, the Act fell short of environmentalists' expectations: it barred civil courts from entertaining environmental complaints, it freed heads of firms and industries from assuming full responsibility, and it denied citizens the right to sue the government for ignoring its duties.

Other Environmental Legislation

Other laws and regulations related to environmental protection include the Hazardous Waste Management and Handling Rules 1989, the Wildlife Protection Act 1972, the National Environment Tribunal Act 1995, the National Environment Appellate Authority Act 1997, the Municipal Solid Waste (Management and Handling) Rules 2000, and the Public Interest Litigation (PIL) Act. Among these, PIL has been effectively used by many NGOs and other activists. Under PIL, courts are empowered to grant relief and prevent any activity that may endanger humans and damage the environment. PIL can be brought by any person who may not even be an aggrieved party to a dispute; and, while dealing with such a petition, courts do not have to follow the detailed procedural formalities if the problem is of national interest or may immediately harm people or the environment. But the main purpose of PIL is to compel government and its agencies to carry out their statutory or constitutional obligation when the public feels that there exists ground for the denial or violation of human rights or persons in misery (Shastri 2002: 52). As discussed in a later section of this chapter, PILs have become major instruments for highlighting environmental crises and the misery of people affected by pollution.

Establishing Institutional Mechanism

The year 1972 marked a watershed in the history of environmental management in India. Prior to 1972, environmental concerns such as sewage disposal, sanitation, and public health were dealt with by different ministries of the Government of India, and each pursued these objectives in the absence of a proper coordination system established at the federal or the intergovernmental level. But when the 24th UN General Assembly decided to convene a conference on the Human Environment in 1972, and requested a report from each member country on the state of the environment, India took immediate steps by setting up a National Committee on Environmental Planning and Coordination (NCEPC) on April 12, 1972 to bring about greater coherence and coordination in environmental policies and programs and to integrate environmental concerns in the process of planning for economic development. Over the years, the NCEPC acted mostly as an advisory body in matters relating to environmental protection and improvement (see Dwivedi 1997). However, in 1976, India became one of the first countries in the world to amend its constitution to provide a national commitment to environmental protection and improvement; under Article 48A of the Constitution Amendment, it was declared that "The state shall endeavour to protect and improve the environment and to safeguard the forests and wildlife of the country" (India 1976). Despite the constitutional amendment, due to electoral politics it was not until November 1980 that a separate Department of Environment was established. Thereafter, the Government of India took the environmental challenge more seriously, for example, by separating budgetary allocation in the Five Year Plans exclusively for environment and development, and by emphasizing sound ecological principles in land-use planning, agriculture, forestry management, wildlife preservation, control of water and air pollution, protection of marine and coastal zones, fisheries, renewable resources and energy, and human settlements. In January 1985, the Department of Environment was changed to the Ministry of Environment and Forests (MEF), and a senior minister with the rank of cabinet minister was appointed to upgrade the status of the environmental agency. Finally, the passage of The Environmental Protection Act of 1986 gave ample authority to enhance the policy and institutional base of the agency. By the end of financial year 2005-06, the MEF had a complement of 1,069 employees and a budget of Rupees 1100.18 crores (equivalent to US $245 million) (India 2006a: 219, 229).

AN ASSESSMENT OF INDIA'S ENVIRONMENTAL LAWS

India's environmental laws appear adequate on the surface, but the infrastructure to support them in terms of human resources and institutional capacity is highly inadequate. The negative implications of this include the following: "chemicals already banned or obsolete in other countries are still being produced in India. In other cases, relatively dirty industries or processes which find themselves under considerable economic and environmental pressure in developed nations, have been installed in India, exacerbating the environmental problems associated with industrial sources" (India 1993: 186). This inadequacy refers primarily to human-resource and financial limitations. The concern regarding inadequate expertise on the part of environmental officers is understandable: at any given time in India, there are a large number of development projects at the planning or implementation stage, but there are not enough trained professionals and technical experts to execute them. This has often led to delays in the clearance of projects. Worse, projects have been cleared without being properly assessed, and consequently they are not environmentally sound. The EPA, the Water Act, and the Air Act use primarily regulation-enforcement mechanisms rather than effluent-charge strategies as a means of securing compliance. While higher courts have taken an active interest in protecting the environment, such actions are limited only when cases reach the courts. Instead, there is a need to strengthen the regulatory mechanism to monitor pollution-generating activities and then to prosecute offending industries and businesses. The government of India has already acknowledged through its planning document, the Tenth Five Year Plan 2002-2007, that "the country has adopted almost all environmental protection Acts and rules enforced in developed countries. But environmental degradation continues despite the existence of a long-standing policy, and legal-cum-institutional framework for environmental protection. The need for reducing the gap between principle and practice cannot be over-emphasised" (India 2002: 1075).

Reliance on conventional regulatory mechanisms makes the environment an exclusively governmental problem, with little incentive for pollution prevention to emanate from industry and an insignificant ability to integrate citizen concerns. Due to this gap, environmental protection depends on voluntary and self-enforcement by polluting industries and private enterprises. This weakness (inadequate mechanism for enforcement and compliance) was acknowledged by the federal government in its Environment Action Programme published in 1993: "Despite legal mechanisms for environmental management, only about 50 per cent of the large/medium scale industries have provided complete/partial emission/effluent control systems and many of these do not achieve [the] stipulated standards. Further, the small-scale industries (SSIs) have not yet been subjected to rigorous pollution control" (India 1993: 185). What are these institutional weaknesses or impediments in enforcing environmental legislation? These are examined below.

IMPEDIMENTS TO EFFECTIVE ENFORCEMENT OF ENVIRONMENTAL LAWS AND REGULATIONS

The case of ineffective enforcement and weak regulatory mechanism for environmental laws and institutions in India was highlighted as early as in 1980, when the N.D. Tiwari Committee, appointed by then-prime minister Indira Gandhi, lamented the major problem faced by the government concerning poor enforcement and compliance of environmental laws and regulations. The Committee reported that "many of the existing laws are updated (in a limited sense) versions of earlier ones which predate the independence era.... The implementing and monitoring machinery of many of these laws are lacking in scientific and technical expertise, as well as other infrastructural resources required to assess and prevent the possibility of adverse environmental impacts" (India 1980: 27-28). Even after more than 20 years, the situation does not appear to have improved much, as India's Planning Commission noted (India 2002: 1075; see above). But how can the gap identified by the Commission be reduced? I have selected the following two issues to illustrate this problem: an inadequate system of compliance and enforcement, and the prevailing atmosphere of corruption.

The Inadequacy of Compliance and Enforcement Mechanism

While over two dozen environmental laws exist, a solid, forceful, and comprehensive administrative network has yet to be established (Dwivedi and Khator 1995: 47-70). More often than not, administrative guidelines are established without a mechanism to follow up on them. The laws governing water pollution, air pollution, and environmental protection are prime examples of this paradox. The administrative machinery created for the enforcement of these three pollution-control laws has failed on several accounts. Among other things, clear-cut lines of accountability have not been defined, and over-centralization of legislative power has crippled innovation. Ironically, under the Indian federal system, the responsibility for environmental compliance and enforcement is delegated only to state governments, but the authority to manage it is still centrally controlled. It is also evident that if enforcement agencies are not adequately administering the law, then the public must use judicial recourse to force both the administrative machinery, and the industrial sector, to pull their weight. The Ratlam Pollution case is a good example in which the Supreme Court of India upheld the concerns of citizens who complained of unhygienic and unsanitary conditions from the dumping of raw municipal and industrial wastes. The case illustrated how an activist court can transform weak legislation into a powerful tool to protect the environment (Shastri 2002: 52-53). Judicial activism acts as a powerful ally for the environment.

Corruption

One of the reasons for the ineffectiveness of India's mechanisms of environmental enforcement and compliance is related to the country's political and administrative culture. Specifically, enforcement and compliance are hampered by political and bureaucratic corruption and the ability of various interests to "capture" government officials. The main hurdle in reducing corruption is the emergence of the popular belief that corruption in public institutions has become a way of life. Bureaucrats are seen as being at the beck and call of whichever politicians happen to be in power. It is therefore not surprising that the public views with cynicism any claim by a political leader that his or her party will conduct an ethical administration. The public in India hold the view that the bulk of the Five Year Plan budget outlay is eaten up by corrupt politicians, officials and their favourite contractors, and because some environmental inspectors demand financial or other considerations from an industry, they are not surprised by the lack of vigour in the application of environmental rules and standards. Of course environmental inspectors succumb to bribes partly because they are poorly paid, partly because of the political culture prevailing in the nation, and also partly because the punishment is not severe enough to deter either them or the polluter. In other words, both inspectors and polluters have an incentive to cheat. Sometimes, when an inspector takes a hard-nosed approach to control pollution, the industry may be able to influence his or her senior officers to get him or her transferred or otherwise harmed. Even when a surprise inspection is carried out, the factory or the industrial unit in question may be able to secure a warning in advance (Dwivedi and Mishra 2005).

ISSUES OF INSTITUTIONAL WEAKNESSES

Three aspects help illustrate the specific nature of some institutional weaknesses concerning environmental management in India. The first one relates to a comprehensive, national environmental policy. Writing about the absence of a coherent national environmental policy for India, I lamented in 1997 that the country needed an environmental action plan along with a national environmental policy that would delineate appropriate enforcement and administrative mechanisms and that would establish "explicit links between the environment and sustainable development" (Dwivedi 1997: 198). The absence of such a policy created jurisdictional fights among government departments and ministries, it continued the weak accountability system existing at the federal and state government levels, and it did not stimulate partnership with different stakeholders, including international development partners. Finally, on May 18, 2006, such a policy, the National Environment Policy 2006, was approved by the Government of India (India 2006b). The policy is intended to "mainstream environmental concerns in all development activities. It briefly describes the key environmental challenges currently and prospectively facing the country, the objectives of environment policy, normative principles underlying policy action, strategic themes for intervention, broad indications of the legislative and institutional development needed to accomplish the strategic themes, and mechanisms for implementation and review" (India 2006b: 2-3). It took more than three decades for the government to usher in such a national policy. It is too early to comment on its effectiveness, but it does fill a major gap in environmental policy planning.

The second aspect refers to the inadequate administrative capacity that generates weak compliance. The Indian administrative system is a good example of several impediments working simultaneously: the effective wielding of power and political influence by public leaders and businessmen; a rigid hierarchical structure; the suppression of junior officers and the blunting of their initiatives; and an overconcentration of decision-making power at the top, which prevents those in the lower echelons from realizing their full potential. An additional problem is the existence of an "administrative trap"—a situation in which policies may exist, but their administration and enforcement are mostly inadequate. For India, the argument that there is a lack of adequate legal mandate can hardly be substantiated, given that there are sufficient environmental laws and regulations to cover various environmental issues to a reasonable extent. (However, there are some areas of environmental protection, such as solid-waste management and garbage dumps, nuclear waste, noise pollution, wetlands, mangroves, and deserts that clearly are in need of better control and regulation.) The overall impact of this arrangement is that it has produced an environmental management system that is inherently weak and ineffective, and enjoys little or no public support. Compliance requires voluntary obedience to the law and mutual trust between the regulators and the governed; in the absence of this, the goals of the National Environmental Policy cannot possibly be realized.

Finally, weak governance creates ineffective management practices. With the ever-growing role of governing institutions, the state's capacity to deliver such functions has declined over time "due to administrative cynicism, rising indiscipline, and a growing perception that the political and bureaucratic elite views the State as an arena where public office is to be used for private ends. In almost all States, people perceive bureaucracy as wooden, disinterested in public welfare, and corrupt" (India 2002: 22). This has led to weak governance, which manifests itself in poor and sometimes non-existent service delivery, excessive regulation, and uncoordinated as well as wasteful public expenditure. A change is required to ensure that critical social, economic, educational, environmental, and related developmental services are provided by reshaping the government administration and by bringing a new work culture that takes their responsibility and duty seriously rather than always asking for more privileges and benefits without concomitant productivity and work ethic. It has been suggested that poverty alleviation requires a significant outlay of funds. However, past experiences indicate that a shortage of funds is not the only reason for not reducing poverty or maintaining environmental quality. Instead, poor governance is the culprit.

These three impediments have continued, partly due to the absence of grassroots support (including the apathy of both the public and political parties, and disinterested business groups who are able to influence party leaders and senior bureaucrats), and partly due to the uncoordinated and sometimes politically motivated civil society groups. The situation has changed since the 1990s, however, when NGOs started asserting their rights to question not only the inadequacy of the environmental regulatory system but also the effects of mega-developmental projects, an aspect to which I now turn.

THE ROLE OF CIVIL SOCIETY

By the time various environmentalist organizations or environmental non-governmental organizations (ENGOs) established their standing in India, some religion-based groups had already started the movement for protecting the environment. In this section, I discuss three fields of civil society activities: the role of ENGOs, religion-based civil society groups, and the emergence of environmental justice.

ENGOs in India

There has been a mushrooming of ENGOs in India, especially since the late 1980s when many governmental agencies decided to extend support either through direct funding—granting intervener status when conducting environmental impact assessments, contracting them for research, approving funds for organizing conferences and seminars—or through supporting their environmental education and awareness programs. In addition, foreign governments through their international aid agencies have supported these ENGOs. As of March 2006, there were 49 ENGOs registered with the MEF, although the number is much larger if one includes those operating at the provincial and local government levels. For example, in 1989, according to WWF-India, there were 908 ENGOs and voluntary organizations registered for work on environment-related causes (Prasad 1991). By 2006, however, their number had increased to approximately 34,000 (IndianNGOs.com 2006). Nevertheless, until the late 1970s, the role of NGOs in the area of environmental issues was not recognized by governmental agencies as important enough to seek their views in the policy formulation process. However, as a result of rising public awareness and with the help of the judiciary,[2] by the early 1990s the role of ENGOs had begun to be acknowledged by the federal government by permitting them to have *locus standi* in environmental cases as well as in related development activities. A federal policy statement also recognized the need to work with NGOs "on environmental surveillance and monitoring, transmitting development in science and appropriate technology to the people at large" (India 1992: 35). In addition, the role of women's organizations related to environmental protection has been encouraged. Indeed, the Chipko movement (discussed below) is one of the earliest and more successful NGOs, and it was spearheaded by women.

One of the most celebrated examples of public participation and NGO activity is the Sardar Sarovar Dam controversy, in which the Narmada Bachao Andolan (movement to save Narmada Valley), under the leadership of Medha Patkar, has played a key role. But the people of India have another weapon with which to tackle bureaucratic hurdles and insensitivity on the part of governmental machinery, and they have also been very successful with it: the Gandhian instrument of *satyagraha*.[3] For example, Van Samitis (the forest committees that are located in the Palamau district of Bihar province), the Van Suraksha Samiti (the forest protection committee in Tehri Garhwal, Uttaranchal province), and the Kerala Sastra Sahitya Parishad (Kerala province), which took the cause of saving Silent Valley from inundation, are some specific examples of how local-based NGOs have been successful in opposing certain governmental development projects that could have brought deforestation or flooded the fragile environment.[4]

There is an erroneous perception in India that there are only two groups that ought to be recognized as the main stakeholders in environmental protection and conservation: governmental institutions and environmental NGOs. Nevertheless, India has a significant number of areas in which tribal people become easy victims of various developmental projects encroaching in their areas. These people are resisting changes as activities causing deforestation, and mining and dams constructed for generating hydro power, are affecting their health, indigenous culture and, evidently, their natural environment. For example, in the Chhotanagpur region of Bihar two steel plants and some chemical industries have been set up because of the abundance of mineral deposits. Large-scale mining operations there have not been careful enough with respect to their industrial effluents and wastes, and the water of the Subarnarekha River now carries six milligrams of cyanide per litre of water, which is extremely dangerous (Upadhyaya 1991). Moreover, one should not exclude cultural and religious organizations from playing a role in environmental protection and helping to create and strengthen an environmentally conscious society. People are likely to accept policies and programs if they can be culturally and spiritually supported. This dimension is examined next.

Religion-based Environmental Movements

The earliest religion-based movement is known as *Chipko*. In March 1973, in the town of Gopeshwar in Chamoli district, in the Himalayas (Uttaranchal Province), villagers formed a human chain and encircled earmarked trees to keep them from being felled for a factory producing sports equipment. The movement's leader, Chandi Prasad Bhatt, declared, "let them know we will not allow the felling of a single tree. When their men raise their axes, we will embrace the trees to protect them" (Shephard 1987: 69). This incident, embracing (or hugging) trees by making a human chain, gave this movement the name of *Chipko*. A year later, the movement acquired a special significance: when the men had gone to a district office to receive compensation, leaving only women, children, and older people, the loggers returned to begin felling trees, and thirty women marched into the forest to protest against logging by hugging trees. The women declared that the forest was like their mother, and before anyone dared to cut a single tree, they had to shoot them (James 2000). After heated confrontation, loggers turned back without felling a single tree. Those thirty women used the Gandhian philosophy of *Satyagraha* to fight against an ecological abuse. Since then, the *Chipko Andolan* (movement) has grown from a grassroots eco-development movement to a worldwide phenomenon. *Chipko* is, in one sense, an eco-feminist movement that affirms the spiritual value of nature; in another sense, it is an eco-feminist movement to protect nature from the greed of men. Moreover, these women specifically saw how men would not mind destroying nature in order to get money, while the women had to walk miles in search of firewood, fodder, and pasture for their cows and goats. The movement, which began as the traditional rights of peasant women to their forest resources, has gradually expanded to include issues of wider ecological concerns, such as the protection of Himalayan ecology and protesting against the construction of Tehri Dam, as well as taking the message to other parts of the world. The *Chipko* movement became a crusade for the preservation of earth and the acknowledgement of the feminine as the creative principle of the cosmos (James 2000).

A few years after the *Chipko* movement emerged, another religiously inspired movement was formed in 1983 in Karnataka, South India, when 163 men, women, and children hugged the trees and forced the loggers to leave. Later, a similar movement started in another province, Kerala, called *Kerala Shastra Sahithya Parishad* (KSSP). The movement successfully saved a genetically rich rainforest in Kerala Silent Valley from being submerged due to the construction of a hydro dam (Bishnoi 2002). The Silent Valley, which is also a National Park, is the first biosphere reserve in India, and perhaps the only remaining undisturbed tropical rainforest in peninsular India.

The Emergence of Environmental Justice

Indian ENGOs have been singularly effective in mobilizing public support for controlling certain polluting industries and constraining governmental planning to build mega-dams. The country's highly committed and extremely active ENGOs are regarded as the true enforcement mechanism to implement environmental regulations by acting as "catalysts for numerous public interest litigations that have resulted, *inter alia*, in closure orders for tens of thousands of polluting industrial units" (Mushkat 2004: 66). These organizations have also started a special advocacy system by launching "people's tribunals" that can hear cases filed by communities affected by environmental pollution. The best example of how ENGOs were able to influence certain mega-dam projects is the Narmada Valley Projects. The Narmada River, the largest westward flowing river in western India, is considered one of the sacred rivers, equal in status with Ganga. The state of Gujarat, which is a water-scarce province, wanted a series of dams to be built on this river so that its water could be diverted to drought-prone northern areas. A developmental project aims for 30 major, 135 medium, and about 3,000 minor dams in the Narmada valley (Fisher 2000: 403; McCully 1996). The main dam, perhaps the largest in Asia, is called Sardar Sarovar and will affect about 100,000 people (mostly tribal and rural) who have been and still are being displaced from their ancestral surroundings. The project has been financed by the World Bank. From the beginning, environmental groups, led by the Gandhian activists Baba Amte and Medha Patkar, have spearheaded the movement to save Narmada from further damage. The movement includes tribal people and eco-feminists. The Narmada valley project illustrates how one may justify the redistribution of natural resources from low-resource-use populations to high-resource populations. From the perspective of the industrialized public of Gujarat, Narmada's water either gets wasted or remains unused. On the other hand, tribal and rural communities from areas where dams are being built consider that they are being forced to accept a change in their traditional space, that they are being marginalized, and that they are being uprooted so that someone else, living thousands of miles away, may enjoy many modern facilities at their cost. The controversy combines social justice with environmental protection by highlighting the needs of poor and tribal people against powerful urban elites and industrial magnates.

In addition to this example, there have been other cases of popular victories in environmental issues. One relates to the Bhopal industrial tragedy, discussed earlier. Because the owner of the plant was located in the United States, it took more than ten years to settle the case of compensation for people affected by that industrial tragedy. The case was decided by the Supreme Court of India in 1989, with a US $470-million fine being imposed on the corporation (Shastri 2002: 3). The Supreme Court of India decided that Union Carbide was responsible for paying compensation to all those injured and to the families of the dead, as well as to those children whose congenital defects were traceable to inherited MIC toxicity (Shastri 2002). A principle of absolute liability was established against polluters.

A third example is the Taj Mahal case. International concern was generated in 1993 when the public noticed deterioration of the famous Taj Mahal in Agra, India. NGOs represented by M.C. Mehta, a lawyer and environmental activist, using the provision of public interest litigation under section 19 of the Environmental Protection Act (1986), took the case to the Supreme Court, which eventually ordered the closure of 212 polluting industrial units in and around Agra. The ruling stated that wilful violators of air-emission standards were contributing directly to the deterioration of this historic seventeenth-century monument. As noted by Mehta, 511 factories had no pollution-control devices (Mehta 1994). However, due to the intervention of the Supreme Court, the provincial government was compelled to force those factories to install air-pollution-control devices.

A fourth example refers to the Ganga Pollution cases. In 1985, Mehta filed a writ petition under Article 32 of the Indian Constitution stating that neither the concerned government agencies nor industry were taking adequate steps to mitigate the flow of pollution into the Ganga. The petition asked the Court to order governmental authorities and tanneries at Jajmau near Kanpur city to halt the discharge of effluents into the river until appropriate treatment facilities had been installed. As a result, 29 tanneries were ordered closed for failing to provide even primary treatment of their waste. However, between 1987 and 1993, in the state of

Uttar Pradesh alone, 860 industries were identified as major water-polluting cases. Out of these 860, 191 were declared in September 1993 as "wilful defaulters" and were subsequently ordered closed by the Supreme Court of India (Dwivedi 1997: 107). Of the remaining "water polluting" industries, 493 installed effluent treatment plants. With the Ganga Action Plan, there have been improvements in water quality, although there are stretches where BOD (Biological Oxygen Demand) exceeds the permissible limits of water quality criteria (India 2001a: 18).

Notwithstanding their healthy growth and some successes related to environmental issues, NGOs have yet to be accepted by governmental machinery as equal partners, simply because some of them have aligned themselves with the existing adversarial political process. At the same time, ENGOs are generally distrustful of the actions of environmental ministry or department officials: from their viewpoint, governmental machinery is inflexible and unimaginative, and most government officers do not show commitment to the cause of environmental conservation and protection.

INTERNATIONAL COOPERATION FOR SUSTAINABLE DEVELOPMENT AND ENVIRONMENTAL PROTECTION

Among all developing countries, India has been at the forefront of international cooperation for sustainable development and environmental protection. India's involvement with international cooperation in issues concerning the environment, forestry, and wildlife started in conjunction with the 1972 Stockholm Conference. The Indian government not only took the lead to prepare an extensive report on the state of its environment, but it also created a National Committee on Environmental Planning and Coordination in April 1972. Then-prime minister Indira Gandhi, in her inaugural address at the Committee, remarked, "As our development progresses perhaps we also shall have environmental problems akin to those in the developed countries. But we can learn from their experience ... Concern with economic and social development need not be a choice between poverty and pollution" (Gandhi 1972: 1). India has tried to balance that choice by not only creating adequate legal mechanisms but also by being cognizant of its role in protecting the international environmental regime and leading the fight for international environmental sustainability through international treaties, conventions, protocols, and agreements. Table 5.2 lists India's cooperation in various international environmental decisions.

To coordinate international environmental activities, India has established an International Cooperation and Sustainable Division within its Ministry of Environment and Forests (MEF). This division is responsible for keeping liaison with the following international agencies: UNEP, the Commission on Sustainable Development, the Global Environmental Facility, the South Asia Cooperative Environment Programme, and the European Union. While the division serves as a nodal agency for various international activities, its major emphasis is on the Ozone Layer Protection Programme through implementation of the provisions of the Montreal Protocol and the UN Framework Convention on Climate Change (UNFCCC). Among these two, the UNFCCC aims to bring the level of greenhouse gases (GHG) back to the pre-industrial era. India has committed itself to cut its GHG emission to the 1990 level. The Kyoto Protocol, which was ratified by a requisite number of states, has a provision whereby industrialized nations may get credit for investing in developing nations in projects that would reduce GHG emissions from drilling at offshore stations. India and the Netherlands have, as a result, entered into negotiations where the Netherlands will assist India in transferring emission-reduction technology. In addition, India committed itself in 2006 to freezing the production and consumption of CFCs at the 1999 level (India 2006a: 44).

With respect to various international meetings, such as the Earth Summit of 1992 and the World Summit of 2002, India not only prepared its "country report" (while many developing countries simply prepared briefing notes) but also showed progress made in controlling pollution, specifically with respect to Agenda 21. Under the joint sponsorship of the World Bank, UNDP, and the UNEP, the GEF[5] has been created to provide grants and low-interest loans to developing countries to relieve pressures on global ecosystems. Each year, eight to 12 projects are considered and approved for implementation on such subjects as biodiversity, climate change, and land degradation (India 2006a: 43). As these projects relate to India, it shows the commitment of the nation not only to implement certain international projects but also to set an example to its neighbours. For example, in 2003, India hosted the eighth UN Framework Convention on Climate Change, where 4,300 delegates from all over the world took part. During this meeting, India helped to create a Special Climate Change Fund with guidelines for bringing down emissions of greenhouse gases. Further, under the Montreal Protocol, India agreed to freeze CFC production and consumption at 1999 levels and to freeze halon production at January 2001 levels. Indications from government sources are that these targets will be met by the 2015 deadline. Finally, India is cognizant of the fact that environmental quality among its neighbours must be improved if it wants its domestic quality of life upgraded. Toward reaching this objective, India has taken serious interest in the functioning of the South Asia Cooperative Environment Programme (SACEP).

Part of the reason why India takes a special interest in international environmental issues is because of its culture, which exhorts Indians to treat the planet earth as Mother Earth, a concept derived from its Vedic culture called *Vasudhaiv Kutumbkam*. According to this notion, we are all members of the same extended family of our Mother Earth. As such, India would like to see especially all the developing nations reaching a better level of environmental quality; and to reach that goal, India sometimes takes a stand that is not favoured by many Western nations.

TABLE 5.2
INTERNATIONAL TREATIES / CONVENTIONS / DECLARATIONS SIGNED BY INDIA

Agenda 21
Rio Declaration
Stockholm 1972
Convention on Biological Diversity
Convention to Combat Desertification
Stockholm Convention on Persistent Organic Pollutants (POPs) (Signed in May, 2002 but to be ratified by India)
Prior Informed Consent (PIC), Rotterdam Convention (For certain Hazardous Chemicals in International Trade) (Ratified on May 24, 2005)
Cartagena Protocol on Biosafety (India ratified it on January 17, 2003)
The Basel Convention on the Control of Transboundary Movements of Hazardous Wastes
Kyoto Protocol to the United Nations Framework Convention on Climate Change
Intergovernmental Panel on Climate Change (IPCC)
Convention on International Trade in Endangered Species of Wild Fauna and Flora (CITES)
World Trade Agreement
Helsinki Protocol on the Reduction of Sulphur Emissions or their Transboundary Fluxes by at least 30 per cent
Sofia Protocol to LRTAP concerning the Control of Emissions of Nitrogen Oxides or their Transboundary Fluxes (NOx Protocol)
Geneva Protocol to LRTAP concerning the Control of Emissions of Volatile Organic Compounds or their Transboundary Fluxes (VOCs Protocol)
United Nations Conference on the Human Environment
United Nations Conference on Environment and Development (UNCED)
Framework Convention on Climate Change (FCCC)
Convention Secretariats of the UNEP

Source: India 2006a.

CONCLUSION

In this chapter I have briefly outlined the efforts made by the Government of India and its MEF to forge various policy and legal means to address the minimization of risks to humans and the environment. As we have seen, it was not until the late 1980s that the political system took serious notice of the environmental issues facing the nation. Since January 1985, the MEF has worked hard to reach its legislative and administrative goals. If there has been any failure in reaching these goals, the fault cannot be laid solely at the door of the ministry and its officers. Instead, other key political and institutional players—such as political parties, powerful ministries, state and local governments, business and industrial groups, and even international development agencies that have their own environmental priorities—must share the blame. Nevertheless, the federal planning strategy, as well as its overall industrial and agricultural development policies, has been unable to entrench a substantive policy for the greening of India. This is what the Government of India, through its National Environment Policy statement, acknowledged in 2006: "[the] key environmental challenges that the country faces relate to the nexus of environmental degradation with poverty in its many dimensions, and economic growth" (India 2006b: 3). It is also clear that environmental degradation is a "major causal factor in enhancing and perpetuating poverty ... when degradation impacts soil, quantity and quality of fresh water, air quality, forests, and fisheries" (India 2006b: 4). It is astounding that similar views were expressed 25 years ago by Indira Gandhi during the 1972 Stockholm Conference. Does this mean that during the past 25 years the situation has not improved much?

There is no doubt that responding to various environmental challenges and facing an atmosphere of increased international economic competitiveness and liberalization of the economy, the federal government has been impelled to inject more resources into the environmental sector. For example, between the Eighth Five Year Plan (1992-97) and the Tenth Five Year Plan (2002-2007), budget allocation for the environment and forestry sector increased from 6,750 million rupees to 59,450 million, approximately an eightfold increase (India 2002: 1076; Dwivedi 1997: 67). Nevertheless, the onslaught on natural resources continues while pollution generated from industrial and municipal activities remains unabated. If Indian leaders do not wish to see the vital natural heritage of their country squandered, they should consider strengthening the financial, administrative and regulatory capability of the Ministry of Environment and Forests and its counterparts at the state level. In the absence of adequate safeguards for the environment built-in at all stages of planning, policy-making, allocation of resources, and implementation, all administrative steps taken toward vigorous monitoring and ameliorative action will not bear fruit. Most importantly, unless the public is convinced that protecting the environment and controlling pollution are critical to safeguarding their personal health and private property, all legislative, administrative, and enforcement measures will remain ineffective.

Ultimately, by reducing pollution and conserving the natural heritage, which are the biggest challenges facing the nation, its prosperity will be enhanced and its environmental problems kept under control a generation earlier than expected. As it has been stated, "there is no reason why India cannot aspire to prosperity on a broad basis and a substantial improvement in the environment and the health of her people" (Dyson et al. 2004: 282). As its population reaches 1.5 billion, India is poised to become a great world power with less poverty, better health and education, and improved environmental quality. On a long-term basis, by moving toward sustainability in every field and notwithstanding its growing population, India may aspire to achieving a good balance of human and environmental well-being. Such a step will not only benefit its people, both rich and poor, but it will also make a contribution regionally and internationally by reducing pollution that spreads beyond its borders. India's response to environmental challenges is being watched by nations the world over: if the country is able to prevent the deterioration of its environment and alleviate poverty, it will be an example for others to follow.

Since the late 1980s, there has been a steadily growing awareness among the people in India about the ecological crisis facing their country. That awareness has led to an impressive growth not only of regulatory and administrative institutions established to deal with the problems of pollution and environmental conservation at both the national and state levels, but also of numerous ENGOs that are now playing a more prominent role than in the past. Nevertheless, the mounting pressures of population, expanding urbanization, and growing poverty have led to the ecologically unsustainable exploitation of natural resources that is threatening the fragile balance between ecology and humanity in India. The task is no doubt stupendous, for the hitherto unimaginative developmental strategies and largely imitative developmental schemes would have to be replaced by environmentally sound sustainable development. The challenge before policy-makers in India is to pursue a pattern of development that is in harmony with enhanced environmental quality. That challenge requires a spirit of partnership with varying stakeholders, such as all levels of government (including Panchayats), NGOs, the business community, political parties, unions, and international development partners in working together for environmental protection, conservation, and sustainability.

Moreover, the challenge within the nation's borders is critical and complex because responsibility for environmental quality is as numerous and fragmented across areas and levels of government as are the sources of environmental degradation. This has added to the level of public frustration at government complication, confusion, and delay. It is thus essential that, in addition to upgrading the administration of regulatory and enforcement mechanisms, the country searches for better ways to preserve and enhance environmental quality and to foster harmony between emergent societal needs and the environment. Finally, it should be noted that many current environmental concerns transcend international boundaries; they represent not simply an Indian challenge, but a global one. The nation will have to look ahead to a future in harmony with environmental imperatives from its rich cultural heritage and the traditional conservation ethos. In effect, a return to the traditional ways of doing things should not be regarded as a step backward but as one guided by the wealth of experience and concern about the environment.

Notes

1 During the monsoon season, with heavy downpours, floods occur that wash away the top soil along with nutrients used by farmers, causing the growth of algae.

2 The role of the judiciary is elaborated below.

3 *Satyagraha* literally means persistence and endurance for the truth; it is a non-violent struggle against the oppression and heavy-handedness of government power, and thus excludes the use of violence against others. The concept was initially used by Mahatma Gandhi in South Africa to mark the non-violent and passive resistance of the Indians against apartheid. Mahatma Gandhi believed that the pursuit of truth and the fight for one's rights does not entail inflicting violence on opponents (Mahatma Gandhi, 1961).

4 In many places, the affected public and NGOs have banded together to fight against forest contractors and government agencies creating Van Samitis (forest committees). In some cases, like Kerala Sastra Sahitya Parishad, which was created as a literary society, such NGOs have taken up the fight against the provincial forest department to highlight the government's failure to save large tracts of forests around Silent Valley, an ecological paradise and a rainforest believed to be 50 million years old (Prabhakaran 2004: 52).

5 The GEF is an organization created by the UN and the World Bank to assist developing countries with respect to projects for controlling pollution. It stands for Global Environmental Facility.

References

Baviskar, A. 1995. *In the Belly of the River: Tribal Conflicts over Development in the Narmada Valley*. New Delhi: Oxford University Press.

Baxi, Upendra, and Thomas Paul. 1986. *Mass Disasters and Multinational Liability—Bhopal Case*. New Delhi: Indian Law Institute.

Bishnoi, Vandana. 2002. "Movements for Tree Preservation." In Krishna Ram Bishnoi and Narsi Ram Bishnoi (eds.), *Religion and Environment*. Hisar, India: Guru Jambheshwar University. 322-27.

Dwivedi, O.P. 1997. *India's Environmental Policies, Programmes and Stewardship*. London: Macmillan Press.

Dwivedi, O.P., and Renu Khator. 1995. "India's Environmental Policy, Programs and Politics." In O.P. Dwivedi and D.K. Vajpeyi (eds.), *Environmental Policies in the Third World*. Westport, CT: Greenwood Press. 47-70.

Dwivedi, O.P., and D.S. Mishra. 2005. "A Good Governance Model for India: Search from Within." *Indian Journal of Public Administration* 51(4): 719-58.

Dyson, Tim, Robert Cassen, and Leela Visaria, eds. 2004. *Twenty-First Century India*. New Delhi: Oxford University Press.

Fisher, William F. 2000. "Sacred Rivers, Sacred Dams: Competing Visions of Social Justice and Sustainable Development along the Narmada." In Christopher K. Chapple and Mary E. Tucker (eds.), *Hinduism and Ecology*. Cambridge, MA: Harvard University Press: 401-21.

Gandhi, Indira. 1972. Address at the UN Conference on Human Environment. Stockholm, Sweden. [Agenda Notes for the National Committee on Environmental Planning and Coordination, published by New Delhi: Department of Science and Technology, July].

India. 1972 (April 12). Inaugural Function. Proceedings of the National Committee on Environmental Planning and Coordination. New Delhi: Department of Science and Technology.

—. 1974. *Water (Prevention and Control of Pollution) Act.* [also known as Water Act, 1974]. New Delhi: Government of India Publications Division.

—. 1976. *The Constitution of India, The Forty-second Amendment Act*. Article 48A. New Delhi: Government of India Publications Division.

—. 1980 (Sept. 15). *Report of the Committee for Recommending Legislative Measures and Administrative Machinery for Ensuring Environmental Protection* (N.D. Tiwari Committee). New Delhi: Department of Science and Technology.

—. 1981. *Air (Prevention and Control of Pollution) Act* [also known as Air Act]. New Delhi: Government of India.

—. 1986. *The Environment (Protection) Act.* New Delhi: Government of India.

—. 1992. *National Conservation Strategy and Policy Statement on Environment and Development*. New Delhi: Ministry of Environment and Forests.

—. 1993. *Environment Action Programme India*. New Delhi: Ministry of Environment and Forests.

—. 2001a. *Annual Report 2000-2001.* New Delhi: Ministry of Environment and Forests. http://envfor.nic.in/report/0001/chap05.html. Accessed on September 10, 2006.

—. 2001b. *A Reference Manual.* New Delhi: Ministry of Information and Broadcasting.

—. 2002. *Tenth Five Year Plan: 2002-2007* (Chapter IX B: Forests and Environment). New Delhi: Publications Division, Government of India.

—. 2006a. *Annual Report 2005-2006.* New Delhi: Ministry of Environment and Forests. http://envfor.nic.in/report/0506.html. Accessed on Aug. 1, 2006.

—. 2006b. *National Environmental Policy 2006*. New Delhi: Ministry of Environment and Forests, 2006. http://enfor.nic.in/nep/nep2006e.pdf. Accessed on Aug. 1, 2006.

IndianNGOs.com. 2006. Database. http://indianngos.com. Accessed on Aug. 9, 2006.

James, George A. 2000. "Ethical and Religious Dimensions of Chipko Resistance." In Christopher K. Chapple and Mary E. Tucker (eds.), *Hinduism and Ecology*. Cambridge, MA: Harvard University Press. 498-530.

Khator, Renu. 1991. *Environment, Development and Politics in India*. Lanham, MD: University Press of America.

Mahatma Gandhi. 1961. *Non-Violent Resistance*. New York: Schocken Books.

McCully, P. 1996. *Silenced Rivers: The Ecology and Politics of Large Dams*. London: Zed Books.

Mehta, M.C. 1992. "Environmental Cases: What the Judiciary Can Do." *The Hindu Survey of the Environment*. Madras: The Hindu. 163.

—. 1994. "Taj Trapezium: A Wonder Under Smog." *The Hindu Survey of the Environment*. Madras: The Hindu. 59-63.

Mushkat, Roda. 2004. *International Environmental Law and Asian Values*. Vancouver: University of British Columbia Press.

Prabhakaran, G. 2004. "Another Battle in the Offing." *The Hindu Survey of the Environment*. Chennai: The Hindu. 51-53.

Prasad, M.K. 1991. "Non-governmental Organizations: Creating Awareness." *The Hindu Survey of the Environment 1991*. Madras: The Hindu. 43-47.

Shastri, S.C. 2002. *Environmental Law*. Lucknow, India: Eastern Book Company.

Shephard, Mark. 1987. *Gandhi Today: The Story of Mahatma Gandhi's Successors.* John Cabin, MD: Sevcn Locks Press.

Thapar, Valmik. 2005. "Sariska: Tigers' Tragedy." *The Hindu Survey of the Environment (2005)*. Chennai: The Hindu. 41-49.

Upadhyaya, Ramesh. 1991. "Bihar: Big Problems of Chhotanagpur." *The Hindu Survey of the Environment, 1991.* Madras: The Hindu. 61.

UNICEF. 2006. *Progress for Children: A Report Card on Nutrition.* New York: UNICEF Division of Communication.

Venkataramani, G. 1992. "Pesticides: Harm Far Outweighs Use." *The Hindu Survey of the Environment.* Madras: The Hindu. 129-35.

CHAPTER 6

DONORS, PATRONS, AND CIVIL SOCIETY

The Global-Local Dynamics of Environmental Politics in Post-war Lebanon

Paul Kingston

INTRODUCTION[1]

Debates about the environmental effects of globalization in the global periphery, or Global South, increasingly revolve around questions of management and governance. The issue is not to reverse globalization—itself a utopic impossibility—but, as Richard Sandbrook argues with respect to the larger globalization debates, to "civilize" it (2003). The targets of such civilizing projects have usually been the agents of global capitalism—multinational companies, governments of the developed world, and international economic institutions. Exponents of "political ecology" would add as an additional target the governments of the developing world, weak if not complicit in harnessing the destructive capacities of economic globalization for their own narrowly focused goals (Bryant and Bailey 1997). In this chapter I suggest that a third set of actors should also be included in this problematic—namely the agents of this global environmental civilization, from the international non-governmental organizations (NGOs) and scientific experts that carry out global environmental work to the multilateral institutions that promote and finance them. Operating under the legitimizing banner of promoting sustainable development and global biodiversity, these global environmental institutions have penetrated deep into the local ecologies of countries in the developing world, setting off processes that have led to significant—and often counterproductive—restructurings of local environmental governance in the Global South.

This chapter presents a case study of the local governance implications of such intervention. The setting for the case study is Lebanon, a Middle East state that entered into a reconstruction program in the 1990s after over 15 years of civil conflict. This re-establishment of the political process in Lebanon coincided with the creation of the Global Environment Facility (GEF) at the United Nations Conference on Environment and Development in Rio de Janeiro, Brazil, in 1992. The GEF, in conjunction with the United Nations Development Program (UNDP) and the World Conservation Union (IUCN), were mandated with the task of promoting the protection of biodiversity through the creation of a global network of protected areas or "biosphere reserves," three of which were eventually designated for Lebanon.

The chapter will take the following form. First, it surveys emerging critical literature on the globalization of the environmental conservation movement, focusing in particular on recent ethnographic work on IUCN itself and its emphasis on the promotion of "grass-roots" and "community" conservation techniques. After a brief overview of the political context of post-war Lebanon, the paper will then dive into the story that surrounds the creation of Lebanon's "biosphere reserves," one that had serious—and, as we shall see, disempowering—implications for environmental governance at both the local and national levels. The chapter will conclude with some broader reflections on the dilemmas faced by global environmentalists in promoting more broad-based environmental governance mechanisms in the global periphery.

ENVIRONMENTAL CONSERVATION IN GLOBAL PERSPECTIVE—A CRITICAL PERSPECTIVE

There is a great paradox to the rise of the global conservation movement. On the one hand, accompanying this rise has been a parallel increase in the power of international environmental actors to intervene in the local ecologies of the Global South, leading to a restructuring of "local human-ecological relations." On the other hand, despite the resources and discursive power at their disposal that have fuelled these interventions, these institutions and organizations have not been able to direct and mould such "restructurings" in ways that have corresponded to their emerging governance goals associated with notions of "community-based conservation." It is this paradox that this chapter seeks to explore.

Recent ethnographic work by Ken MacDonald on the work of IUCN (2003; 2005) reflects on the growing global interventionist power of conservation institutions. IUCN describes itself as "the world's oldest and largest global conservation body." It is principally known as a "knowledge-based" organization that generates, integrates, and disseminates information about conservation and the equitable use of natural resources. This knowledge is produced by various commissions that exist within IUCN's federated structure, one of which is the World Commission on Protected Areas. MacDonald describes its mission as being "to influence, encourage, and assist societies throughout the world to conserve the integrity and diversity of nature and to ensure that any use of natural resources is equitable and ecologically sustainable" (2003: 5).

The global reach of the IUCN has been transformed significantly with the establishment of the GEF after the 1992 UN Conference on Environment and Development (UNCED). This has had two major effects. First, it has created stronger institutional linkages between the IUCN and the more powerful array of international development institutions that fund environmental activities. Indeed, MacDonald argues that IUCN has increasingly begun to resemble a standard development NGO, supporting government- and donor-approved projects and tailoring its appeals to the funding requirements of donors and the willingness of governments to approve their work. This has not only "radically altered the operational structure of IUCN" with its "knowledge" commissions taking on much less importance (2003: 12), but it has also transformed the IUCN into "a node in an organizational hierarchy that flows from Government and inter-governmental agencies through IUCN to the locales where projects are implemented" (MacDonald 2003: 21). In short, it has become a part of the global "network of power" (MacDonald 2005: 269).

This emerging power that flows from its growing international institutional linkages, with the resultant access to greater financial resources, has also been accompanied by a rising discursive power that has served to legitimize its right to intervene in environmental affairs in the Global South. MacDonald argues that the rise of discourse around "global biodiversity," buttressed by the notion that "human survival relies upon the maintenance of biodiversity," has facilitated the incorporation of local ecologies into the mandate of international ecology organizations such as IUCN, creating "a global image of an interconnected web of life." Indeed, MacDonald suggests that it has given such concerns an "ontological status" that is seemingly "non-partisan" (2005: 269). As MacDonald concludes, "through the representation of environmental problems as territorially transcendent, transnational NGOs have assumed an enhanced political significance and have come to play a crucial role in the transnational arenas" (2005: 270).

This enhanced power and legitimacy to intervene in local ecologies, however, has not necessarily translated into power to control and mould the effects of such interventions, particularly as we shall see with respect to politics and governance. On the one hand, as represented by a series of strategies released in the 1980s and 1990s, IUCN has been active in promoting community-based conservation techniques (Adams 1993). This seeming movement away from old colonial techniques of conservation was driven by a number of factors, the most important being that most of the world's biodiversity lies in areas inhabited by communities of poor in the developing world. The new goal was not to separate nature from people but to adopt an approach that recognized their inextricable integration and interdependence. Further strengthening this re-orientation was the more pragmatic argument that conservation strategies worked best when local communities exhibited "a robust and durable interest in conservation" (Western and Wright 1994: 429).

In reality, however, community-based conservation has been difficult to operationalize for a variety of reasons. Indeed, one of the conclusions of a book devoted entirely to an analysis of community-based conservation efforts is that only in a few cases were local communities actively involved in defining and designing conservation problems and solutions (Western and Wright 1994). At best, community-based conservation was characterized as being pursued "with" as opposed to "by" local communities (Little 1994: 36). MacDonald goes so far as to describe efforts at community-based conservation as rhetorical "tokenism" (2005: 272). In certain cases, this has been dictated by the fact that "biologically concentrated hot spots" tended to be located in remote areas where problems of community exclusion were minimalized. Exacerbating this trend has been the difficulty in finding internally cohesive communities through which conservation governance mechanisms can work effectively, a reality that MacDonald argues tends to act—or be used—as a further justification for the intervention of an external agency. Given IUCN's position within the broader development apparatus, community-based conservation techniques also fall prey to the institutional demands within the donor community for efficiency and effective delivery that can often run at cross-purposes to the more timely and inefficient processes associated with community decision-making (see Kingston forthcoming; Ferguson, 1990). The result, as articulated by MacDonald, is the emergence of "new forms of governmentality" that have little relation to community-based conservation techniques. On the contrary, he describes them as representing "a regime of distanced practice" in which "the ability to direct environmental use is being distanced from those who live with the immediate consequences" (2003: 24). As MacDonald concludes, "the decisions that affect localized and particularized human-environmental relations are increasingly being made in institutions and places distant from those locales and in the absence of an appreciation of local context" (2003: 25); in short, "local ecologies are becoming subject to institutional agents of globalization" (2005: 261).

IUCN, GRASSROOTS CONSERVATION, AND THE CREATION OF "ENVIRONMENTAL MONOPOLIES" IN POST-WAR LEBANON

IUCN sent its first mission to Lebanon to scout out the possibilities of funding "biosphere reserves" in 1993, four years after the signing of the Ta'if Accord that eventually led to the end of Lebanon's 15-year civil war one year later. The mission lasted three months and found a terrain rich in possibilities of funding. First, Lebanon has a diverse array of ecosystems in coastal plains, mountains, and drylands that give rise to a similarly wide array of biodiversity. It is said that the forests of Ehdin, for example, have more species of trees than all of the United Kingdom. Second, IUCN also found a country whose ecological heritage was under serious threat, the result of centuries of Ottoman and French neglect, the unregulated nature of Lebanon's post-war development model, and the degradation and destruction as a result of Lebanon's 16-year-long civil war. The existence of just one minister responsible for the environment who had virtually no staff or budget—in short, no actual ministry—offered little hope that this situation would be turned around on its own. The one ray of hope, however, appeared to be Lebanon's emerging coterie of environmental NGOs. The Friends of Nature (FON), headed by Ricardo Haber, a botanist and former professor at Lebanon's American University of Beirut (AUB), for example, had already played a crucial role in convincing the fledgling post-war Lebanese state to set up "paper" reserves in two locations—Ehdin and the Palm Islands off the coast of the city of Tripoli in the north of the country; and the Society for the Protection of Nature in Lebanon (SPNL), headed by Assad Serhal, another major NGO player in the early post-war years, had played a key role in facilitating the visit of the IUCN mission to Lebanon in the first place.

Impressed by the activities of Lebanon's NGOs and encouraged by the commitment of the Lebanese state to set up an actual Ministry of the Environment (MOE), the IUCN, in conjunction with the GEF and the UNDP, agreed in 1995 to formulate a US $2.5-million project over a five-year period to help establish three biodiversity reserves: Barouk, in the Shouf mountains; Ehdin, in the Zghorta region; and the Palm

Islands, off the coast of Tripoli. What is interesting about the project proposal was its governance framework, advocating what it called "a multidisciplinary approach to management." On the one hand, this meant building up the administrative capacity of the soon-to-be established Ministry of Environment through the creation of a Department of Protection and Wildlife (PAW). The key strategic actors, however, were to be NGOs, who were given responsibility for managing the actual reserves: the Arz as-Shouf Cedar Society (ASCS) in the case of the Barouk reserve, the Friends of Horsh Ehdin (FOHE) in the case of the Ehdin reserve, and the Environmental Protection Committee (EPC) in the case of the Palm Island reserve. Indeed, IUCN's project document wrote of tapping into "a new phenomenon of emerging cooperation" between the state and the "increasingly vocal NGOs." The project document also wrote of bringing the various local communities affected by the reserves alongside in the long term, through a variety of mechanisms that included employment on the reserve, support for local development outside of the reserve through the creation of a revolving credit fund, and education and public-awareness campaigns that would also hold out the carrot of eventual eco-tourism revenue. Finally, to top off this ambitious set of goals, it was hoped that this project would contribute to the process of "national reconciliation" in Lebanon, providing "a stimulus for people to get together again" after the long civil-war period. The project was accepted by the Lebanese Government and officially started in November 1996.

Through its inclusion of the state, NGOs, and local communities, IUCN hoped to promote a sustainable and institutionalized system of conservation management. However, while the reserves have now existed for over a decade, management teams hired, uniforms for guards bought, and Toyota Land Cruisers acquired, it is my contention that the project has not contributed to the emergence of a sustainable and institutionalized system of management. To understand why this is the case, one needs to understand the broader political context of post-war Lebanon—in particular, the battles between political elites, at both the local and national level, to re-establish their political hegemony after the long civil-war period.

Administratively, Lebanon is a republic and democracy in which deputies are directly elected to an Assembly for four-year terms and who, in turn, elect a president for a six-year term. Underpinning this republican democratic political system, however, is an informal power-sharing agreement (the National Pact of 1943), which (in the pre-war period) guaranteed electoral representation of its eighteen different religious communities on a 6:5 basis between Christians and Muslims, and which distributed executive power between the three main religious communities: the Maronite Christians to whom the office of President is reserved, the Sunni Muslims to whom is reserved the office of Prime Minister, and the Shi'a Muslims whose claim is to the office of Speaker of the Assembly. The reforms agreed to in Ta'if in 1989 altered the confessional balance of power in the Assembly from a pre-war ratio of 6:5 in favour of the Christian population to a post-war ratio of 5:5, and have also created a more equitable balance of power within the executive offices of the state. It is this combination of religious power-sharing with republicanism that has led many to describe Lebanon as a paradigmatic case of consociational (or confessional) democracy.

In this political battle between Lebanon's republican and confessional/sectarian roots, between its formal and informal political institutions, it has been the latter that have proven to be most powerful. This power of sectarian leaders, networks, and communities in Lebanon has been underpinned by three principal factors. First, the decentralized and confessionally based electoral system has entrenched the power of local elites and has made it virtually impossible to build cross-cutting, national, and ideologically based political parties and coalitions (Richani 1998). Politics at the national level, therefore, is driven by the primary logic of protecting local and, in the case of rural areas, territorial bases of power.

Second, these local and particularistic networks of power are strengthened by their contact with, and access to, resources of the state. Be it through the allocation of state-sector employment, the opportunity to win bureaucratic favours, or ability to manipulate government regulations and laws, access to the resources and influence generated by the state is of vital importance in maintaining and expanding local support, done through the time-honoured tradition of patron-clientelism (Khalaf 1987; Hudson 1985; Salam 1998). Consisting of inequitable but consensual exchanges between patrons and clients in which the latter offer their support and continued socio-economic and political subordination in exchange for access to resources and favours, patron-clientelism has had an ambiguous effect on Lebanese politics, helping to stabilize an extremely fragmented polity while, at the same time, perpetuating a long-standing "system of inequality" and militating against the establishment of a more stable and institutionalized political order (Johnson 1986).

Finally, of crucial importance in maintaining and expanding elite clientelist power in Lebanon has been their ability not only to capture public resources, but also to accumulate private ones. In addition to being called a consociational democracy, Lebanon has also been called "a merchant republic" characterized by its open economic policies and a minimalist state dominated by a powerful commercial-financial bourgeoisie (Johnson 1986). Whether by political default or design, this absence of economic regulation, combined with the lack of enforcement capacity for those regulations that exist, has facilitated the vast accumulation of resources by Lebanon's elite classes. The result has been the further perpetuation of socio-economic inequality in Lebanon upon which patron-clientelism depends and is sustained (Gilsenan 1977; Scott 1977; Eisenstadt and Roniger 1980; Chabal and Daloz 1999).

Lebanon's post-war period has seen the reconsolidation and expansion of these particularistic and clientelist sectarian networks. The reconstitution of the Lebanese state, for example, has renewed opportunities for access to state-generated surplus—especially with the emergence of Lebanon's grandiose and ambitious post-war reconstruction program (Denoeux and Springborg 1998). Further broadening the scope of elite activity and accumulation have been the increasing connections between Lebanon and the global economic arena, connections strengthened by particular linkages on a regional level, such as those between the former prime minister Rafiq al-Hariri and Saudi Arabia and the Shi'a community and Iran, on a global level with the numerous Lebanese diaspora communities abroad, as well as with global criminal networks established during the war. Finally, selective elites have also strengthened their positions in Lebanon by way of their connections with Syria; in Lebanon, most roads lead to Damascus (al-Khazen 2003).

However, while all these factors have served to increase the power of elites in Lebanon, leading not unexpectedly to the rise of massive levels and quite open displays of corruption (see Malik 1997; Leenders forthcoming-a; forthcoming-b), the stability of elite power in Lebanon remains fragile. Unable to overcome the resilient hostility built up over years of civil war and forced to share the political spotlight with selected wartime militia leaders exonerated by the blanket wartime amnesty and Syrian support, post-war Lebanese governments have been fractious affairs ultimately held together until their withdrawal in 2005 by the self-appointed brokering role of the Syrians. The point is this: with the elite structure of power being in a fragile state of reconsolidation, competition between the various expanding elite patron-client networks has been intense.

It is within this context of intra-elite competition for resources and hegemony that the protected area project must be analyzed. The protected area project was of interest to some of Lebanon's elite class for a number of interrelated reasons. First, despite the fact that the amount of foreign capital was limited—with the proposed US $2.5 million being spread out over a five-year period and between three reserves with a significant proportion of the capital being spent on foreign technical advisers—the project nonetheless represented real, if local, opportunities for patronage, especially in the distribution of reserve-related employment. Moreover, given the very nature of the project, it also provided additional opportunities to the affected *zu'ama* (local elites) to consolidate their hold over land. This was not only of symbolic value; it also offered real opportunities for accumulation in the future if and when the post-war reconstruction program began to pay dividends in the areas of tourism and eco-tourism.

These benefits, however, would not materialize if the project was managed according to the principles outlined in the project document. The idea of building up a more robust and nationally oriented conservation administration represented on the ground by public-minded NGOs, held accountable by a local MOE-appointed committee, and backstopped by an emergent Protected Area and Wildlife Department within the MOE, placed too many formal constraints on the more informal and particularistic workings of Lebanon's clientelist system. Therefore, the affected elites moved in a variety of ways to weaken the national, legal, and conservation-minded framework of the project—preventing the creation of a department of Protected Areas and Wildlife, interfering in and ensuring the appointment of a more pliable Project Director, and limiting the resources available for boundary delineation, essential in placing the protected areas on a solid legal footing—all for the purpose of ensuring that the proposed reserves would operate first and foremost on the basis of political loyalty.

This insistence on "compliance rather than conservation" as the framework for the project is perhaps best exemplified by an examination of the politics that surrounded the local management of the reserves. The clearest example of this lies with the Barouk Reserve in the Shouf Mountains—the location of some of the few remaining wild Lebanese cedar trees. The Shouf region is populated predominantly by the Druze of Lebanon, a confession that continues to be represented at the national level by the Jumblatt clan. Its head, Walid Jumblatt, is said to have had an interest in environmental questions and, during the war, had directed significant resources from his canton administration, the Administration of the Mountain, toward maintaining the environmental heritage and integrity of the area. In the early post-war world, he had also toyed with the idea of creating a Lebanese "green" party (Kingston 2001). However, though environmentally minded and, hence, supportive in principle of the IUCN initiative, Jumblatt was not interested in the establishment of the kind of autonomous system of governance proposed by the IUCN in "his" territory. He therefore worked quietly through his own connections within the Lebanese state as well as with the UNDP to undermine the governance provisions of the project. First, Jumblatt ensured that the NGO appointed to manage the reserve be his own—the Arz as-Shouf Society, over which he presided as chairman of the board. Second, he insisted that the MOE-appointed local committee, created in the initial reserve legislation in 1991 to oversee the activities of the NGOs in the two initial reserves (Horsh Ehdin and the Palm Islands), not be included as part of the governance structure of the Barouk reserve, effectively removing a local level of accountability that could in theory have challenged Jumblatt's overarching control. By acquiescing to these governance changes, it certainly seemed that IUCN, in the name of conservation, was willing to give up on its other stated goal of promoting grassroots participation, especially given Jumblatt's role as one of Lebanon's more prominent militia leaders during the civil war.

Perhaps the most interesting example of *zu'ama* struggles to ensure that development projects operated on the basis first and foremost of "compliance" rather than conservation was the case of Horsh Ehdin. Located in the Zghorta region of Lebanon, the politics surrounding the creation and management of Horsh Ehdin became enmeshed with the broader and historically rooted feudal rivalries of the region in which the Franjieh clan sought to maintain its predominant though not hegemonic control (Johnson 2001). The development of the reserve at Ehdin posed two challenges to the head of the Franjieh clan, Sulieman. First, the NGO earmarked by The World Conservation Union (IIJCN) to manage the reserve–the Friends of Horsh Ehdin (FHE)–was set up by Ricardo Haber, an independent-minded environmentalist with little respect for the prevailing clientelist political dynamics of the country. Further complicating the situation was the prior existence of the MOE-appointed local committee created in 1991 before IUCN's entry into the country—filled essentially with environmentalists and Haber loyalists—that could act to support the activities of the NGO in the face of Franjieh pressure. The stakes were particularly high for Franjieh in that there was an emerging interest in developing the areas around and above Horsh Ehdin into a tourism resort, a project that would generate significant opposition from environmentalists (Haddad 2000).

Hence, faced with symbolic, political, and potentially material challenges to his regional regime of compliance, Sulieman sought to "clip the wings" of Haber and his environmentalist followers. Using his own clients within the FHE, Sulieman worked to have the independent members of the NGO expelled; he forced the NGO to hire a management team on the basis of loyalty rather than competence, leading to a situation on the ground of virtual inactivity: he engineered the appointment of more compliant members to the local MOE-appointed local committee; and, perhaps most significantly, he blocked Haber's impending appointment as project manager for the IUCN "biosphere reserve project" in Lebanon as a whole. All along the road, the process was acrimonious, characterized by severe political infighting and stalemate and, at one point, there was even discussion of withholding external funding for this particular reserve. In the end, however, Franjieh's efforts prevailed and, with the political battles for local control complete, the reserve has now begun to show some signs of technical progress on the ground.

What is the result? Some environmental work has gone ahead, with the technical groundwork for biodiversity preservation being laid in the short term. Whether or not these reserves will operate on the basis of grass-roots conservation techniques backed up by a sustainable conservation-management system, however, is more problematic. With the exception of small income-generating projects around the Barouk reserve, local communities, especially goat herders in the Horsh Ehdin area, have had little involvement in the reserves and see them as areas from where they are now excluded. Moreover, NGOs and local committees, designed to be important agents for the promotion of more grassroots conservation governance, have been penetrated by local political leaders in ways that ensure that they will operate, first and foremost, on the basis of political loyalty. In effect, therefore, rather than promoting "grassroots *in situ* conservation," the IUCN project has unwittingly contributed to the consolidation of "environmental monopolies" controlled by local Lebanese political elites.[2]

LEBANON'S "GREEN" CIVIL SOCIETY AND THE NEGATIVE EFFECTS OF THE IUCN INITIATIVE

The most disturbing consequence of the IUCN "grassroots" conservation initiative in Lebanon, however, is the effect that it has had on the "green" component of Lebanese civil society as a whole, and it is to this last story that we now turn. Lebanon's "green" movement is very modest. Before the war, it was virtually nonexistent save for the efforts of a few individuals. With the beginning of the postwar period, however, its numbers and public profile have been steadily rising, sparked by the significant environmental problems with which the country has been confronted, problems that range from the degradation of the coastline, the regulation of waste management that includes dealing with the illegal dumping of toxic waste in the country from the West during the civil-war period, air pollution emanating from both cars and industrial enterprises, and uncontrolled quarrying that is, in effect, carving out the Lebanese mountains in order to provide building material for Lebanon's reconstruction program (Kingston 2001). In response to these problems, university campuses in the country have increasingly taken up the environmental issue, regular columns have begun to appear in Lebanese newspapers on environmental issues, and this has resulted in the proliferation of NGOs interested in promoting environmental awareness. In 1993, this rising interest also resulted in the creation of a national coordinating committee for environment-related NGOs called the Lebanese Environmental Forum (LEF), and this, in turn, has helped to attract the attention of a number of foreign donors, the most outspoken being Greenpeace, which has now established a local office in Lebanon—the first in the Arab world.

However, Lebanon's "greens" are also a divided lot. On the one hand are those whose focus is primarily "conservationist," located by and large within the LEF. On the other are those more attracted to the broader agenda of "sustainable human development." This latter group has not paid much attention to the IUCN project of creating "biosphere reserves" and looks upon the idea of reserves as "an old way of thinking" that is too nature-focused, especially since most biodiversity destruction occurs outside their confines. Their focus is more on issues such as toxic and industrial waste dumping, rising consumerism, waste disposal and recycling, the destructive effects of quarrying, land reclamation, and unregulated development and, more broadly, the basic provision of social services—issues that they see as being directly linked to Lebanon's model of post-war reconstruction and development and its congruence with the broader neo-liberal agendas of globalization (Masri 1999). This latter tendency within Lebanon's "greens" organized itself into an informal network called "the environmental meeting place" (or *multaqa al-bi'a*) consisting of citizens and communities immediately affected by these issues, scientists interested in being more socially and policy active, and redefined leftists looking for "a new framework for political action" within which to challenge the emerging globalist agenda of Lebanon's economic and political elites.

However, these divisions, caused in part by clear differences in ideological approach, have been exacerbated by "the hostile context of power" in which they have found themselves (Torgeson 1999). Lebanon's political and economic elites, for example, who have been used to a free hand to pursue economic ventures and who are eager to take unfettered advantage of the opportunities offered by Lebanon's reconstruction program, have not looked with favour upon the emergence of a network of NGOs pushing for a broader and more radical agenda on environmental sustainability. Indeed, many are involved in the very activities that Lebanon's "greens" are campaigning against—notably the Interior Minister in the late 1990s, Michel Murr, who was both responsible for regulating the activities of quarrying in the country while having substantial interests in the business at the same time (Leenders forthcoming-a). Having sensed the existence of divisions within the environmental movement, affected elites have clearly tried to drive an even larger wedge between them and thus debilitate the influence of their advocacy work at the national level.

The story revolves around the activities of the Lebanese Environmental Forum. As stated, the LEF was created in 1993 to be a national forum for promoting awareness about the environment; its main purpose was to be advocacy. However, it has not performed that function and has been notably muted or, indeed, silent in discussions of the country's most serious environmental problems. In part, this is a generational issue with many of LEF's members being of an older, more cautious generation: "it's like an archive!" commented one of LEF's original foreign backers (interview with Samir Farah, Beirut, 1999). Moreover, many of LEF's members have taken this approach in order to manoeuvre themselves into a position of acquiring favour from those in authority; in short, LEF has become a mechanism used by its members to promote their personal interests. It is clear, for example, that many of the members of LEF are either involved or interested in becoming involved in the management of a "biosphere reserve." However, it is also clear that LEF has been penetrated by political elites themselves in order to regulate and enforce an implicit clientelist bargain—one that sanctions the LEF as a distributive mechanism for its members in exchange for its effective silence on environmental matters that relate to elite processes of accumulation. The most notable example of this elite clientelist penetration is the membership in the LEF of an NGO called *amwaj al-bi'a* (the environmental wave). Run by Randa Berri, the wife of the Shi'a speaker of the Parliament (Nabbi Bern), *amwaj al-bi'a* has virtually no membership to speak of, controlled the position of secretary of the LEF (including its files) throughout the 1990s, and was eventually awarded its own "biosphere reserve" on the sand beaches near the southern city of Tyre—ironically, the site of one of Lebanon's first post-war environmental campaigns against the extraction of sands by companies controlled by Bern's own financial empire.

The consequences of this clientelist bargain are twofold. First, the LEF has been neutralized as a forum for serious environmental advocacy. Its work has increasingly reverted to tourism promotional activities, painting a picture of Lebanon as a country of immense natural beauty in ways that completely ignore the serious environmental problems that lie not so far beneath the surface (see LEF 2001-2002). Moreover, as increasing numbers of its progressive original members have left, as foreign donors have withdrawn their initial support, and as some of its remaining members have been convicted of corruption, the LEF's initial promise as a national forum for advocacy on environmental issues has been seriously eroded. However, perhaps the most serious consequence of the LEF's fall from grace is its continued existence, one that crowds out other more serious environmental networks seeking to have a more institutionalized and national profile. In the absence of an effective intermediary institution through which civil society activists in the field of the environment can operate—a problem endemic to Lebanon's civil society movement as a whole (Kingston 2001)—Lebanon's more progressive network of environmentalists will have a difficult time translating their numerous grassroots successes into effective and sustained pressure for "sustainable human development" policy frameworks at the national political level.

CONCLUDING REFLECTIONS

I would like to conclude with three main points. First, it is clear that the introduction of finance from abroad to promote the creation of biosphere reserves in Lebanon has had negative repercussions for civil society organizations involved in environmental advocacy at both the local and national level. Indeed, it is startling how broad the affects of this seemingly small project have been, leading in effect to the "greening," and hence legitimization, of patron-client networks that are inimical to the promotion of more broad-based governance in the country. All this confirms MacDonald's conclusions emphasizing the significant restructuring effects of global environmentalism on local human-ecological, as well as society-state, relations in the Global South.

In the Lebanese context, this raises the important question of whether no financial and technical assistance from the GEF/UNDP/IUCN group might have been better than some. It is clear, for example, that local efforts to protect biodiversity had already been initiated, ones that had resulted in the creation of two protected areas before the IUCN mission had even arrived in the country. This was part of a much wider trend in the early post-war period in Lebanon in which local environmental actors, filling the void created by the effective absence of a Lebanese state, emerged as autonomous actors in an environmental field that they essentially created as they went along—promoting recycling campaigns, organizing advocacy campaigns against local polluting industries, and acting to protect local "green" and heritage spaces against unregulated commercial encroachment, especially with respect to quarrying (Karam 2004). Interventions from abroad, while often designed to assist such initiatives, often ended up pre-empting and disempowering them. As a recent study of donor interventions in the field of governance promotion concluded, "in many Third World situations, there is [already] an intensive and continuous exploration of new modes of structuring relations between state and society ... such pressures are usually problematic and difficult but, above all, require space and occasionally some cautious support. It remains uncertain whether the introduction of external models of 'universal' good governance as conditionalities is helpful in such situations" (Doornbus 1995: 387; see also Kingston forthcoming).

Perhaps the most important question, therefore, is why international donor and technical assistance agencies are not more sensitive to the local power dynamics of the localities in which they work. MacDonald suggests that this insensitivity is rooted in resilient but hidden colonial-type assumptions about the incapacity of local actors to act as effective agents of modernization, be it in the environmental or other fields of activity—ones that work simultaneously to justify international intervention (2005: 268). Such assumptions, in turn, result in downplaying the importance of doing serious research into the nature of local contexts. In his ethnographic work on IUCN, for example, MacDonald writes of the severe under-representation of social scientists in the secretariat of IUCN, epitomized by the marginalization of its Social Policy Group in the late 1990s, only a few years after its actual creation in the first place (2003: 18). This in turn affects the ability of organizations like the IUCN to "learn" from its experiences and to act in a more "reflexive" manner in terms of its own governance (2003: 21). The more IUCN becomes integrated into an emerging global configuration of development institutions, what MacDonald has called the "regime of distanced practice," however, the less likely it is that such institutional reflexivity will be developed. The paradoxical lament, as this case study of IUCN's involvement in post-war Lebanon shows, is that "distanced practice" seems antithetical to the achievement of IUCN's more community-oriented conservation goals.

Notes

1 Much of the material in this article is based upon interviews with Lebanese active in the environmental movement during a six-month period from January to June 1999. Names have for the most part been excluded for reasons of confidentiality.

2 When a UNDP employee with responsibilities for overseeing the implantation of protected area projects in the Arab region as a whole was contacted about this counter-productive outcome in the Lebanese cases, that employee refused to comment, asking that he not be quoted as "having either agreed or disagreed" with the contents of my paper" (Email communication with author, Oct. 12, 2002).

References

Adams, Bill. 1993. "Sustainable Development and the Greening of Development Theory." In Franz Schuurman (ed.), *Beyond the Impasse: New Directions in Development Theory*. London: Zed Books.

Al-Khazen, Farid. 2003. "Political Parties in Post-war Lebanon: Parties in Search of Partisans." *Middle East Journal* 57,4: 605-24.

Bryant, Raymond, and Sinead Bailey. 1997. *Third World Political Ecology*. New York: Routledge.

Chabal, Patrick, and Jean-Pascal Daloz. 1999. *Africa Works: Disorder as Political Instrument*. Bloomington: Indiana University Press.

Denoeux, Guilain, and Robert Springborg. 1998. "Hariri's Lebanon." *Middle East Policy*. VI, October 1998.

Doornbus, Michael. 1995. "State Formation Processes Under External Supervision: Reflections on 'Good Governance.'" In Olav Stokke (ed.), *Aid and Political Conditionality*. London: Frank Cass.

Eisenstadt, Shuel, and Louis Roniger. 1980. "Patron-Client Relations as a Model of Structuring Social Exchange." *Comparative Study of Society and History*. Vol. 22, No.7 (January): 42-77.

Ferguson, James. 1990. *The Anti-Politics Machine: 'Development,' Depoliticization, and Bureaucratic Power in Lesotho*. Minneapolis: University of Minnesota Press.

Gilsenan, Michael. 1977. "Against Patron-Client Relations." In Ernest Gellner and John Waterbury (eds.), *Patrons and Clients in the Mediterranean Societies*. London: Duckworths.

Haddad, R. 2000 (May 19). "Ehden in Danger of Development." *The Daily Star* [Beirut].

Harik, Judith. 2005. "Coping with Crisis: Druze Civil Organizations During the Lebanese Civil War." *The Druze: Realities and Perceptions*. London: Druze Heritage Foundation.

Hudson, Michael. 1985. *The Precarious Republic*. Boulder, CO: Westview Press.

Johnson, Michael. 1986. *Class and Clients in Beirut: The Sunni Muslim Community and the Lebanese State, 1840-1985*. London: Ithaca Press.

—. 2001. *All Honourable Men: The Social Origins of War in Lebanon*. London: I.B. Taurus.

Karam, Karam. 2004. *Revendiquer, Mobiliser, Participer: Les associations civiles dans le Liban de l'après-guerre*. Thèse doctoral, Science Politique, L'Université Paul Cézanne—Aix Marseille III.

Khalaf, Samir. 1977. "Changing Forms of Political Patronage in Lebanon." In Ernest Gellner and John Waterbury (eds.), *Patrons and Clients in the Mediterranean Societies*. London: Duckworths.

—. 1987. *Lebanon's Predicament*. New York: Columbia University Press.

Kingston, Paul. 2001. "Patrons, Clients and Civil Society: Environmental Politics in Post-war Lebanon." *Arab Studies Quarterly* 23(1): 55-72.

—. forthcoming. "The Anti-Politics of Peacebuilding: Governance Promotion and Local Political Processes in Post-war Lebanon." Unpublished paper.

Leenders, Reinoud. forthcoming-a. "Lebanon's Political Economy: After Syria, an Economic Ta'if?"

—. forthcoming-b. *Divided We Rule: Reconstruction, Institutions Building, and Corruption in Post-war Lebanon*. Cambridge: Cambridge University Press.

LEF (Lebanese Environmental Forum). 2001-2002. "Think Green, Act Green: The Lebanese Environmental Forum." In *Cedar Wings* (In-flight magazine of Middle East Airlines).

Little, Peter D. 1994. "The Link Between Local Participation and Improved Conservation: A Review of Issues and Experiences." In David Western and Michael Wright (eds.), *Natural Connections*. Washington, DC: Island Press.

MacDonald, Ken. 2003 (Feb. 6). IUCN: *A History of Constraint*. Text of Address to Permanent Workshop of the Centre for Philosophy of Law, Higher Institute for Philosophy of the Catholic University of Louvain (UCL), Belgium.

—. 2005. "Global Hunting Grounds: Power, Scale, and Ecology in the Negotiation of Conservation." *Cultural Geographies* 12: 259-91.

Malik, Habib. 1997. *Between Damascus and Jerusalem*. Washington, DC: Washington Institute for Near East Policy.

Masri, Rana. 1999. "Development—At What Price? A Review of Lebanese Authorities' Management of the Environment." *Arab Studies Quarterly* 21(1): 117-34.

Richani, Nazih. 1998. *Dilemmas of Democracy and Political Parties in Sectarian Societies: The Case of the Progressive Socialist Party of Lebanon, 1949-1996*. New York: St. Martin's Press.

Roniger, Louis. 1994. "The Comparative Study of Clientelism and the Changing Nature of Civil Society in the Contemporary World" and "Conclusions: The Transformation of Clientelism and Civil Society." In Louis Roniger and Ayse Gunes-Ayata (eds.), *Democracy, Clientelism, and Civil Society*. Boulder, CO: Lynne Rienner Publishers.

Salam, Nawaf. 1998. *La Condition Libanais*. Beyrouth: Dan an-Nahar.

Sandbrook, Richard, ed. 2003. *Civilizing Globalisation: A Survival Guide*. New York: SUNY Press.

Scott, James. 1977. "Patronage or Exploitation?" In Ernest Gellner and John Waterbury (eds.), *Patrons and Clients in the Mediterranean Societies*. London: Duckworths.

Torgeson, David. 1999. *The Promise of Green Politics: Environmentalism and the Public Sphere*. Durham, NC: Duke University Press.

Western, David, and Michael Wright. 1994. *Natural Connections*. Washington, DC: Island Press.

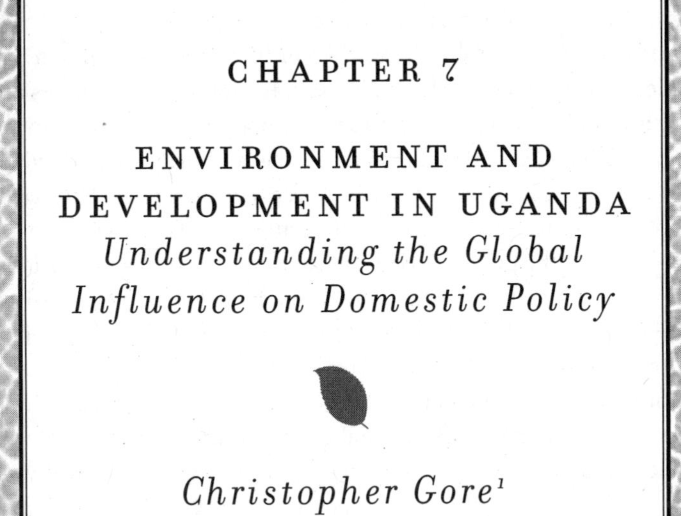

CHAPTER 7

ENVIRONMENT AND DEVELOPMENT IN UGANDA

Understanding the Global Influence on Domestic Policy

Christopher Gore[1]

INTRODUCTION

In 2004, Dr. Wangari Maathai became the first African woman to win the Nobel Peace Prize. This was a landmark event for several reasons, but perhaps most notably because the Nobel Committee rewarded Maathai for her ongoing efforts to link the quality of the environment to political participation and representation. In her acceptance speech, based on her experience initiating and leading the Green Belt Movement in Kenya, Maathai articulated the need to galvanize civil society and grassroots movements in order to catalyze environmental improvements. At the same time, she expressed the need for governments "to recognize the role that social movements play in building a critical mass of responsible citizens, who help maintain checks and balances in society." While few would question the validity of Maathai's statement, her proposal is much more complex than it first appears, given the significant global dimension to environmental issues in Africa.

As is the case in all parts of the world, conflicts over environmental quality, environmental service provision, and access to and use of resources are inherently political; that is, conflicts over the environment are most often about who has authority, who is "in authority," and how that authority is exercised to allocate resources and make decisions over resource use. On the surface, Maathai's suggestion that governments need to recognize the critical role of citizens in environmental issues positions the politics of the environment as a two-way relationship. We know, however, as Maathai certainly would too, that understanding environmental management according to a two-way domestic interaction omits the international influence on environmental issues and resources in Africa. It is in this context that the environmental challenges of globalization start to be revealed.

Popular knowledge of environmental issues in Africa typically centres on two things: degradation of the environmental resource base due to poverty and harsh climatic conditions; or conflict over minerals, oil, and timber. While interpretations and responses to these conditions are often misguided (see Leach and Mearns 1996; Fairhead and Leach 1996), both clearly have significant and complex global economic dimensions that often implicate foreign corporations or individuals in the exacerbation or production of conflict while natural resources are extracted for international markets (see Le Billon 2001a; 2001b; Duffield 2001). No less significant is another economic dimension of globalization, Foreign Direct Investment (FDI). With the exception of investment in South Africa, the largest percentage of foreign direct investment in Africa is tied to oil (Asiedu 2006).[2] These natural resource-based economic dimensions of globalization reinforce what might be considered a "dueling environmental imperative" in Africa.

On the one hand, almost all sub-Saharan countries are highly dependent on the international sale of primary products (coffee, tea, oil, minerals, fish, timber) for foreign-exchange earnings. Hence, there is a strong desire to make use of the international comparative advantage offered by a rich natural resource base to assist in poverty reduction and economic development. On the other hand, for years African governments have been working to establish, implement, and maintain systems of environmental management that minimize negative social, economic, and environmental outcomes from resource extraction. On the surface, this contest between extraction and management would seem to be consistent with countries throughout the Global South. What makes this "duel" unique in Africa is that international actors are often driving the promotion and implementation of both imperatives; bilateral and multilateral agencies such as the World Bank are at once promoting better domestic environmental management regimes while at the same time promoting economic policies that are driven by market-liberal norms.

Donors promote international trade and foreign investment as solutions to African economic woes, while also promoting stronger environmental management *and* transparency *and* participation in environmental policy and decision-making. While international trade and foreign investment need not conflict with strong environmental management and greater citizen participation in decision-making, given the urgency for development and FDI communicated by African governments and donors, they often do. Moreover, African governments are often antagonistic to domestic and international civil society organizations wishing to debate or challenge government and/or donor policies relating to the extraction and use of environmental resources or in changes to environmental management regimes. The outcome of this complex picture is that globalization, vis-à-vis dominant economic models and actors, places civil society in a bind: donors argue that civil society should assist in managing environmental resources and holding government accountable (De Coninck 2004), while national governments, who are implementing donor-designed reforms, are often antagonistic to civil society organizations that challenge their policies and decisions, labelling them as anti-development, "enemies of the state" or "economic saboteurs." As a result, with time, resources, and the capacity of domestic civil society organizations (CSOs) often limited, and opportunities to influence domestic policy also limited, CSOs that do wish to challenge national decisions and policies relating to resources and the environment reach out to international NGOs and networks to help support their efforts.[3] Hence, while many national governments in Africa have made great strides in addressing environmental concerns, including articulating the right to a healthy environment in their constitutions, a tension over environmental management often persists and manifests itself implicitly and explicitly in the processes of domestic policy and decision-making. To understand the influence of globalization on environmental issues in Africa, then, it is important to go beyond the duel between resource extraction and environmental protection. It is important to understand how global actors work through and in domestic decision-making, and how the actors and ideas inherent to globalization become embedded in the domestic processes where decisions over resource extraction, policy, law, protection, conservation, and management are scrutinized.

Hence, one of the central arguments of this chapter is that it is in the institutional milieu of domestic policy- and decision-making processes where we see the multi-faceted impact of globalization play out most poignantly.

It is in the domestic policy system where national governments are in theory expected to mediate between the norms and goals of international interests (firms, NGOs, and donors) and rights and knowledge of civil society, while also communicating to domestic citizens the expected outcomes and benefits to be gained from these processes. What emerges from this daunting task is that at a time when systems of environmental management and policy-making in Africa are rapidly evolving and maturing, globalization and the attendant actors, ideas, processes, and demands it carries place an enormous strain on the capacity of governments and to some degree, I would suggest, undermine them. To make this point, I will discuss environmental policy-making in the East African country of Uganda specifically, with a particular focus on energy and electricity. One of the issues that will come to light with respect to the theme of this book is that while international civil society organizations are present and influential in Uganda, to a greater extent it is domestic civil society organizations that are working, often slowly but importantly, to challenge and question international norms and actors.

This chapter is organized into three subsequent sections. In the second section, I describe and explain the character of politics in contemporary Uganda, along with its current political system and ongoing reforms. This is followed by a discussion of environmental management and policy in Uganda, emphasizing the role of international and domestic actors in environmental management and policy-making. This section highlights the dominant influence of international donors and norms on environmental policy in Uganda, but also the increasing role domestic civil society is trying to play in environmental decision-making, even though opportunities to participate are largely due to donor requirements. In the third section, I examine the principal case for this chapter: energy and electricity. Uganda has one of the lowest levels of access to electricity in the world (4 per cent of the total population) and for years has been trying to increase the amount of electricity generated in the country by building a new hydroelectric dam on the Nile, the Bujagali dam. Despite the demand for electricity, and the endorsement provided by the World Bank and the national government, it has been delayed on numerous occasions and construction has still not begun. While the World Bank communicates that the delays are a result of international and domestic financial conditions, I argue that one cannot understand delays in dam construction without understanding the domestic policy process, in particular how international firms, international donors, the national government, and domestic and international environmental NGOs (ENGOs) have converged. In following this story, we see that domestic civil society, despite reaching out for international support, has had some but limited influence on environmental affairs due to the power of the World Bank's policy agenda and the commitment of the national government to that agenda. Whether one agrees with the construction of large dams or the privatization of electricity services, this case reveals the complex tension surrounding domestic and international interests and environmental norms in Uganda. The final section of this chapter concludes with a reflection on the international influence on domestic environmental politics and policy in Uganda, particularly given that donors act as dominant authorities in shaping environmental policy and reform in the country.

POLITICAL AND ECONOMIC CONTEXT

Uganda gained independence from the United Kingdom on October 9, 1962. Not long afterwards, however, Uganda suffered through almost fifteen years of civil conflict and some would say much terror, particularly under the notorious reign of General Idi Amin (1971-79). Four brief regimes followed Amin, including Milton Obote's second term as president (Obote II). In 1986, the degree of stability in Uganda markedly changed with Yoweri Museveni's ascendance to power under the banner of the National Resistance Movement and following a protracted "bush-war." Museveni, however, inherited a country whose economy was in ruin, and whose infrastructure, social services, and political and administrative institutions were dilapidated (Kiyaga-Nsubuga 2004: 89). In response, in 1987 the government announced its Economic Recovery Programme. External support for the program marked the beginning of President Museveni's ongoing relationship with the World Bank and International Monetary Fund (IMF) under the auspices of a "series of consecutive and sometimes overlapping structural adjustment loans" (Dijkstra and Kees van Donge 2001: 842).

Since the late 1980s, there is a sense that Uganda's "relative" political stability and prudent macroeconomic reforms have created "one of the better investment climates and the most liberal trade regime[s] in the region" (World Bank 2005).[4] Other praise reinforces this view, with suggestions that maintaining economic discipline while continuing to place anti-poverty and sector policies in the country's macroeconomic framework has the potential to produce great long-term stability (Kasekende and Atingi-Ego 1999; Muhumuza 2002). While some have questioned the veracity or authenticity of the country's economic success (Tangri and Mwenda 2001), the near 6-per-cent annual economic growth in the last two decades generally produces few critiques in relation to macroeconomic performance. General achievements in increasing access to primary education, HIV/AIDS prevention, and poverty reduction in the 1990s reinforce these economic accomplishments. But at the same time, in recent years the durability of Uganda's economic success in the 1990s has been questioned, as the percentage of the population living below the poverty line between the period 1999-2000 and 2002-2003 increased from 34 per cent to 38 per cent (Hickey 2005: 997). Moreover, the legacy of the country's political regime has been increasingly scrutinized. Scholars have raised concern over the character of President Museveni's rule, the role of international donors in Uganda, and the legacy of Uganda's reform strategies and political system on state-society relations and approaches to policy-making.

Contemporary Political Context

Museveni came into power under the banner of the National Resistance Movement (NRM). The Movement system, as it came to be known, could be neutrally described as a "no-party democracy"—a system where the President and legislative representatives were elected by popular vote but not permitted to stand as representatives of political parties. The argument in support of the no-party system was that political parties had historically exacerbated ethnic and religious cleavages leading to conflict. Therefore, while political parties did exist, they were not permitted to hold mass meetings or campaign on party platforms. While some argued that the NRM effectively operated as a single-party state, others suggest that it is more accurate to characterize Uganda's no-party political system as a "hegemonic party system" where "political supremacy was exercised by a single organization, with smaller opposition groups not able ... to put up any significant challenge" (Hickey 2005: 98). Others, however, push the character of the no-party system and Museveni's rule further. Well-known Africanist scholar Joel Barkan, for example, remarked that after the 2001 presidential elections, the movement morphed into an old-style one-party state. Barkan suggests that political institutions that once held great promise for evolving into meaningful institutions of countervailing power to the Executive "have come under varying degrees of pressure to support Museveni" (Barkan 2005: 4).

In early 2006, the Constitution was amended to allow parties to function as political organizations and for candidates to campaign on party platforms. This followed several years of donors publicly expressing their desire for Uganda to open its political system to political parties, but not pushing for it. In addition, in 2006 the Constitution was also amended to remove term limits for the presidency, thus allowing Museveni to stand for a third term as leader of the National Resistance Movement in the March 2006 national elections. Museveni won the election with 59 per cent of the popular vote.[5]

What is striking about the recent change in Uganda's political system is the length of time the no-party system was able to persist without strong donor pressure for change (Harrison 2001; Dijkstra and Kees van Donge 2001; Muhumuza 2002). Several prominent scholars of East African politics note that donors seemed to hold Uganda to a different standard than other African states (Tripp 2004), particularly Kenya, where administrative reform and demands for multipartyism have been constantly reinforced through the threat or execution of the withdrawal of financial support (Harrison 2001: 659). Of all bilateral and multilateral donors operating in Uganda, one, however, figures most prominently—the World Bank. In our interview, the outspoken yet controversial Ugandan journalist Andrew Mwenda described the relationship between the Bank and Uganda as a "marriage of convenience": "The World Bank needs Uganda as much as Uganda needs the World Bank."[6] Similarly, a bilateral donor representative with whom I discussed the World Bank's role in energy-sector reforms confidentially yet light-heartedly remarked, "people refer to Uganda as the Pearl of Africa, but some people say that Uganda is the Pearl of the World Bank." Others have argued that Uganda's preferred status is due in part to its serving as a "fertile ground on which to test new approaches" relating to poverty alleviation and economic development (De Coninck 2004: 60). Why are these perspectives important to introduce when considering globalization and environmental management in Uganda?

Close to 50 per cent of Uganda's national budget is financed by donors, with the World Bank the largest (East African 2005; USAID 2005).[7] The World Bank has been lending to Uganda since 1961. In the period between 1991/92-1993/94 the World Bank required 86 policy reforms as part of its conditional lending (Harrison 2001: 668). Today, if one consults the World Bank Project Database, one learns that between 2000 and 2005, the World Bank was financing, and the Government of Uganda administering, 57 different projects.[8] What is important to note when considering these numbers is that they do not factor in the range of other donor-countries with projects in Uganda. These would include Denmark, Norway, Sweden, the UK, the United States, and Germany, to name a few. From one perspective, the simple message from the volume of donor activity in Uganda is that there is a strong propensity for donors to support many reforms at the same time (see Therkildsen 2000), thus raising concerns about the government's capacity to administer so many projects at once, particularly in the absence of evidence that the projects and reforms produce the intended outcomes. More directly in relation to the theme of this book and argument of this chapter, the high presence of donors in Uganda clearly indicates that it is difficult to understand environmental management and policy in Uganda without understanding the role and norms of donors in domestic policy-making given that the policy-making process is an important channel through which donors may influence reform (Therkildsen 2000: 65). In Uganda, to date this has certainly been the case.

When the National Resistance Movement first took power, Museveni initially took a stand against International Financial Institutions (IFIS) (Kjaer 2004: 396), stressing "their complicity with the Obote II regime and their role as agents for external intervention" (Harrison 2001: 662). Imposing policies running counter to liberal-economic orthodoxy—price and foreign exchange controls to curb inflation—enforced his position (see Kjaer 2004). These initiatives, however, did not produce the desired results and inflation accelerated, leading to a shift in Museveni's approach to the economy that sided with pro-market reformers (Kjaer 2004: 396). The decision to move the economy toward a market-liberal framework has since raised questions for researchers about an apparent compromise by Museveni: in exchange for donor acceptance of the president's "no-party" political system, Uganda would permit, amongst other things, a dramatic reduction in the state's economic controls. According to Andrew Mwenda, "Museveni gave donors almost complete control of the economic policy making process, and in return the donors allowed him a free hand to pursue his preferred political and security machinations like banning political party activities in the country and pursuit of military adventures at home and in the region" (2005).[9] Not surprisingly, Museveni would fervently disagree with Mwenda, having said, "We did not adopt market economics as a consequence of pressure, but because we were convinced it was the correct thing to do for our country. If we had not been convinced, we would not have accepted it" (Kjaer 2004: 396). Goran Hyden has also recently commented on Museveni's decision to reject his earlier intellectual leanings toward socialism (2006). Hyden argues that what is clear is that Museveni "initially conceded considerable autonomy to economists and technocrats who were free to design neoliberal economic policies that reduced state involvement and encouraged private investments" (2006: 131). Whether or not a formal compromise was agreed to, we will never know. What remains significant is that the influence of donors in economic reforms continues to be strong and has also manifested itself in the manner in which policy-making evolves and also how civil society groups participate in policy-making.

The National Resistance Movement's ascendancy to power in Uganda ushered in a new era for civil society organizations (CSOs). According to John De Coninck, in the early going, "a space was provided ... for the emergence of indigenous civil society organizations ... *so long as they had no political agenda*" (2004: 57, emphasis added). De Coninck continues:

> This era of growth for civil society organizations ... accelerated as the World Bank and other donors forced fiscal orthodoxy upon government. Seen as ideologically preferable to state delivery, CSOs were considered less corrupt and closer to the people. This was the heyday of NGOs. Generously funded, they could act with impunity and without reference to government policies.... This era also established two important dimensions of civil society in Uganda: firstly, the lasting association—even equation—of "civil society" with NGOs ... secondly, the tendency for NGO growth to be driven by the availability of donor funding rather than the need to provide a direct answer to specific locally rooted social or political imperatives. (De Coninck 2004: 57–58)

De Coninck also explains, however, that as the state started to reassert itself at the local level, national poverty and adjustment policies were introduced that started to redefine the role of the state: the "bonanza years for NGOs were over. The latitude for their involvement in service delivery was narrowing while donors were reconsidering the funding of such activities through NGOs" (2004: 60). A signpost for the change in NGO activity was the 1989 introduction of the Non-Governmental Organization Registration Statute, which required all NGOs to register with the National Board for Non-Governmental Organizations based in the Ministry of Internal Affairs. "In addition to approving applications, the Board was to guide and monitor organizations in carrying out their activities and could revoke a certificate of registration if the organization failed to operate in accordance with its constitution and if such revocation was considered in the public interest" (Tripp 2000: 61). The National Board's membership and regulations suggested a high degree of suspicion surrounding NGOs, but the NRM remained accommodating for the most part, largely due to its inability to enforce rules and monitor activities. The Board also had weak coordination and planning capacity (see Tripp 2000: 61-62; Dicklitch 2001).[10] As a result, NGO accommodation was a default outcome rather than an expressed position (Tripp 2000: 61). Moreover, because of weak internal coordination and planning, "NGO agendas tended to be donor-driven and Ugandans consequently had little negotiating power" (Tripp 2000: 62). Despite the early rise in number and influence of NGOs, the de-registration of a handful of them, along with the delay and near denial of registration for others, reinforced the position that "political" activities would not be accepted (Dicklitch 2001: 35). Hence, even though NGOs were tolerated, the NRM was ready and willing to limit their autonomy if there was "the slightest possibility that they might prove to be too much of a challenge" (Tripp 2000: 63).

One of the important outcomes of these conditions is that most Ugandan CSOs have by choice and necessity worked with the state and donors, rather than challenged their activities and policy choices.[11] As Brock has noted, "the Movement system of government, with its rhetorical and structural focus on inclusion and decentralisation, has subsequently shaped a political landscape where the dividing line between state and non-state actors is blurred" (2004: 95). The role of donors in this blurring has been central. In the mid-1980s and early 1990s, donors were promoting and supporting NGOs as important agents of poverty reduction and development activities. While this has not changed, and the number of NGOs in Uganda has steadily increased since the early 1990s,[12] concerns with corruption, accountability, poverty, and the desire to see a multiparty system led donors to promote a new role for NGOs: holding government accountable (De Coninck 2004: 62). This current situation, as noted in the introduction, has placed Ugandan CSOs in an extraordinarily difficult situation: donors increasingly want NGOs to monitor government activities, thus encouraging them to be engaged in political activities that are clearly at odds with a government intolerant of challenge. This is despite the fact that both NGOs and civil servants acknowledge that most Ugandan NGOs lack the capacity and expertise necessary to monitor compliance and ensure accountability—an observation with significant implications for environmental policy debates in Uganda, as well as for explaining the absence of a recognizable environmental movement in the country.

Environmental Context: Issues, NGOs, and Policy-making

Comparatively, Uganda's land mass is relatively small at just under 200,000 square kilometres; this is slightly smaller than the state of Oregon, about one fifth the size of the province of Ontario, and about four-fifths the size of the United Kingdom. The very southern portion of the country lies astride the Equator and sits on the northern shore of Lake Victoria. Uganda is bordered by Sudan to the north, Kenya to the east, the Democratic Republic of Congo to the west, and Rwanda and Tanzania to the south. Driven by two long and two short rainy seasons, this location provides a temperate, moist, and favourable climate for agricultural production, which remains the most important sector of the economy (Urban Harvest–Kampala 2004: 6).

Uganda's total population is currently about 28 million. With most of the population living in rural areas (87%) and relying on natural resources and agricultural production for their livelihoods, concerns surrounding land degradation and soil erosion have historically been prominent and remain dominant environmental concerns (Tumushabe 1999; NEMA 2001; 2005). Concerns with land degradation are intensified by Uganda's high population growth rate (3.3%) and by the fact that forest resources (charcoal and firewood) are the dominant source of energy for over 95 per cent of the population. Recent reports suggest that access to water resources and biomass in many Ugandan districts is becoming an increasing problem, exacerbating the gendered dimension of household responsibilities as women and children have to walk further to meet daily needs (Monitor 2006). Therefore, though we must be very careful not to recapitulate the highly problematic and overly simplified argument that there is a direct causal relationship between poverty and environmental degradation, the combination of a dominantly rural and agriculturally dependent population, a high population growth rate, and poverty does explain the presence of (but not the reason for) some of Uganda's most pressing environmental issues: soil erosion, land scarcity, unsustainable harvesting of fish and other natural resources, officially sanctioned encroachment into protected areas, and poor administration of land-related issues (NEMA 2005). Despite these challenges, in its "State of the Environment Report, 2004-2005," the National Environmental Management Authority (NEMA) suggested that the country had also made several notable achievements in environmental management in the past decade, including improvements in dealing with several environmentally related illnesses like malaria and sleeping sickness, increases in permanent housing, stabilization of biodiversity loss, and the presence of enabling policy framework for environmental and natural resources management (2005). While this last point is true, the origins of this policy framework and the challenges in its implementation speak directly to concerns over the global influence on domestic policy.

Beginning in the late 1980s, a number of countries in sub-Saharan Africa embarked on strategic policy and planning exercises to address mounting concerns over environmental issues. Although in some cases international organizations such as the World Conservation Union (IUCN) had earlier sponsored the development of national conservation strategies and a handful of countries had independently undertaken national environmental reviews, it was a World Bank drive for countries to develop National Environment Action Plans (NEAPs) that came to dominate (Wood 1997). This in turn meant that in many African countries the formal processes that governments undertook to inventory national environmental conditions and to create new policies, laws, and administrative units were largely a result of Bank requirements and less so due to the urging of domestic civil society. The Bank's requirements, in conjunction with a president that recognized the significant role of science, technology, and agriculture, all played an important part in Uganda's future environmental reforms.

One of the first changes relating to environmental management in Uganda came in 1991 with the establishment of the Uganda National Council for Science and Technology. The establishment of this institution followed President Museveni's public acknowledgement of the role of science and technology in development (Tumushabe 1999: 73). The year 1991 also marked the beginning of Uganda's formal NEAP process. The inventory of national environmental issues along with the review of environmental policies and laws lasted for three years, culminating in the creation of a new National Environmental Policy in 1994. In 1995, three new pieces of legislation emerged in Uganda with significant environmental influence: the Water Act, the National Environment Act, and the National Constitution. Significantly, as the supreme legal document, the Constitution articulated the right to a healthy environment for all citizens of Uganda. Other environmental legislation that followed included the Wildlife Act (1996); a host of new regulations relating to such things as environmental impact assessment, waste management, wetlands, ozone, and water resources between 1998 and 2003; the Animal Breeding Act (2001); the Mining Act (2003); and the National Forestry and Tree Planting Act (2003). In many circumstances, the emergence of a new piece of legislation also coincided with the creation of a new donor-funded regulatory authority. For example, the National Environment Act established the creation of the NEMA. NEMA is responsible for the management, coordination, and supervision of environmental issues in the country. Since its creation, the World Bank has wholly funded this institution. On December 31, 2007, World Bank funding for NEMA is supposed to end, and the Government of Uganda is expected to fund NEMA exclusively. There is, however, some concern about whether the government will continue to support NEMA to the extent it has been supported by the Bank. In 2007, the national government was supposed to fund a portion of NEMA's national budget, but by mid-2007, this money had reportedly still not been disbursed. Other regulatory authorities also funded by donors include the Uganda Wildlife Authority (an amalgamation of the National Parks and Game departments) and the National Forest Authority. Thus, to what extent can the emergence of new legislation in Uganda be attributed to donor requirements as opposed to domestic obligations to international treaties or pressure from domestic civil society? According to personal observation and conversations with ENGOs in Uganda, the environmental policy and regulation that began to emerge in the early 1990s was largely a result of donor requirements. By 1992, the Bank was insisting that projects it funded comply with its environmental guidelines. But these requirements had little impact as there was little compliance, the Bank could not enforce them, and domestically there was also no enforcement (Kenneth Kakuru, lawyer, Kakuru and Co. Advocates, Uganda, personal communication). At the same time, in the early 1990s, Uganda had embraced IMF and World Bank economic reforms, which promoted privatization. Thus it made sense domestically to develop new legislation to meet enforcement requirements and also to scrutinize new private-sector development. The new laws were certainly in keeping with international trends and obligations following the 1992 Rio Summit. However, when trying to understand the global influence on domestic environmental policy in Uganda, it is clear that donors have historically been and continue to be central instigators of change.

None of the above information is presented to suggest a lack of domestic commitment to environmental issues. Indeed, Uganda's commitment to environmental concerns has received cautious praise from researchers (Mugabe and Tumushabe 1999; Okoth-Ogendo 1999; Tumushabe 1999; Wood 1997; World Bank 2001). This praise has been variously attributed to Uganda's senior political commitment to the sustainable use of environmental resources, including the President's; NEMA's role in environmental management; and Uganda's system of decentralization that rests authority and responsibility for environmental management at subnational levels. (With respect to decentralization and environmental management in Uganda, the theoretical benefits of subsidiarity have been mixed to date. Limited local capacity, jurisdictional conflicts, and the misappropriation of local funds remain prominent concerns.) These characteristics are complemented by the financial and logistical support the national government receives from donors, but also the approximately 400 NGOs working in the area of environmental management in the country (World Bank 2000).

A majority of ENGOs are based out of the capital city, Kampala. Most NGOs, however, restrict their activities to five or fewer districts, and only seven of 1,777 NGOs surveyed in 2002 operated nationally (see Barr et al. 2005). With respect to financial resources, for the Ugandan NGO sector as a whole, grants received from international NGOs account for nearly half of total funding in 2001, with grants from bilateral donors the next most important source (Barr et al. 2005: 664).

These survey results also show that over 50 per cent of NGOs indicated that they were engaged in activities relating to natural resources or the environment. In my experience, most ENGOs rely on a small number of permanent employees (five to ten) and depend greatly on volunteers and contract consultants. Importantly, there is also a very limited permanent presence of international ENGOs in Uganda.

One exception is the World Conservation Union (IUCN). Given that NGOs play a dominant role in driving environmentalism and environmental movements globally, one can ask how "environmentalism" might be characterized in Uganda, and whether there is a recognizable environmental movement; that is, is there evidence of a collective challenge by organizations and individuals with common purposes and solidarity relating to the environment in Uganda (see Tarrow 1994; Doyle and Mceachern 2001)? The answer to this question is not easy to discern.

There are a handful of ENGOs in Uganda that are increasingly playing a prominent role in advocating for greater access to information, more opportunities to participate in policy development, and openly challenging donor and government projects that raise significant environmental and economic concerns. The National Association of Professional Environmentalists (NAPE), Advocates Coalition for Development and Environment (ACODE), Greenwatch, and Uganda Wildlife Society (UWS) are examples of Kampala-based NGOs that have raised significant concerns about national and local policies and development plans, and have gained national media attention along with the ire of the national government for their activities. To date, however, owing to the small size and limited capacity of NGOs, NGOs need for non-domestic financial support, and the dominant rural character of the country, in terms of both population and poverty, localized and poverty-oriented environmental concerns have largely been the focus of ENGO activities. This has meant that, although ENGOs regularly meet, there are few environmental issues in Uganda that have invoked a recognizable national "movement" that is sustained and focused on one collective goal. Despite this, ENGOs remain important domestic actors who increasingly gain national and international attention, particularly for their role in drawing attention to how decisions surrounding environmental resources are made in the country. For example, during past elections, elected officials have frequently used environmental resources, particularly forests, to leverage political support by encouraging constituents to cut trees illegally. Incidents like these, combined with high-level corruption within some environmental bureaucracies,[13] and unsuccessful historic reform efforts have reinforced the trend in Uganda and elsewhere in Africa to institute regulatory or "independent authorities" to oversee environmental resources. This trend has an important connection to NGO advocacy in Uganda, for donors have not only been responsible for establishing new institutions like NEMA, the Forest Authority, and Wildlife Authority, but have also defined how and when NGOs get to participate in policy debates and reform.

Some refer to the creation of "independent authorities" as an "executive agency model" (Therkildsen 2000). The intent of this model is to leave government with policy-making responsibilities while revenue-generating and service-provision activities are done at an arm's length from government. However, according to a senior civil servant in Uganda's Ministry of Public Service, "there is not much known about how they [independent authorities] are doing in Uganda."[14] Using the example of the Uganda Revenue Authority (URA), it was explained that the revenue the URA had collected had decreased while its operating costs had increased. URA also recently registered a 5-billion shilling loss for one financial quarter (Monitor 2006). Therefore, despite it being established in 1991, "it's not known whether the URA is adding any value."[15] In addition, despite the intent of separating policy functions from service delivery and management functions, some regulatory authorities like NEMA have in fact been given specific responsibility for policy development. As a result of this situation, there are functional overlaps and problems of accountability and responsibility, for roles are not clear or consistent. In one interview, a donor representative suggested that NEMA was "a bit schizophrenic" because it crossed so many roles.[16] Hence, one argument is that the executive agency model in Uganda is "not very well thought through." Nonetheless, some recent donor-led reform efforts are receiving cautious praise from NGOs for the greater opportunity to participate. The forest-reform process (1999-2003) stands out in this regard as it is perceived to be one of the best consultative processes relating to the environment experienced in Uganda in recent years. NGOs point to the regional workshops, working groups, and opportunities to comment as important opportunities to contribute to the reform process. But the forestry reform process also highlights two other significant issues surrounding NGO participation in environmental policy-making in Uganda.

First, participation does not mean influence. For all but a small number, environmental NGOs in Uganda are constantly engaged or consumed with what one author describes as a "politics of presence" rather than a "politics of influence" (Brock 2004: 103). That is, ENGOs are consumed with trying to be invited to different donor-sponsored information sessions or workshops that donors and government promote as consultation. Interviews with Ugandan ENGOs confirm this trend, with representatives explaining that while workshops and policy-review sessions are by no means ideal, they are often the only opportunities available to participate in policy development. The second important observation is that whether one speaks to NGOs or civil servants, when asked where the motivation for reform comes from, the answer is always the same: donors. This was certainly true for forestry, the environment as a whole, and, as we shall see, the energy sector. Hence, for some observers, "Donors on the Ugandan policy scene are ... not just funders but actors who contribute to various policy processes and are also very aware of the power that they wield in shaping policy" (Ssewakiryanga 2004: 83). Others push this observation further: "rather than conceptualizing donor power as a strong external force on the state, it would be more useful to conceive of donors as *part of the state itself*. This is not just because so much of the budgeting process is contingent on the receipt of donor finance, but also because of the way programmes and even specific policies are designed and executed" (Harrison 2001: 669).

The fact that NGOs are looking to be present at policy fora is not surprising given that the majority of their funding comes from international sources, which are often "preoccupied with accountability to their donors" (De Coninck 2004:

63). Abigail Barr et al., in the conclusion to their recent paper on the Ugandan NGO sector, argue that NGOs are quite entrepreneurial, are led by educated individuals interested in attracting international aid, are generally enhancing the well-being of their beneficiaries, and are generally well perceived in the country (2005: 676). However, the authors also note that Ugandan NGOs are also operating as "subcontractors for international donors" (2005: 676). Being present at workshops and policy fora, therefore, provides important opportunities to demonstrate engagement in issues, to promote individual work, and also to network and seek out additional funding opportunities. Hence, on one level the "politics of presence" is ideologically in keeping with donor preferences and international trends surrounding citizen participation and access to information in environmental decision-making. Here we should recall Principle 10 of the 1992 Rio Declaration, which states that "Environmental issues are best handled with participation of all concerned citizens, at the relevant level," and that citizens should have access to information held by public authorities regarding the environment, and that opportunities should be provided for citizens to participate in decision-making. But pragmatically, the Ugandan NGO desire to participate in policy fora and workshops is also simply a necessity; these fora often serve as the only policy game in the proverbial town. Here, the bind in which Ugandan environmental NGOs function emerges more explicitly.

There exist few formal opportunities to influence environmental policy and decision-making in Uganda. The opportunities that do exist are relatively limited with respect to opportunities for dialogue and debate, are generally better understood as information sessions, and are contingent on donor requirements for participation and the level of controversy surrounding a sector or issue. For example, NGOs will be invited to participate in a health or gender policy forum, while something relating to defence will be closed (ACODE 2002). Moreover, the government's strong aversion to public criticism and NGO dependence on external funding for support leaves few NGOs willing or able to challenge government and donor reforms publicly, for fear that it will ultimately undermine their very own existence. When the volume of reforms being simultaneously undertaken is added to this scenario, along with a desire for reforms to produce quick results that meet human needs and encourage and support international investment, the enormity of the domestic policy task begins to be revealed. The case of the energy-sector reform and dam-construction effort in Uganda helps bring this complex global-domestic policy picture further into light.

ENERGY AND ELECTRICITY IN UGANDA

Uganda has an installed capacity to generate 300 MW of electricity, largely derived from two hydroelectric facilities—the 1954 Owen Falls Dam (now renamed Nalubaale) and the 1993 Owen Falls extension (now renamed Kira). Stemming from the protracted civil conflict in the country, following the completion of the Nalubaale dam, no new generation facilities have been constructed in the country. In 2006, because of recent drought in the region, low water levels on Lake Victoria decreased hydroelectric generating capacity to just 135 MW. As a comparison, Canada's most populous province, Ontario, has a population of 12.5 million (less than half of Uganda) and a peak electricity generation capacity of 22,000 MW! Demand for electricity in Uganda is also growing at about 30 MW per year. In the interim, large diesel generators have been brought into the country to meet electricity demand, while rolling blackouts continue, and demand for biomass (firewood and charcoal), already the primary source of energy for Ugandans, increases. The situation in the country is so extreme that projections of seven-per-cent economic growth for the country have been reduced to 4.5 per cent, largely due to power shortages (East African 2006a), and the capital, Kampala, has been sarcastically described as "generator city" given the constant hum of generators (East African 2006b). In 2002, only four per cent of Uganda's total population of close to 25 million people were connected to the electricity network. In rural areas, the figure was closer to one per cent of the population. In 2005, while Uganda's population had increased to nearly 27 million, the percentage of people connected to electricity had not changed. Today, it is not inflammatory to suggest that Uganda's electricity sector is functioning in crisis mode. Given this scenario, it would come as little surprise that both reform to the energy sector and new sources of electricity generation have been government and donor priorities. Both of these initiatives were intimately connected and driven by international actors.

Electricity-Sector Reform and the Bujagali Dam

In 1986, a consulting report assessed Uganda's electrical system and identified three sites for future hydroelectric development. In 1990, the Canadian consulting firm Acres International completed another survey of the hydroelectric potential in Uganda. The Acres report first recommended the expansion of the Nalubaale dam. This was followed by a recommendation to build a large new 200 to 250 MW facility at Bujagali Falls, 12 kilometres downstream from Nalubaale. Another dam at Karuma Falls in northern Uganda should then follow. The report suggested that these facilities were the "least cost electricity generation options" for Uganda and helped lay the groundwork for future energy investments in the country.[17] Shortly after World Bank support for the dam expansion, Bujagali Falls was promoted and pursued.

As the story of the Bujagali Dam unfolds, Nitin Madhvani, head of the South Africa-based Madhvani Group of companies (and also the largest private investor in Uganda) contacted US-based AES International to gauge their interest in building Bujagali. In 1995, AES representatives visited Uganda. In the same year, Museveni signed a Memorandum of Understanding (MOU) with AES and Madhvani, giving the companies first right of refusal to build a dam at Bujagali. Neither company had ever constructed a dam before. Together, AES and the Madhvani group established AES Nile Power (AESNP).[18] At the same time, other Independent Power Producers (IPPs) were in the country, including Enron. Another company, Norpak Ltd., a subsidiary of Norwegian-based utility company, Agder Energi AS, was given the right to develop the second most-favoured site, Karuma Falls, located in northern Uganda. How was their right assured? According to a consultant working on Norpak's behalf, this was the result of a "gentleman's agreements that we are number two." Hence, no competitive bidding process was undertaken for either of the sites. At this point, however, the institutional and legal conditions in Uganda were not amenable to international firms generating electricity in the country. Since the Uganda Electricity Board's (UEB) creation, it was the only company legally permitted to generate, transmit, and distribute electricity in the country. While parliamentary debate suggests there was some controversy over whether the Electricity Act had to be amended in order to permit new companies to generate electricity, an amendment to the legislation was pursued. Complementing this legal reform was internal resistance to change within the UEB. Nonetheless, a slowdown in project development did not arise. Rather, at the same that time private firms were developing their hydroelectric projects, they were also working with government to amend the domestic institutional and legal frameworks responsible for the approval and oversight of the projects. For example, the national environmental authority (NEMA) was developing its own legislation for environmental assessment procedures while AES was planning Bujagali—a project that would eventually be scrutinized using the EA legislation. According to NEMA officials, while AES did not dictate the review procedures, they certainly put a lot of pressure on the organization. Therefore, while AES was pursuing the largest private-sector investment in Africa (approximately $550 million), and Norpak was waiting to start construction on Karuma, three principal government agencies relating to energy issues—Forestry, Energy, and Environment—were all undergoing significant reform and change at the same time.

By 1999 a new Electricity Act had been promulgated and, by 2001, the UEB had been split into three separate companies, with distribution and generation intended for privatization. Meanwhile, AESNP was slowly gaining the approvals necessary for the construction of the Bujagali dam. But the period between 1998 and 2001 was also a period when domestic civil society organizations began to question seriously the rationale behind the construction of the dam. In 1998, the one-year-old National Association of Professional Environmentalists (NAPE), a domestic environmental NGO with about 50 registered members in 2004, became engaged with the Bujagali dam issue. They visited the site where the dam was to be built and spoke with some of the residents who were going to be resettled—approximately 100 households and 8,700 residents. NAPE suggested that despite the general perception that the local community was in support of the project, those that opposed it were "suppressed." The reality is very hard to determine, given that by this point AESNP was already engaged in environmental and community assessments, thus bringing out community expectations of resettlement and compensation. This same year, another ENGO, the Uganda Wildlife Society (UWS), organized a workshop to discuss the project, as there was concern circulating that Parliament was being pressured to approve the Power Purchase Agreement (PPA) — the 30-year contract stipulating the guaranteed price and volume of electricity that the government, under the auspices of the Uganda Electricity Transmission Company, would have to buy from AESNP — before the environmental assessment of the project was complete and approved. One year later, the Berkeley, California-based International Rivers Network (IRN) became engaged in the Bujagali project. NAPE, in its advocacy efforts, approached IRN for assistance owing to its prominent global role in questioning the need for large hydroelectric dams and the protection of river systems. IRN began working with NAPE and one other Ugandan NGO, the Save the Bujagali Crusade. Despite the fifty or more members affiliated with NAPE, both organizations were very small and received no direct budgetary support from outside the country, and their Bujagali campaigns were largely driven by two or three individuals. The original concerns of both domestic organizations were largely linked to the dam's impact on the natural environment, particularly fisheries and agriculture, but in a short period, and with the advice and support of IRN, the central concerns that came to dominate were procedural, economic, social, and political.

First, the purpose and benefits of the project were never accurately communicated. While the rationale for the project was pitched as a poverty-reducing measure, the reality, as interview data with NGOs, donors, and government officials confirm, was not to supply electricity to individuals but to fuel industrialization. Thus the "poverty impact" of the project rested on a trickle-down theory of economic development. This model, in and of itself, is not the problem. The problem was that Ugandan citizens, perhaps naively but honestly, were led to believe that the dam was going to benefit them directly, a point again confirmed by my conversations with lay citizens. Second, the general assumption was that the cost of electricity would decrease with the completion of the dam. The reality, however, again not communicated, was that the price of electricity would increase dramatically in order to pay for the cost of the project, the infrastructure and the high return AESNP was to receive on its investment. On more than one occasion I was blankly told, even by those supportive of the project, that "AES is not bringing anything to Uganda and that all this talk about a $550 million investment is rubbish — AES will make a lot of money in Uganda." Furthermore, the details of the Power Purchase Agreement (PPA) were not publicly disclosed until after the Ugandan NGO, Greenwatch, challenged AESNP and the government's argument that the PPA was proprietary, and the Ugandan High Court ruled in favour of making the agreement public. The World Bank's economic analysis of the project was also not disclosed. Finally, as the above information would suggest, there were serious concerns about access to information and transparency, along with the character and opportunity to participate in the dam's assessment and review process. Here a striking debate remains. On one hand, project proponents suggest that the Bujagali review process was extremely open and the consultation processes were widespread. On the other hand, the issue is not how many times consultation took place, but the quality of participation. Indeed, in 1999, one of the central public meetings that took place in the town of Jinja, near the headwaters of the Nile, was a raucous event described as "havoc," where proponents and opponents of the project charged each other with buying off and paying off participants.

The height of conflict over the dam came about in 2001. AESNP began compensating dam-affected communities prior to World Bank financial approval, and prior to AESNP having the necessary $115 million equity it needed. Owing to these and some of the issues outlined above, after writing to World Bank management about concerns and receiving an unsatisfactory response, by mid-year seven individuals—one from NAPE, two from Save the Bujagali Crusade—filed a complaint with the International Finance Corporation's (IFC) Compliance Officer and the World Bank Inspection Panel. (The Inspection Panel is the independent body established in 1994 by the World Bank Executive Board owing to the protracted controversy with its financing of the Sardar Sarovar Dam in India.) The request for inspection focused on several Bank operational policies the NGOs felt had been contravened. The concerns largely centred on environmental and economic considerations.[19] The request was made five months prior to the date when the Bank Board was expected to vote on the project for approval. The Bank management's response to the issues raised in the Inspection Panel request was deemed unsatisfactory by the NGOs. Subsequently, the Inspection Panel recommended a full investigation of the allegations, which the Bank's Board of Executive Directors approved. Nonetheless, despite the initiation of the investigation, on December 28, 2001, the Bank approved the partial risk guarantee for the project, in effect muting whatever findings would result from the Inspection Panel's investigation. Hence, by the end of 2001 it was assumed that the project would go ahead. Less than a month later, on January 24, 2002, a groundbreaking ceremony was held.

During remarks at the ceremony a senior representative of AES described Bujagali as a "perfect infrastructure project for Africa." Along with other speakers, Syda Bbumba, Uganda's Minister of Energy and Minerals Development enthusiastically noted how President Museveni's leadership, vision, and persistence were central to the success of the project. Bbumba also thanked the US ambassador, who, she explained, had "done nothing in the last four months but lobby for support of the AES project." President Museveni was the last to speak to the crowd of dignitaries and citizens. His reflections were not marked by celebration, however. Museveni began by stating that he was "not happy at all" and was "ashamed." He said he did not want to talk about how happy Ugandans were but just wanted to get on with the project. He suggested that a project that should have taken two years to launch had by that point taken seven. Moreover, he said he did not accept any person's thanks;

indeed, people were foolish for thanking him: "Do you thank people for feeding their children?" he asked rhetorically. After describing Uganda's great potential for producing thousands of megawatts of electricity, and noting the serious electricity deficit in the country, he suggested that the process leading to the construction of Bujagali was a "circus," which had led to embarrassment and undermined their interests: "this is an occasion of shame and repentance," he said. Museveni finished by critiquing the World Bank, suggesting that it "needed to stop listening to so many people" and instead "talk to people in the Third World." The Bank "listens to a lot of nonsense" and is "too squeamish and too sensitive to shallow opinions of those who aren't supportive of transformation." Indeed, two days after the ceremony, Museveni stated, "Those who delay industrial projects are enemies and ... I am going to open war on them" (New Vision 2002). Despite the fanfare surrounding the project, four years later, in 2006, the physical construction of the Bujagali dam had still not advanced beyond the placement of the foundation stone. In August 2003, estimating a financial loss of US $75 million, and amidst investigations into corruption surrounding construction contracts for the project, AES withdrew from its protracted ten-year effort to construct the dam.[20] Why, in a country where so many conditions seemed aligned to initiate a project, has so much frustration evolved and so many problems emerged? The President was the project's largest champion; the World Bank, Uganda's largest international donor, supported the project; compared to other large international dam projects, Bujagali was comparatively benign as there were few resettlement concerns, and the physical construction would not be difficult; the most active and vocal opposition to the dam was generally isolated to two small domestic ENGOs and one international NGO; and electricity is desperately needed domestically and desired regionally. Moreover, what does the case of the Bujagali dam say about globalization and environmental management?

In an October 2005 Project Completion Note, the World Bank suggests that three events led to the cancellation of the Bujagali project: 1) withdrawal of export credit agency support due to the high level of perceived country and business risk in January 2002; 2) ongoing investigations and allegations of corruption involving one of the engineering/procurement/construction contractors; and 3) the deterioration of the private sponsor's (AESNP) global financial situation. Together, the basic reading of the problem is financial. This fact is not in dispute and deals with an essential component of economic globalization. Uganda's desire to construct Bujagali was in large part dependent on favourable international financial conditions. When Enron collapsed, global investments and confidence in higher-risk energy undertakings decreased, and therefore confidence in AES's international activities, combined with the higher-risk Ugandan market, proved hugely problematic for the completion of Bujagali. But the economic dimension to globalization is intimately connected to the global norms and actors that influence domestic policy and politics.

Uganda continues to be on the receiving end of several dominant global economic theories about what developing economies need for success: regulatory authorities, privatization, foreign direct investment, competition. As noted earlier, while these norms have not yet produced the benefits anticipated, they are driven by international donors and are the grounds by which much economic and administrative reform in Uganda continues. Donors are also actively engaged in the promotion of dominant global norms relating to participation, transparency, accountability, and liberal democracy. These norms are simultaneously promoted and implemented in projects, programs, and reform procedures. But here it is important to reiterate the country context in which these norms are directed: poverty levels are high; service provision is low and in some cases desperate as with electricity; state-civil society relations have historically been uneasy and remain so particularly when highly coveted projects or reforms are at stake; and national and local institutional capacity to manage transformations, while much stronger than African bureaucracies are given credit for, are still maturing and often weak. When one adds a more vigilant and internationally connected domestic civil society (which donors promote) into this mix, as well as international firms, particularly energy-related, who are aggressively in search of new markets and opportunities, it would have been shocking if the problems with and conflict over Bujagali *had not* emerged. Globalization has a multi-faceted impact on domestic politics, and to ignore this undermines the complexity of the relationship between environment and development on the ground.

National governments have the unenviable task of trying to mediate successful policy interventions from both international expert bodies and marginalized groups (Forsyth et al. 1998: 38). This situation is further complicated by the evolving and maturing bureaucratic systems in African countries that are trying, and being forced to learn (at an historically unprecedented pace), to respond to a more vigilant, informed, and globally connected citizenry, while also trying to administer a litany of reforms. But to suggest that problems in project implementation or reform in Uganda are simply an outcome of international financial conditions as the World Bank has, or to suggest that domestic and international NGOs are responsible for undermining Bujagali as other observers have done (Mallaby 2004), fails to heed the interplay and conflict between the social and political change taking place in Uganda at the same time that economic and environmental norms and reforms are being promoted and instituted. Certainly, international and domestic civil society do challenge the path of development sometimes preferred and desired by governments, but to date only to a limited degree. This will change, and is changing, particularly as a result of international norms and donor policies that suggest that civil society's participation in development policies is needed and beneficial. Hence, it must be recognized and clearly stated that globalization, vis-à-vis actors and dominant norms—most centrally those carried and promoted by donors—are complicit in the increasing challenge of executing policies, projects, and reforms. And, therefore, when things do not go as planned, and national governments become frustrated with the process of reform or the influence of stronger, more participatory environmental management regimes, international actors—firms, donors, and civil society—must recognize their complicity in making the developmental effort more difficult.

CONCLUSION

In this chapter I have demonstrated how global actors and norms have played a central role in the evolution of Uganda's policies and practices generally, and in relation to the environment and energy sector specifically. The most dominant of these influences have been donors, along with the norms they promote. While the economic and environmental outcomes of these norms have received cautious praise, the more profound global influence in Uganda may not yet have fully emerged. That is, the degree to which global actors and norms are shaping state-society relations and the processes with which they interact. In this regard, the global influence on environmental management in Uganda is direct and indirect.

Directly, Uganda is seen as a potentially lucrative market for energy provision and expansion. International firms have clearly shown an interest in Uganda's future energy market, and because of that have had a direct impact on shaping laws, institutions, and policies.[21] Bilateral and multilateral donors, as agents of globalization, also have a direct influence on the shape of the legal and institutional environment. Whether one believes that donors should be understood as external to, or as a part of, the state in Uganda, their influence in promoting liberal-economic and environmental norms cannot be challenged. For the environmental management regime generally, or for forestry and energy specifically, donors have defined both the character of the reforms, and the character of debate over reforms—how and who gets to participate. It is in this regard that we see the more indirect, but perhaps more long-term, influence of globalization.

Poverty alleviation and economic development should remain central priorities in Uganda. However, it must be recognized that the international actors that promote global norms, and the pace at which these norms are implemented in Uganda, while potentially increasing the rate of economic growth and social benefit, have dramatic, pervasive, and unproven impacts on state-society relations and policy-making practices in the country. This observation is not to suggest that development should be slowed down, but more that the manner in which these global forces work through domestic environmental policy and politics needs more scrutiny. Domestic and international civil society organizations are also influencing environmental management practices in Uganda, but not as one would expect or to the degree that critics of liberal economic policies and practices might hope. Domestically, NGO capacity and competency are growing, and international civil society is playing a supportive, but not dominant, role. Perhaps more significantly, NGOs clearly understand that they are immersed in a difficult political context that is highly influenced by global actors. Domestic NGOs recognize that donors are playing a central role in shaping the character of domestic state-society relations, the processes and opportunities available for participation, and ultimately, the character and outcome of decisions regarding the environment.

In this regard, it must also be said that suggestions that NGOs are out for themselves is inconsistent both with my experience and with the research of others (see, for example, Barr et al. 2005). Domestic NGOs understand the need for development in Uganda and are not "anti-development" or "economic saboteurs," as has been suggested. With respect to electricity, most NGO representatives I spoke with clearly said that they were not by principle against large dams, and the Bujagali dam speaks to this issue. NGOs were aggravated by the lack of opportunity to debate the merits of the project, the failure for donors and government to communicate the rationale for the project, and the apparent privileging of international investment over domestic concerns. Ugandan NGOs are looking to have access to information and they want to participate in a debate over projects, policies, and reforms.

To be blunt, the processes and practices of development are messy, and the presence and dominance of global norms and actors are making it messier. Globalization produces a vexing domestic situation in countries like Uganda because international and domestic actors and norms converge, and disputes between speed of reform and process of reform are heightened. This is unlikely to change in the near future, and it will likely be intensified as NGOs become more active, competent, politically astute, and internationally connected. As a result, it would seem to be extremely important for development practitioners and researchers to be very attentive to the manner through which global norms and actors are shaping domestic politics, and equally, for donors to scrutinize carefully the volume and substance of reforms they dictate, and the impact these have on state-society relations.

Notes

1 This research was carried out with the aid of a Doctoral Research Award from the International Development Research Centre (IDRC), Ottawa, Canada. Information on the Centre is available on the Internet at www.idrc.ca. Additional funding for this research was provided by the C.B. Macpherson Dissertation Fellowship, Department of Political Science, University of Toronto; the Labatt Fellowship, Centre for Environment, University of Toronto; and the School of Graduate Studies Travel Grant, University of Toronto.

2 Elizabeth Asiedu writes, "When it comes to foreign direct investment (FDI) in Sub-Saharan Africa (SSA), the common perception is that FDI is largely driven by natural resources and market size. This perception seems to be consistent with data: the three largest recipients of FDI are Angola, Nigeria and South Africa – from 2000 to 2002, these countries absorbed about 65 per cent of FDI flows to the region" (2006: 63). Asiedu goes on to explain that South Africa has a large local market and contributes about 46 per cent of SSA's total GDP. Angola and Nigeria are oil-producing countries, with oil accounting for 90 per cent of total exports. SSA's total breakdown of FDI is: 36 per cent to South Africa, 16 per cent to Nigeria, 13 per cent to Angola, and 19 per cent to the remaining 45 countries in the region (Asiedu 2006: 63).

3 Perhaps the most famous example of this occurrence is the Movement for Survival of Ogoni People (MOSOP), which originated in the Niger Delta Region of Nigeria. MOSOP was made famous by the international attention it gained in speaking out against human rights and environmental abuses in the

Niger Delta, which linked Royal Dutch Shell and the then-military regime in power. The figure heavily responsible for MOSOP's international recognition in the mid- to late-1990s was Ken Saro-Wiwa, later executed by Sani Abacha's regime (1993-1998) for his actions. The lesser-known story of MOSOP, however, is that MOSOP was riddled with internal power struggles, and for many years it had failed to gain international attention, in part due to the complexity of issues it articulated, such as ethnicity, political independence, and greater oil revenues. It was only after "framing" and "pitching" its message more simply as a case of human rights and environmental abuses that international NGOs could better advocate on MOSOP's behalf (see Bob 2001; Watts 1997; Hunt 2006).

4 The word "relative" is used because Uganda has been widely accused of producing instability in other countries, most notably the Democratic Republic of Congo and Sudan. Domestically, the devastating instability in Northern Uganda surrounding the conflict with the Lord's Resistance Army (LRA) certainly raises serious concerns about a suggestion that Uganda has been a "stable" country.

5 The leader of the Forum for Democratic Change (FDC), Dr. Kizza Besigye, received the next highest percentage of votes at 38%. This represents an approximate 10-per-cent reduction in popular support for Museveni, compared to 2001, and a 10-per-cent increase in support for Besigye. What is noteworthy about these results is that in the poorest and most unstable region of the country, the North, Besigye had overwhelming support. During the "bush war" Dr. Besigye was Museveni's personal physician and a member of the National Resistance Army.

6 Interview, Jan. 17, 2003, Kampala, Uganda.

7 A recent press report notes that external donors "finance up to about 48 percent of the country's annual budget" (Monitor 2006).

8 This number includes projects that were approved during the 2000-05 period or that would close during this period.

9 These remarks were written in response to the November 2005 arrest of Col. Kizza Besigye. Besigye returned to Uganda to stand as the Forum for Democratic Change (FDC) presidential candidate, after self-imposed exile for four years. At the time Mwenda wrote this commentary, Besigye was remanded in Luzira Prison on counts of treason and rape in Uganda's High Court, as well as on counts of terrorism and illegal possession of firearms before the military General Court Martial. Mwenda's discussion about the historic relationship between Museveni and donors stems from his contention that the state of political affairs in Uganda, is a result of donor complicity.

10 The National Board was made up of members of nine ministries and one member from the Office of the Prime Minister, the Internal Security Organization (ISO), and the External Security Organization (ESO).

11 Some notable exceptions to this observation are the Uganda Debt Network and the Uganda Law Society.

12 Barr et al. report that as of December 2000, approximately 3,500 NGOs were registered with the NGO Registration Board. "However, not all of these are operational ... of the 1,777 registered NGOs with headquarters in Kampala, only 451 or 25% could be traced. In contrast, in the rural districts, 41% could be traced" (2005: 662).

13 In 1999, the top eight people in the Forest Department were indicted for corruption. In the end, all were given a "slap on the wrist," suggesting that if these people went down they were going to take bigger fish down with them—the "forces of darkness were raging," according to Dr. Mike Harrison, Forest Development Advisor in Uganda's Forest Sector Coordination Unit (Interview, Dr. Mike Harrison, March 28, 2002).

14 Interview, May 23, 2003, Kampala, Uganda.

15 Interview, senior civil servant, Ministry of Public Service, Kampala, Uganda, May 23, 2003.

16 This observation was reinforced by the fact that at the district level, District Environment Officers are supposed to be overseeing regulation but are also planting trees; hence, they are seen as a regulating agency but also doing operational work.

17 It is noteworthy that, in 2002, Acres International was charged with bribery in relation to the massive Lesotho Highlands Water Project. Acres was found to have made over $2 million in payments to project officials, and was subsequently sanctioned by the World Bank in 2004, halting Acres' ability to bid on World Bank-related contracts for three years. Other engineering firms have also been found guilty of bribery, including Lahmeyer International. Acres is the first international company to be found guilty of bribery charges.

18 AESNP was a consortium established in 1994 between the US-based corporation AES and the Uganda-based Madhvani Group. Estimates of the total cost of Bujagali have varied and are currently being revised. Expected costs when AES was leading the project were $582 million. AES was to provide $115 million of equity while AESNP would raise $464 million for debt financing. According to the financing plan, the International Finance Corporation (IFC) and the African Development Bank (AfDB) would lend the project company $60 million and $55 million respectively, while Export Credit Agencies (ECAs) and the World Bank's International Development Association (IDA) would guarantee the remaining $349 million of commercial debt (Esty and Sesia Jr. 2004: 6).

19 For further details of Panel procedures please see World Bank 2003.

20 At the same time that AES withdrew from Uganda, it also suspended a $2.5-billion investment in thermal electric power facilities in Brazil.

21 It is notable that several international companies are currently drilling exploratory oil wells in Northern Uganda. In May 2006, tests confirmed that one drilling site did have oil of a quality and quantity that could potentially be commercially viable.

References

ACODE (Advocates Coalition for Environment and Development). 2002. *Consolidating Environmental Democracy in Uganda Through Access to Justice, Information and Participation*. Kampala: ACODE.

Asiedu, Elizabeth. 2006. "Foreign Direct Investment in Africa: The Role of Natural Resources, Market Size, Government Policy, Institutions and Political Instability." *The World Economy* 29(1): 63-77.

Barkan, Joel D. 2005 (June 2). "Uganda: An African 'Success' Past its Prime?" Unpublished paper of summary remarks presented at the Woodrow Wilson International Center for Scholars.

Barr, Abigail, Marcel Fafchamps, and Trudy Owens. 2005. "The Governance of Non-Governmental Organizations in Uganda." *World Development* 33(4): 657-79.

Bob, Clifford. 2001. "Marketing Rebellion: Insurgent Groups, International Media, and NGO Support." *International Politics* 38: 311-34.

Brock, Karen. 2004. "Ugandan Civil Society in the Policy Process: Challenging Orthodox Narratives." In Karen Brock, Rosemary McGee, and John Gaventa (eds.), *Unpacking Policy: Knowledge, Actors and Spaces in Poverty Reduction in Uganda and Nigeria*. Kampala: Fountain Publishers. 94-112.

De Coninck, John. 2004. "The State, Civil Society and Development Policy in Uganda: Where Are We Coming From?" in Karen Brock, Rosemary McGee, and John Gaventa (eds.), *Unpacking Policy: Knowledge, Actors and Spaces in Poverty Reduction in Uganda and Nigeria*. Kampala: Fountain Publishers. 51-73.

Dicklitch, Susan. 2001. "NGOs and Democratization in Transitional Societies: Lessons from Uganda." *International Politics* 38: 27-46.

Dijkstra, A. Geske, and Jan Kees van Donge. 2001. "What Does the 'Show Case' Show? Evidence of and Lessons from Adjustment in Uganda." *World Development* 29(5): 841-63.

Doyle, Timothy, and Doug Mccachern. 2001. *Environment and Politics*. 2nd ed. New York: Routledge.

Duffield, Mark. 2001. *Global Governance and the New Wars*. London: Zed Books.

East African. 2005 (June 13). "Donors to Meet 40 percent of Uganda's Budget." Retrieved from: http://allafrica.com/stories/printable/200506140943.html.

—. 2006a (May 9). "Government's Power Plan to Cost $4.4 Billion." Retrieved from http://allafrica.com/stories/printable/200605090145.html.

—. 2006b (May 30). "Uganda: In Generator City, Birds Don't Sing, And You Have No E-Mail." Retrieved from: http://allafrica.com/stories/200605300449.html..

Esty, Benjamin C., and Aldo Sesia, Jr. 2004. *International Rivers Network and Bujagali Dam Project* (A). Cambridge, MA: Harvard Business School. N9-204-083.

Fairhead, James, and Melissa Leach. 1996. *Misreading the African Landscape*. Cambridge: Cambridge University Press, 1996.

Forsyth, T., M. Leach, and I. Scoones. 1998. *Poverty and Environment: Priorities for Research and Policy. An Overview Study*. Prepared for the United Nations Development Programme and European Commission. Falmer, Sussex: Institute of Development Studies.

Harrison, Graham. 2001. "Post-Conditionality Politics and Administrative Reform: Reflections on the Cases of Uganda and Tanzania." *Development and Change* 32(4): 657-79.

Hickey, Sam. 2005. "The Politics of Staying Poor: Exploring the Political Space for Poverty Reduction in Uganda." *World Development* 33(6): 995-1009.

Hunt, J. Timothy. 2006. *The Politics of Bones: Dr. Owens Wiwa and the Struggle for Nigeria's Oil*. Toronto: McClelland and Stewart.

Hyden, Goran. 2006. *African Politics in Comparative Perspective*. Cambridge: Cambridge University Press.

Kasekende, Louis A., and Michael Atingi-Ego. 1999. "Impact of Liberalization on Key Markets in Sub-Saharan Africa: The Case of Uganda." *Journal of International Development* 11: 411-36.

Kiyaga-Nsubuga, John. 2004. "Uganda: The Politics of 'Consolidation' under Museveni's Regime, 1996-2003." In Tasier Ali and Robert O. Matthews (eds.), *Durable Peace: Challenges for Peacebuilding in Africa*. Toronto: University of Toronto Press. 86-112.

Kjaer, Anne Mette. 2004. "'Old Brooms Can Sweep Too!' An Overview of Rulers and Public Sector Reforms in Uganda, Tanzania and Kenya." *Journal of Modern African Studies* 42(3): 389-413.

Le Billon, Philippe. 2001a (June). "The Political Ecology of War: Natural Resources and Armed Conflicts." *Political Geography* 20(5): 561-84.

—. 2001b (January). "Angola's Political Economy of War: The Role of Oil and Diamonds, 1975-2000." *African Affairs* 100(398): 55-80.

Leach, Melissa, and Robin Mearns. 1996. "Environmental Change and Policy: Challenging Received Wisdom in Africa." In Melissa Leach and Robin Mearns (eds.), *The Lie of the Land: Challenging Received Wisdom on the African Environment*. Oxford: James Currey. 1-33.

Mallaby, Sebastian. 2004. "NGOs: Fighting Poverty, Hurting the Poor." *Foreign Policy* (Sept./Oct.): 50-58.

Monitor. 2006 (March 24). "Economy to Boom Despite Power Crisis." Retrieved from: http://allafrica.com/stories/printable/200603230725.html.

Mugabe, J., and G.W. Tumushabe. 1999. "Environmental Governance: Conceptual and Emerging Issues." In H.W.O Okoth-Ogendo and G.W. Tumushabe (eds.), *Governing the Environment: Political Change and Natural Resources Management in Eastern and Southern Africa*. Nairobi: African Centre for Technology Studies. 1-28.

Muhumuza, William. 2002. "The Paradox of Pursuing Anti-Poverty Strategies under Structural Adjustment Reforms in Uganda." *The Journal of Social, Political, and Economic Studies* 27(3): 271-306.

Mwenda, Andrew. 2005 (Dec. 11). "Without Mincing Words." *Monitor Online*.

NEMA (National Environmental Management Authority). 2001. *State of the Environment Report*. Kampala: Government of Uganda.

—. 2005. State of the Environment Report. Kampala: Government of Uganda.

New Vision. 2002 (Jan. 26). "Uganda: Museveni Warns Economic Saboteurs."

Okoth-Ogendo, H.W.O. 1999. "The Juridical Framework of Environmental Governance." In H.W.O Okoth-Ogendo and G.W. Tumushabe (eds.), *Governing the Environment. Political Change and Natural Resources Management in Eastern and Southern Africa*. Nairobi: African Centre for Technology Studies. 41-62.

Ssewakiryanga, Richard. 2004. "The Corporatist State, the Parallel State, and Prospects for Representative and Accountable Policy." In Karen Brock, Rosemary McGee, and John Gaventa (eds.), *Unpacking Policy: Knowledge, Actors and Spaces in Poverty Reduction in Uganda and Nigeria*. Kampala: Fountain Publishers. 74-93.

Tangri, Roger, and Andrew Mwenda. 2001. "Corruption and Cronyism in Uganda's Privatization in the 1990s." *African Affairs* 100(398): 117-33.

Tarrow, Sidney. 1994. *Power in Movement: Social Movements, Collective Action and Politics*. Cambridge: Cambridge University Press, 1994.

Therkildsen, Ole. 2000. "Public Sector Reform in a Poor, Aid-dependent Country, Tanzania." *Public Administration and Development* 20(1): 61-71.

Tripp, Aili Mari. 2000. *Women and Politics in Uganda*. Madison: University of Wisconsin Press, 2000.

—. 2004. "The Changing Face of Authoritarianism in Africa: The Case of Uganda." *AfricaToday* 50(3): 3-26.

Tumushabe, Godber W. 1999. "Environmental Governance, Political Change and Constitutional Development in Uganda." In H.W.O Okoth-Ogendo and G.W. Tumushabe (eds.), *Governing the Environment: Political Change and Natural Resources Management in Eastern and Southern Africa*. Nairobi: African Centre for Technology Studies. 63-88.

Urban Harvest - Kampala. 2004 (May). *Strengthening Urban and Peri Urban Agriculture in Kampala*. Formal survey (Draft Final Report). Lucy Aliguma, CIAT. Centro International de Agricultura Tropical (CIAT) and Urban Harvest.

USAID (UNITED STATES AGENCY FOR INTERNATIONAL DEVELOPMENT). 2005. "Uganda". Retrieved from: http://www.usaid.gov/policy/budget/cbj2005/afr/ug.html.

Watts, Michael. 1997. "Black Gold, White Heat: State Violence, Local Resistance and the National Question in Nigeria." In Steve Pile and Michael Keith (eds.), *Geographies of Resistance*. London: Routledge. 33-67.

Wood, A., ed. 1997. *Strategies for Sustainability: Africa*. London: IUCN and Earthscan.

World Bank. 2000. *Project Information Document. Uganda-Second Environmental Management and Capacity Building Project*. Report No: PID10088. Washington, DC: Africa Regional Office.

—. 2001. *Project Appraisal Document on a Proposed Credit in the Amount of* SDR *17.1 Million* (US *22.0 Million Equivalent) to the Republic of Uganda for a Second Environmental Management and Capacity Building Project*. Report No: 21343-UG. Washington, DC: Africa Regional Office.

—. 2003. *Accountability at the World Bank. The Inspection Panel 10 Years On*. Washington, DC: World Bank.

—. 2005. *Uganda Country Brief*. Retrieved from: http://web.worldbank.org/WBSITE/EXTERNAL/COUNTRIES/AFRICAEXT/UGANDAEXTN/0,,menuPK:374947~pagePK:141132~piPK:141107~theSitePK:374864,00.html.

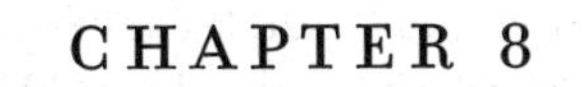

CHAPTER 8

ENVIRONMENTAL MANAGEMENT IN WEST AFRICA

The Case of Nigeria

Godwin Onu

INTRODUCTION

West Africa is made up of nine Francophone countries (Benin, Burkina Faso, Niger, Mali, Ivory Coast, Mauritania, Senegal, Guinea, and Togo), five Anglophone countries (Ghana, Liberia, Nigeria, Sierra Leone, and Gambia), and three Lusophone countries (Guinea Bissau, Cape Verde, and Sao Tome and Principe). The extremes of its geographical location range from the desert Air Tenere in the northeast of Niger (0-100 mm of rainfall per year), to the Guinea forest of Kindia in the west (4,000 mm of rainfall per year); from the few oases of Taoudennie in the northern part of Mali, to the mangrove swamps of Port Harcourt in Nigeria; from the sand dunes of Bir Moghrein in Mauritania to the evergreen forest of Kakum national park of Cape on the Ghanaian coast.

West Africa's environment is inhabited by over 56 species of large mammals and 1,300 bird species: from the bare-headed rock fowl of Sierra Leone to the Sahelian ostrich; from the oryx and the addax antelope living in the most severe deserts, to the reserves of the Hadeija Nguru valley wetland of Nigeria (Zeba 1998: 1). Specifically, according to Zeba, among the nine Francophone West African nations, the vegetation found consists mainly of savannah with abundant grass and tree species such as acacia and commiphora, bordered in the south by sparse isoberlina forests, and in the north by grassy steppes and deserted landscapes. But these semi-natural woodlands in this arid and semi-arid region are overexploited for wood fuel (in particular charcoal for urban areas) and submitted to intense pressure from grazing and farming activities resulting from population growth (Zeba 1998: 14). With respect to the original dense forest cover, Zeba notes that human activities have reduced the cover to less than 11 million hectares (less than 10 per cent of woody surface areas), while forest fallow land and savannah extend over some 130 million hectares (90 per cent of woody surface areas). As E.P. Stebbing has stated, "the people of West Africa are living on the edge, not of a volcano, but of a desert whose power is incalculable and whose silent and almost invisible approach must be difficult to estimate. Furthermore, overgrazing has been found to have had similar effects on farmlands like the drought" (1935). Indeed, Pierre Hiernaux found out, in a study of Mali, that changes in the floristic composition in response to drought are often similar to those in response to grazing pressure, which explains the correlation between overgrazing and desertification (1999).

It is against this general context of the deterioration of the environment in West Africa that I examine in this chapter the state of the environment in Nigeria by looking at the challenges confronting Nigerian policy-makers with issues of environmental management. In this chapter, I analyze the role that civil society organizations play in environmental protection, the international demands that have been placed on the country, and what the country's responses have been. The argument that I present here is that efficient and sustainable management of the environment in Nigeria will depend entirely on the extent to which political will by the leadership will be able to change environmental culture in the country through better enforcement of environmental regulations, the involvement of all stakeholders—including international institutions, policy networks, international non-governmental organizations (NGOs), and local actors—and more transparent, accountable, and participatory processes of decision-making at the local level.

THE STATE OF THE ENVIRONMENT IN NIGERIA

Surrounded by Niger, Chad, and Benin, Nigeria is the largest black nation in the world, with a population of over 111 million people in 1990, and it is projected to add 168 million by 2020 (Brown and Kane 1995). In addition to the 64 per cent of the population who live in rural areas, there are about 200 urban settlements, each inhabited by more than 20,000 people (Osuocha 1999). Nigeria is endowed with abundant natural resources, which range from fish, wildlife, timber, medicinal plants, liquid and solid mineral deposits, and water, to ornamental and food crops. In general terms, the environment provides all the life-support systems in the air, on water, and on the land, as well as the materials needed to bring about economic development (UNEP 2006). The dominant mode of farming, especially in the rural areas, is shifting cultivation. But with population explosion, land has become overused and has lost its fertility, thereby affecting crop yield.

Nigeria's diverse climate (ranging from the extremely wet and humid equatorial type to the hot, dry, and arid) accounts for the abundance of plant and animal species in the country. However, due to deforestation and prospecting for mineral and oil resources, some of the biotic resources are being increasingly depleted. Woods such as iroko, mahogany, oak, and walnut are being felled and exported to Europe and North America. This also has a tremendous negative effect on agriculture and agricultural production. In addition, issues of overgrazing, overpopulation, desertification, and other forms of degradation of the environment have been linked to food insecurity. Soil degradation, for instance, has been responsible for flooding, which washes away fertile top soils and affects crop yield. Nigeria's climate and soil structure are all ideal to support enough food production for the survival of its population. Unfortunately, the northern fringes are progressively becoming Sahelian, so persistent crop failures are becoming the norm and livestock are migrating southwards in an irreversible trend. There are also problem areas in the southern and southeastern sectors, where floods, erosion, and oil pollution are disruptive to crops and fish production. If the soils are well managed, they will be capable of supporting agriculture for many generations in spite of their inherent medium to low fertility.

Even though institutional and regulatory measures have been put in place over the last twenty years for better environmental management, Nigerians continue to suffer from what amounts to official irresponsibility and abject neglect of existing provisions by both governments and corporations. The consequences range from flooding arising from the reckless dumping of refuse in drainage areas, to oil pollution due to the lack of implementation of environmental laws. Outside government, and the lack of ethical standards and sustainable environmental behaviour of average Nigerians are to blame. Even when some provisions are made for dumping garbage, the garbage is often allowed to stay for weeks, and in some cases months, or is completely abandoned without proper disposal. In many instances, there are no refuse-treatment plants or hazardous-waste landfills. Subsequently, such refuse becomes mountainous and is set on fire, which causes choking, pollution-filled smoke that in many cases can envelop an entire city.

In most urban centres, refuse can even block major roads through which presidents and governors pass without showing any sign of irritation. Oludare Hakeem Adedeji (2002) lamented such erosion of ethical standards in Nigeria's attitude toward the environment, citing countless examples of blatant disregard for both national and international environmental agreements and common sense. Perhaps more important in this regard are the regulatory ability of the agencies charged with environmental regulation and conflicts among various agencies in the process of enforcement; in many cases the state of environmental ethics is made clear through the lack of enforcement of environmental regulations—as if, in fact, such regulations did not exist.

In the urban centres of Nigeria, residents' inability to treat the environment with a reasonable degree of respect or sense of moral responsibility has been disappointingly overwhelming. Despite the existence of regulatory authorities in cities such as Lagos, Ibadan, and Onitsha, these cities are notoriously known for being among the dirtiest in West Africa. Residents dump refuse inside drains and anywhere that looks a little isolated. Onibokun and Kumuyi have noted the problems of the high rate of urbanization in African countries (Nigeria included), a situation that has led to a rapid accumulation of refuse (1991: 1). They found that an estimated 20 kilograms of solid waste is generated per capita yearly. Given that Nigeria's population is more than 100 million, this amounts to 2.2 million tons per year. Moreover, Onibokun and Kumuyi also found that there has been a rapid increase in the rate of waste generation in many Nigerian cities. In Lagos, for example, an estimated 625,000 tons of waste were generated in 1982 alone. This, according to the Federal Ministry of Housing and Environment, was expected to have risen to 998,000 tons by 2000. Likewise, approximately 258,000 tons of waste were generated in 1982 in Kaduna, and this is estimated to have increased to 431,000 tons by 2000. The authors attribute this environmental disaster to the lack of adequate management services.

Environmental regulatory agencies exist, but they are weak. Refuse disposed of in drains further blocks the already poor drainage systems, leading to flooding. This situation is worsened by an increasing influx of rural people to urban centres, in addition to the government's failure to come to terms with this challenge.

It was projected in 1996 that the number of people living in Nigerian towns and cities would more than double from 40 million to reach about 100 million by 2010. Such urban population growth is likely to exacerbate the consequences of human activity on the urban environment (World Bank 1996: 3). This may spell disaster for urban governance.

This lacklustre attitude to environmental issues has led to the loss of biodiversity, and to deaths, poverty, and unrest in some parts of Nigeria. It has also affected economic development. For instance, between 1976 and 1996, a total of 2,369,470 barrels of crude oil were spilled into the rivers and lands of the Niger Delta. A 1998 study by Agbola and Olurin found that major land-use and land-cover changes have occurred in the Niger Delta as a result of petroleum and gas exploitation activities. These changes have resulted in significant air and water pollution, the loss of biodiversity, habitat fragmentation and loss, and climate change. They have also had devastating effects on ecosystems and have led to the loss of farmlands and fisheries, which are the major source of livelihood for the people. Consequently, food security has become severely threatened. It is also important to note that sulphur and nitrogen compounds emitted into the atmosphere, concentrations of gases, and acidic deposition have further impacts on the local environment. This could cause further acidification of lakes and soils and affect human health, crop productivity, and forest growth biodiversity.

Nigeria's environmental degradation has led to material and relative poverty. This phenomenon facilitates the recruitment of the large numbers of unemployed, who become easy prey for insurgent and terrorist groups as well as for violent civil conflicts. Moreover, continuing war in the West African sub-region has not only slowed economic development but has also damaged the environment. Furthermore, a strange combination of insurgency and terrorism in Nigeria has used oil exploration as a means to create a situation where violent protests become associated with perceived injustice for the poor and needy masses. Environmental problems thus have been relegated to the backburner. For example, regular threats by insurgents in Nigeria to blow up oil wells, and the actual success in accomplishing some of these threats, sent oil markets soaring to an unprecedented US $70-75 per barrel in 2006.

In addition, the elimination of forest cover leaves the land poorly protected and prone to erosion, rills, and gullies. In the process the soil loses its fertility, in turn increasing hunger, starvation, and poverty. Poverty affects savings and savings affect investment. The menaces of erosion and desertification have also ravaged many parts of Nigeria and have left an increasingly larger part of it barren. These have affected agricultural production and have had a consequent impact on the economy. In Anambra state, for instance, many people have become refugees after being chased away by rampaging erosion (Egboka 1993: 1).

The World Bank notes that numerous industries, from pulp to petroleum, dump untreated and often toxic liquids in open gutters, streams, rivers, and lagoons. In addition, emissions from dilapidated motor vehicles and motorcycles that have been adopted as means of transportation, as well as urban solid waste, add to the pollution of air, land, and water. These are major sources of concern. The government's apparent incapacity to manage what Onibokun (1999) has called "monsters" has led directly or indirectly to environmental unrest and insurgency in the Niger Delta region of Nigeria. The consequences have been enormous on oil workers, who are mostly expatriates from across the world, as well as on the economy of Nigeria.

Apart from environmental unrest, there are other problems, such as acidification of lakes and pollution, that can have an impact on human health, crop productivity, forest growth, biodiversity, and man-made materials. They irritate the respiratory tract and may give rise to disease, especially in children and asthmatics, and bronchitis (Stockholm Environmental Institute 2003: 2). In Nigeria, major scientific research has not been done on the corrosive effects of environmental damage on buildings, vehicle bodies, and historical places. However, the United Nations Environmental Programme found that the concentration of carbon dioxide (CO_2) in Nigeria increased from 280 parts per million in 1980 to about 380 in 2006 (UNEP 2006: 2). Such an excessive amount of CO_2 in the atmosphere is not good for anyone in Nigeria or their neighbours.

ENVIRONMENTAL MANAGEMENT IN NIGERIA

Environmental management refers to the process or processes that would enable efficient and sustainable use of human and natural resources for the betterment of all. This assumes enlightened, informed, and capable leadership, and the participation of critical partners or stakeholders having the capacity to do so, within the framework of full transparency and accountability. Furthermore, efficient and productive environmental management requires established structures or institutions tasked with environmental monitoring, policy implementation, regulation, and enforcement. Let us see the extent to which these factors have shaped and created environmental challenges and what the traditional and official policy responses in Nigeria have been by looking at four specific challenges: desertification, flooding, soil erosion, and pollution.

Desertification

The 1997 United Nations Convention to Combat Desertification in developing countries (CCD) defined desertification as land degradation in arid, semi-arid, and dry sub-humid areas resulting from such factors as climatic variations and human activities. Though the natural causes of desertification range from poor physical conditions of soils, vegetation, and topography to climatic variability, as evidenced in periodic droughts in Nigeria, climate variation is perhaps the most important natural cause. There are also cases of villagers burning bushes while clearing land for agriculture, hunters setting fire to vegetation while searching for game, and cattle herdsmen setting fire to dry grass to stimulate the growth of dormant grass buds. Other causes of desertification in Nigeria include the use of wood for building, the arts, fuel, and the fishing industry, and the removal of trees, shrubs, herbaceous plants, and grass cover from the fragile land.

Despite the fact that reliable and current data are not easily available, based on common observation and some surveys it is believed that those areas which are considered dry lands are facing severe desertification problems (UNEP 2006). The visible sign of this phenomenon is the gradual shift in vegetation from grasses, bushes, and occasional trees, to grass and bushes alone, and then, in the final stages, to expansive areas of desert-like sand (Federal Ministry of Environment 2001). Nigeria's Federal Ministry of Environment (FME) estimates that between 50 and 75 per cent of Bauchi, Borno, Gombe, Jigawa, Kano, Katsina, Kebbi, Sokoto, Yobe, and Zamfara states are being affected by desertification. These states, with a population of about 27 million people, account for about 38 per cent of the country's total land area (see Table 8.1). In these regions, the FME notes that population pressure resulting in overgrazing and overexploitation of marginal lands has aggravated desertification and drought. For instance livestock population in Nigeria has been estimated to consist of 16 million cattle, about 13.5 million sheep, some 26 million goats, approximately 2.2 million pigs, and 150 million heads of poultry (Federal Ministry of Environment 2001: 14).

Nigeria also faces additional pressures from cattle farmers from the neighbouring countries of Cameroon, Chad, and Niger. These frontline states are located along two of the Nigerian pastoral corridors that are shared with Nigeria. These corridors carry millions of heads of cattle annually (Federal Ministry of Environment 2001: 14). Also, due to overgrazing, entire villages and major access roads often get buried under sand dunes in the extreme northern parts of Katsina, Sokoto, Jigawa, Borno, and Yobe states. As the FME has stated, "the pressures of the migrating human and livestock populations from these areas are absorbed by pressure point buffer states such as the Federal Capital Territory, Plateau, Adamawa, Taraba, Niger, Kwara and Kaduna states. It is reported that these buffer states have about 10–15 per cent of their land area threatened by desertification. This action leads to an intensified use of fragile and marginal ecosystems resulting in progressive degradation even in years of normal rainfall" (Federal Ministry of Environment 2001: 14).

Deforestation is a major cause of desertification, and it has become a serious problem in the rainforest zone of eastern and western Nigeria. The situation is more pronounced within the Guinea savanna area. In the case of the savanna area of the middle belt, deforestation is combined with bush burning to open up the already sterile soils for further degradation, thereby paving the way for further reduction of fertility, agricultural productivity, and yield. Continuing deforestation and bush burning in northern Nigeria have further increased the rate of desertification. Moreover, when cattle herds move from one grazing area to the other, such migration leads to soil compaction and nutrient decline.

A second cause behind desertification in Nigeria has been drought. According to the FME, the history of the Sudano-Sahelian zone of Nigeria is replete with severe and prolonged droughts, some lasting several years. For example, this zone suffered a prolonged drought between 1911 and 1914. Later droughts faced by the region included those of 1919, 1924, 1935, and 1951–54. The FME notes that the 1983–85 period was the driest time in the twentieth century in this zone, as Lake Chad fell to its lowest level in recorded history (Federal Ministry of Environment 2001: 13). Further evidence seems to suggest that the 1979–85 droughts were a function of tropical anomalies. According to the FME, a series of severe and prolonged droughts as witnessed since the 1950s have made the Sudano-Sahelian environment (already a fragile zone) more vulnerable than ever. It is widely believed that climate change is greatly responsible for the acceleration of desertification in the country.

TABLE 8.1
POPULATION DENSITY OF FRONTLINE STATES OF NIGERIA, 2001

States	*Land area*		*Population*	
	km²	*%*	*Size*	*Density/km²*
Bauchi/Gombe	64,605	6.99	4,294, 413	66
Bornu	70,890	7.67	2,596.598	37
Yobe	45,502	4.93	1,411,481	31
Kano	20,131	2.18	5,632,040	280
Jigawa	23,154	2.51	2,829,929	122
Katsina	24,192	2.62	3,878,344	160
Sokoto/Zamfara	65,735	7.12	4,392,391	67
Kebbi	36,800	3.98	2,062,226	56
Total/Average	351,009	38.00	27,097,422	77

Source: Federal Office for Statistics 2001.

The control of desertification programs is the responsibility of the environmental sector of the federal government of Nigeria. According to the FME, the following are some of the main efforts made to contain drought and desertification in Nigeria: (1) a National Action Programme to Combat Desertification and mitigate the effects of drought towards the implementation of the Convention to Combat Desertification (CCD); (2) a strengthening of national and state institutions involved in drought and desertification control programs; (3) promoting sustainable agricultural practices and the management of water resources, including water harvesting and inter-basin transfers; (4) encouraging individual and community participation in viable "aforestation" and reforestation programs using tested pest- and drought-resistant and/or economic tree species; (5) encouraging the development and adoption of efficient wood stoves and alternative sources of energy; (6) involving local people in the design, implementation, and management of natural-resource conservation programs for combating desertification and ameliorating the effects of drought; (7) intensifying international cooperation and partnership arrangements in the areas of training, research, development, and transfer of affordable and acceptable environmentally sound technology and provision of new and additional technical and financial resources; (8) intensifying cooperation with relevant inter- and non-governmental organizations in combating desertification and mitigating the effects of drought; and (9) establishing, reviewing, and enforcing cattle routes and grazing reserves (FME 2001: 18).

Flooding and Soil Erosion

Flooding in Nigeria has been caused by natural and artificial factors. The unevenly distributed rainfall and climatic variability result in abnormal runoff generation, which channels are unable to hold (Federal Government of Nigeria 1996: 10). Flooding has become a major environmental challenge in Nigeria. This phenomenon occurs in three main forms: coastal flooding (in the low-lying belt of mangrove and freshwater swamps along the coast), river flooding (in the flood plains of the larger rivers), and short-lived snap floods (associated with rivers in the inland areas where sudden heavy rains can change them into destructive torrents within a short period). These floods carry municipal waste directly to rivers, since no storm-sewage treatment system has been put in place in most urban areas such as Lagos, Ibadan, Aba, Warri, and Benin.

Perhaps the latest environmental challenge in both cities and some semi-urban areas is polythene and sachet water bags. The presence of sachet water[1] in urban centres has signalled some kind of private-sector involvement in water provision, especially in light of the apparent failure of the state in providing pipe-borne water. Sachet water is very cheap and within the reach of even a beggar. It is hawked at every nook and cranny of urban centres in Nigeria, with major clusters found around motor parks. These plastic bags—in addition to pieces of metal, decaying food materials, and glass of all kinds—have constituted an eyesore in almost all Nigerian urban centres and have blocked both urban roads and drainage systems. Among the consequences are flooding and erosion, which are the result of the poor environmental behaviour of urban residents. In addition, unlike the majority of modern cities, in most Nigerian cities, urban planning started after residential buildings and markets had been erected everywhere. Sometimes streets end at residential buildings.

Official responses to erosion management have ranged from undertaking shoreline protection projects in Lagos, including replenishing the Bar Beach; carrying out soil-erosion and flood-control measures in Adamawa; and establishing erosion- and flood-control projects in Mararaba in Nasarawa State, Nawfia in Anambra State, Nyanya in the Federal Capital Territory (FCT), Okpanku in Imo State, and Dura Dyke in Benue State. Additional efforts have also been made to plant trees in the savanna region of the country. To prevent bush burning, cover crops have been planted and barriers constructed to stop water from eroding the topsoil in the gully region of the south-eastern region. The construction of dams to store excess water and the demolition of structures along stream and river banks have also proven somewhat effective in cases of flooding. Despite all these measures, government efforts at erosion control in Nigeria have not yielded the expected results. This is because of the poor management of ecological funds made available by both government and development partners for such purposes, and because of the top-down approach adopted in the distribution of these funds.

Pollution in Nigeria

The Federal Environmental Protection Agency has defined pollution as "man-made or man-aided alteration of chemical, physical or biological quality of the environment to the extent that is detrimental to that environment or beyond acceptable limits and 'pollutants' shall be construed accordingly" (1990). This alteration of air, water, or soil materials interferes with human health, the quality of life, or the natural functioning of ecosystems and food security. Pollution can affect the air we breathe, the water we drink, and the land upon which we derive our food and existence. Little regional variation exists in the nature of pollution. In advanced industrial countries of the world, problems of nuclear waste, radioactivity, and emissions from industries pose major challenges. A similar situation exists in developing countries, with some degree of variation.

AIR POLLUTION

The major sources of air pollution in Nigeria are the outdoor flaring of natural gas, the setting alight of heaps of domestic and industrial solid waste, and indoor pollution resulting from the use of wood stoves for cooking. For instance, Nigeria is said to flare more gas than any other country in the world (Efole 2004: 6). This gas flaring is dominant in the Niger Delta region of the country, an area from which most of Nigeria's vast oil reserves are extracted. It is estimated that 45.8 billion kilowatts of heat are discharged into the atmosphere of the Niger Delta alone, flaring 1.8 billion cubic feet of gas every day (Iyayi 2005). According to a report, Nigeria flares 17.2 billion cubic metres of natural gas a year (Global Gas Flaring Reduction Initiative 2002). The UNEP has also found out that the concentration of CO_2 in Nigeria has increased from 280 parts per million in the 1980s to about 380 parts per million today as a by-product of oil production (2006: 2). Air pollution and acid rain from this gas flaring, which has been occurring for decades, not only contaminate the air humans breathe but also affect the ecosystem and in some cases could lead to loss of life through respiratory diseases, especially in the oil-producing areas of the country.

In most cases, the household use of stoves and even wood hardly achieves efficiency in combustion (Smith 1999). Moreover, as it has been noted, they emit a large number of pollutants, including particulate matter, CO, CO_2, methane, nitrogen dioxide, formaldehyde, and benzo[a]pyrene (Ezzati and Kammen 2002). These are associated with a variety of adverse health impacts, including cancer and neurological, reproductive, and developmental effects (Kindzierski 2000). According to Ezzati and Kammen, the emissions from incomplete combustion not only contribute to climate change as greenhouse gases, but they also have major adverse health impacts, including acute respiratory infections, chronic obstructive pulmonary disease, asthma, nasopharyngeal and laryngeal cancer, tuberculosis, perinatal conditions, adverse pregnancy outcomes, and eye irritation (2002). Furthermore, sulphur and nitrogen compounds emitted into the atmosphere can cause the acidification of lakes and soils and have an impact on crop productivity, forest growth biodiversity, and man-made materials.

Another peculiar source of air pollution in Nigeria is emissions from aging vehicles and motorcycles. The presumed benefit of these modes of transportation lies in their ability to provide fast service to the suburbs. The concern about motorcycles is not only because of deaths occasioned by reckless drivers, but also because of the noise and pollution of the environment that most governments of West Africa have been unable to bring under control. The motorcycle exhausts emit air pollution all day long and create health risks for drivers, passengers, and the residents of the streets they ply (Tadegnon 2005). The contributory impacts to respiratory diseases and other ailments such as respiratory infections, cardiovascular diseases, cancer, eye ailments, and irritability have not been fully determined, but they are widely believed to be significant; in addition, there is a shortage of medical facilities to deal with these diseases. Air pollution is therefore an increasingly serious environmental concern in Nigeria that is worsened by increasing urbanization and a lack of adequate policies and enforcement measures to check environmental hazards.

WATER POLLUTION

The key sources of water pollution in Nigeria are oil exploration and the consequent blow-outs that have become a regular feature, ineffectively managed sewage treatment plants, and floods (causing runoff that is accompanied by nitrogen and pesticides). With respect to oil exploration, there appears to be no regard for the environment where oil is exploited. There are regular reports of oilwell blow-outs and oil spillages, oil ballast discharges, and, according to the UNEP (2006), improper disposal of drilling mud from petroleum prospecting. These have unleashed significant damage to ecosystems, destroying the aesthetic values of natural beaches due to unsightly oil slicks. In addition, oil spillages, especially in the Delta areas of Nigeria, have caused extensive damage to marine life, completely eliminated various species, and effected delays in biota (fauna and flora) succession.

But oil pollution is not only a cause for concern for Nigerians alone; it also concerns neighbouring countries and the international community as a whole. In the Delta area, pollution has become endemic and a routine, polluting both farmlands and fishing stocks. This problem started around 1970, when Nigeria had its first significant oil well blow-outs at the Bomu and Obagi wells near Port Harcourt (Okebukola 2001). The regularity with which oil pollution occurs has become a national embarrassment, especially when it has led to insurgency in the Delta region. Between 1976 and 1996, a total of 2,369,470 barrels of crude oil were spilled into the rivers and lands of the Niger Delta. The pollution is still a regular occurrence and no permanent solution has been found.

Water pollution is not limited to rivers and streams; it also affects wells, especially the ones located near abattoirs in Nigeria. A visit to abattoirs across Nigeria will reveal glaring weaknesses in regulation and enforcement of environmental laws and expose the dilapidated places where humans purchase their major source of protein. This is the situation in most of Nigeria's urban centres.

The above discussion about various kinds of pollution confronting Nigeria is supplemented by what laws and institutions have been created by the federal and state governments to fight pollution and defend the environment.

THE ESTABLISHMENT OF THE FEDERAL ENVIRONMENTAL PROTECTION AGENCY

In 1992, the mandate of the Federal Environmental Protection Agency was expanded by a presidential decree to cover conservation of natural resources and biological diversity. A Federal Ministry of Environment (FME) was then created in June 1999 against the backdrop of the National Environmental Policy (NEP) that was launched on November 27, 1998, by Nigeria's federal government. Part of the objective was to have a more holistic and coordinated approach to environmental management and to secure for all Nigerians an adequate environment for their health and well-being, as well as to conserve and use the environment and natural resources for the benefit of the present and future generations. Second, the NEP was designed to restore, maintain, and enhance ecosystems and ecological processes essential for the functioning of the biosphere and for the preservation of biological diversity, and to adopt the principle of optimum sustainability in the use of living natural resources and ecosystems. Third, it was to create public awareness and promote understanding of essential linkages between environment and development and to encourage individual and community participation in environmental improvement efforts. Finally, the FEP was meant to generate a framework of cooperation in good faith with other countries, international organizations, and agencies in order to optimize the use of trans-boundary natural resources and to prevent or reduce trans-boundary environmental pollution.

In order to reach these objectives, some relevant departments and units from other ministries were transferred to the newly created FME (UNEP 2006: 2). These include the Forestry Department (Wildlife, Forestry Monitoring Evaluation and Coordination Unit-FORMECU) of the Federal Ministry of Agriculture; the Forestry Research Institute of Nigeria, also from the Federal Ministry of Agriculture; the Environmental Health and Sanitation Unit of the Federal Ministry of Health; the Oil and Gas Pollution Control Unit of the Department of Petroleum Resources of the Federal Ministry of Petroleum Resources; the Coastal Erosion Unit, Environmental Assessment Division and Sanitation Unit of the Federal Ministry of Works and Housing; and the Soil Erosion and Flood Control Department of the Federal Ministry of Water Resources (UNEP 2006: 3). Following this development, some states have created full-fledged Ministries of Environment to replace their Environmental Protection Agencies. The state and federal ministries regularly meet to harmonize environmental policies and discuss vital issues affecting the environment of all of Nigeria.

In addition, at a national level, several sources of funds for environmental protection activities exist. These include the following: (i) The Ecological Fund: One per cent of the Federation Account is set aside for the amelioration of ecological problems such as soil erosion and flood control, desertification, drought, and general environmental control (refuse, solid waste, water hyacinth, industrial waste). This amount was recently increased to 2 per cent and paid into a Special Ecological Fund. (ii) Funding from Crude Oil Revenues: The government has committed 3 per cent of the revenue accruing from crude oil in the country to tackle some ecological problems through the Oil Minerals Producing Areas Development Commission (OMPADC). (iii) Financial Contributions from NGOs. (iv) Bilateral and Multilateral Financial Assistance: Several environmental financial assistance initiatives from such agencies like the World Bank, UNDP, UNEP, FAO, IUCN, UNICEF and ADB cover such problems as desertification control, capacity building, and so on (Federal Government of Nigeria 1996).

Environmental protection issues have been before the policy-makers of Nigeria for many decades, even during colonial times.[2] When the Federal Environmental Protection Agency (FEPA) was created in 1988, the institution was further strengthened by the introduction of a series of other environmental policy and regulatory decisions.[3] However, it was not until 2005 that Nigeria's effort at environmental protection received a tremendous boost when its first satellite, the SATT I, was launched into orbit by the administration of Olusegun Obasnajo, in power since 1999. The satellite is designed to monitor the environment, especially for early warnings on possible environmental disaster, through remote sensing. The remote sensing makes it possible to monitor the environment more systematically, particularly the degradation of the Sahelian and coastal environments. It can provide information on climate, biomass, animal populations, soils, water, and human activity; and it can also help to track surface temperatures, the vegetative cover, and the productivity of the soil (Debailleul et al. 1997: 3).

Perhaps the most far-reaching initiative by the federal government on environmental management was the enactment in 1988 of decree No. 58 establishing the FEPA. This agency facilitated the establishment of state environmental protection agencies (SEPAS) in the 36 states of the Federation and the Federal Capital Territory (FCT), which were mandated to address all environmental problems (including drought and desertification) at the state level. The FME, in furthering its mandate, has created the Drought and Desertification Amelioration department to strengthen the existing institutional arrangement for more effective coordination of activities by government toward the implementation of the CCD in the country. This will further ensure a sharper focus on rehabilitation and restoration of desertified and near-desertified conditions in the affected areas (Federal Ministry of Environment 2006: 19). The government does this in partnership with NGOs, representative ministries, and research institutes.

In what ways has pollution (both air and water) been managed in Nigeria both traditionally and officially? Water pollution is still a minor problem for the rural population, although the major urban centres do suffer from it partly because of the location of industrial units and also because municipal waste that is partially or not fully treated is released into streams. Because a limited number of industries are located in rural areas, industrial pollution there has never been a subject of major concern. Green vegetation blossoms in almost all parts of rural Nigeria, providing oxygen for healthy living. The rural population is generally aware of the environmental effects on their livelihoods but has little option given their poverty; their immediate needs mean that they have to satisfy their hunger first before worrying about the environment. Sometimes bushes are set on fire for hunting purposes. Streams are natural sources of drinking water and people take them in their natural form. Therefore, in the rural communities youths are organized periodically to keep streams clean, and every morning they sweep village squares and move refuse collected earlier. Most streams have been cleaned in this way. These youths have a long list of rules of "dos" and "don'ts" (such as "do not fight in the stream," "do not defecate," "do not spit around the stream," "do not urinate around the stream," etc.). Citizens who do not obey these rules are fined. Every month, youths organize (and in the Eastern part of Nigeria use masquerades) and go around to defaulters to collect fines. The services from the police or state or local government agents are not sought in this self-regulating enforcement mechanism. This traditional method of enforcement has proved to be very successful in the rural localities, especially when some of the streams are ascribed to gods and goddesses; according to some of their cultures, in addition to physical punishment, the pollution of rivers and creeks could also attract the punishment of the gods. Rivers and streams that fall within this framework have untouched fish supplies and nobody kills them. However, even though these measures have been partly successful, they cannot replace government-operated and -managed environmental regulations and policies.

THE ROLE OF CIVIL SOCIETY IN ENVIRONMENTAL GOVERNANCE

Environmental NGOs (ENGOs) have become active players in environmental governance in Nigeria. The ENGOs act as pressure groups on the government, individuals, and organizations, as well as business, for environmental protection. They engage in monitoring the environment and sometimes publish environmental issues of general concern. In the process they alert the government and even the general public on ending environmental disaster. Nigeria's ENGOs are involved in numerous anti-desertification efforts. The federal government instituted an Arid Zone Afrostation Project (AZAP) in 1976 to tackle the problems of desertification through the establishment of woodlots, shelterbelts, and windbreaks. This was done with the assistance of the World Bank. Many of the programs that are undertaken by the federal government are being managed with the assistance of the European Union, Nigerian NGOs, and Community Based Organizations (CBOs). These organizations are engaged in a variety of action programs, the coordination of mechanisms and partnerships, capacity-building, education and public-awareness activities, and financial resources and mechanisms. Moreover, some of the NGOs in Nigeria are actively participating in the activities of the Global NGO Network on Desertification. In fact, the Nigerian Environmental Study/Action Team (NEST) is the sub-regional focal point of this network for Anglophone West Africa. Other prominent national and international NGOs include the Nigerian Conservation Foundation (NCF), the Forestry Association of Nigeria (FAN), and the International Union for Conservation of Nature (IUCN) (Federal Ministry of Environment 2006: 24).[4] It then becomes obvious that problems of desertification require collective action from all stakeholders to check its encroachment. In addition to these ENGOs, the fight against desertification has involved numerous other stakeholders, such as resource users, CBOs, the private and public sectors, academia, and development partners. Furthermore, Nigeria's FME has also established the National Council on Shelterbelt, Afforestation, Erosion and Coastal Zone Management, with a mandate to prepare and implement an integrated shelterbelt and afforestation action plan as well as erosion and flood-control programs. This includes participatory input from key stakeholders, states, local governments, communities, and the private sector under the supervision of the Office of the Vice-President.

In Nigeria, NGOS such as the National Conservation Fund (NCF) affiliated with World Wildlife Foundation International (WWF), Friends of the Environment (FOE), and the Forestry Association of Nigeria have been operational for the past 15 years by contributing to the development of environmental policies and legislation, action plans, and programs at the national, regional, and local levels. FOE and NCF have been able to attract a considerable inflow of international funds to support their environmental projects in Nigeria. The country has also benefited from the Food and Agricultural Organization (FAO), UNICEF, and the African Development Bank (ADB) for desertification control, capacity-building, and other issues. Perhaps the environmental problems of the Niger Delta Region in Nigeria have generated the most international and local concern, something that has led to the establishment of a plethora of NGOS, primarily to draw national and international attention to the social, political, economic, and environmental problems of the Niger Delta.[5] Nevertheless, even with such impressive growth of NGOS, their impact on forcing governmental agencies to monitor and control pollution remains weak.

INTERNATIONAL POLLUTION CONTROL DEMANDS AND NIGERIA'S RESPONSES

Because environmental issues do not respect national boundaries, international action is needed to control pollution and curb the dumping of waste between countries. It is in this regard that such bodies as The Permanent Inter-State Committee on Drought Control in the Sahel (CILSS), have played prominent roles assisting with the establishment of some inter-regional committees of which Nigeria is a member, given that it is part of the Sahel region. The CILSS, the United Nations Sudano-Sahelian Office (UNSO), and Club du Sahel have all helped in the control of drought and desertification and contributed in coordinating international assistance and the promotion of the adoption of sound strategies and policies to all the countries that fall within the Sahel region (CILSS 1991). While UNSO is attached to the UNDP, the Club du Sahel is made up of over 20 funding agencies under the aegis of the Organization for Economic Cooperation and Development (OECD). UNSO and CILSS were established in October 1973 by the United Nations General Assembly to respond to requests for assistance from CILSS and member governments, while providing aid in UNEP's name to the nine CILSS countries, the six countries of the Intergovernmental Authority on Drought and Development

(IGADD), and other Sudano-Sahelian countries. IGADD is charged with the mandate of providing support to various countries and regional institutions in the planning and design of projects and programs in the area of drought and desertification control (Debailleul et al. 1997: 9). Nigeria is a member of the Niger-Basin Authority and Chad Basin Authority. The two agencies have international status consisting of all countries at the basin of the respective water bodies. They consider issues relating to the allocation, development, and use of water in the respective basins to forestall international water conflicts.

Nigeria has benefited from resources provided by international institutions such as the UNDP and the World Bank. Membership to international bodies in an increasingly globalized world has had a tremendous impact on the environmental policy process not only in Nigeria, as it has committed itself to the implementation of international environmental programs, agreements, and treaties. For instance, the World Bank has assisted several African countries in their transition to environmentally sustainable development through the promotion of national Environmental Support Programmes (ESPS) (World Bank 1997: 4). This effort has helped African states seek environmental planning as a participatory process at the national and local levels, develop environmental content investment lending, strengthen environmental training, public information, and communication, and develop financial instruments for effective financing of sustainable development, among others.

In recognition of the importance of cooperation with other nations to achieve the effective protection of the global environment, the government has, over many years, ensured collaboration with the international community. Such collaborative efforts have resulted in positive contributions to the development of appropriate policies, legislation, action plans, and programs at regional and international levels (Federal Government of Nigeria 1996). The World Bank, through the IDA, has also entered into collaboration with the National Network for Environmental Monitoring to build an environmental data bank for the purpose of gathering information and research on pollution in Nigeria. Included in this arrangement are data on Ozone Depleting Substance (ODS), Lagos Pollution Control. Grants for this purpose came from British Official DA (Development Assistant) (Adegoke 1996: 47). Nigeria has also agreed to some memoranda of understanding with the US Environmental Protection Agency in order to secure technical assistance for staff training and environmental management, and four Nigerian universities have received funds for environmental research and capacity-building through these agreements.

Finally, the FEPA, which has been the anchor for national environmental programs in Nigeria, has received assistance funds for pollution monitoring equipment from Japan. Canada's International Development Research Center (IDRC) has also provided significant assistance to Nigeria to fund programs for environmental sustainability. At the regional level, the New Partnership for Africa's Development (NEPAD) program on environmental sustainability has also provided funds.

CONCLUSION

Nigeria is the most populous black nation in the world and it possibly has the strongest economy in West Africa. The country's environmental issues can have major implications for other West African states and the rest of the world. What happens in Nigeria can potentially have a serious impact on the economic development of the sub-region. It is therefore important to the world community to know about the country's environmental situation.

As I have outlined in this chapter, Nigeria's political leadership has taken action to address the country's environmental challenges over the last twenty years. The government's policy response to environmental pollution was initially triggered by the alarm raised in 1987 following the dumping of toxic wastes in Koko village, in the former Bendel State, supposedly by an Italian firm. In the face of mounting criticism by the public, human rights groups, and the press, the federal government of Nigeria was compelled to establish the FEPA in 1988. Since its creation, however, and as a result of the various commitments the country has taken internationally, a series of institutions and pieces of legislation have been introduced since then with the objective of achieving more sustainable environmental management.

Despite these actions, Nigeria still faces significant environmental challenges. In 2005, the Vision 2010 Committee was set up by the federal government to set benchmarks to achieve sustainable development practices. The committee has identified several important obstacles that must be overcome, however. A first set relates to undermining efforts at sustainable environmental management: the general inability of the agencies responsible for the environment to enforce laws and regulations, particularly with respect to urban planning and development; the prospecting for minerals and adherence to industrial standards; the siting of public and residential space in flood-prone areas; the improper reclamation and conversion of unsettled dump sites to plots for residential quarters, public buildings, and market stalls in ecologically sensitive areas; inappropriate agricultural practices; the destruction of watersheds and opening up of river banks and other areas leading to the silting of river beds and loss of water course. It also identified bush burning for farming and the ever-increasing depletion of young forests for fuel-wood; gas flaring and the resultant problems of ecosystem destabilization, heat stress, acid rain, and acid precipitation-induced destruction of freshwater fisheries and forests in the coastal areas of the country (Nigeria alone accounts for about 28 per cent of the world's total gas flares); and poverty as a cause and consequence of environmental degradation, with the poor scavenging marginal lands to eke out a living.

The commission's intentions will remain on the books and will not be effective unless corruption is tackled. For example, it is alleged that funds allocated for state-run programs get diverted by state and local governments into private uses. (Though there is no empirical back-up for this claim, still newspaper reports and common sense reveal that it is real.) Such corrupt practices have led to youth restlessness, which in turn has resulted in, for example, a disturbingly high degree of insurgency in the Delta areas of Nigeria. I would argue that the unrest that has taken place in the Niger Delta of Nigeria over the years, which has at times resulted in violence, is the result of a feeling of environmental injustice and the appalling manner in which oil is exploited from their territory and its consequent environmental disasters. Despite this unrest and a general feeling of environmental injustice, embezzlement continues, and the political elite of the Delta region has not been affected. Thus, corruption remains the biggest impediment for the present weak state of the environment in Nigeria.

The second most important obstacle is the weak enforcement of environmental laws and regulations as well as the conflicts that emanate from intergovernmental friction. The basic problem is the way Nigerian's federal system operates in which powers are shared between the centre, the states, and local governments. The overlapping nature of power decentralization results in jurisdictional disputes and interstate and intergovernmental conflicts. For example, the Nigerian Constitution has a provision that allows local governments to manage their own public transportation, sewage, and refuse disposal. This means that municipal waste disposal is the responsibility of the local government, and this level of government is empowered to pass by-laws on such issues. But local governments depend on state governments for sufficient funds to manage these programs. On the other hand, state governments believe that funds and personnel for managing such programs should come from the federal government, which has enormous resources derived from the exploitation of oil, which is under federal jurisdiction. As a result, there is a general view held by officials from state and local governments that unless the federal government shares its oil exploration money, FEPA may not succeed. The best example of the results of this situation is the ever-growing heaps of refuse that are seen all over Nigeria's urban centres.

I would argue that effective environmental management in Nigeria requires the integration of human and financial resources and coordinated efforts by the public sector, international institutions, NGOs, and business. But it also requires political will across the various levels of government, something that does not seem to be forthcoming. In 2005, Nigeria launched a satellite to monitor the environment through remote sensing. While this will help to map environmentally sensitive areas, the success of such mapping depends on the extent to which government management and regulatory processes are able to come to terms with such issues as enhancing transparency, accountability, and popular participation, as well as controlling the corruption that has been the bane of Nigeria's economic and political development. It is only in this way that we can discourage such dastardly practices as trading in hazardous wastes and be in the best position to protect and sustain the national and global commons.

Notes

1 This is prepared by putting a small quantity (two small glasses) of water into a small polythene bag and, with the help of machine designed for it, it is sealed and sold cheaply to those who cannot afford the bottled water sold in big shops.

2 Numerous pieces of legislation have been passed in Nigeria since 1956. These include the Oil Pipeline Act, 1956; Forestry Act, 1958; Destruction of Mosquitoes Act, 1958; and Public Health Act, 1958. And since independence in 1960, the following five laws have been enacted: Mineral Oil (Safety) Regulations 1963 (cap. 350 LFN 1990); Oil in Navigable Waters Act (1968 cap. 108 LFN 1990); Endangered Species Act (cap. 108 LFN 1990); Quarries Act (cap. 385 LFN 1990); and Sea Fisheries Act (caps. 404) (Adegoke 1996).

3 These include the National Policy on Environment (1989 and revised in 1999); the National Agenda 21 (1999) (this is based on the Earth Summit of Rio in 1992: Agenda 21, which touches on the various cross-sectoral areas of environmental concern and maps out strategies for addressing them); the National Guidelines and Standards for Environmental Pollution Control in Nigeria (1999); the National Effluent Limitation Regulation, S.I.8 (1991); the Pollution Abatement in Industries and Facilities Generating Wastes Regulations, S.I.9 (1991); the Waste Management Regulations S.I.15 (1991); the Environmental Impact Assessment (EIA) (Decree No. 86 of 1992); the Procedural and Sectoral Guidelines for EIA (Jan. 1999); the Natural Resources Conservation Action Plan; the National Fuel Wood Substitution Programme; the National Guidelines on Waste Disposal Through Underground Injection (1999); the National Guidelines and Standards for Water Quality in Nigeria; the National Guidelines for Environmental Audit in Nigeria (1999); the National Guidelines on Environmental Management Systems in Nigeria (1999); the National Guidelines for Spilled Oil Fingerprinting (1999); and the National Guidelines on Registration of Environmentally Friendly Products and Eco-labeling (1999) (UNEP 2006: 4).

4 Other NGOs that have general concern for the protection of the environment include the Nigerian Field Society, the Young Foresters Club, the Fauna Conservation Society, the Ecological Society of Nigeria, the Horticultural Society of Nigeria, the Nigerian Society for Environmental Management and Planning, the Nigerian Environmental Society, the Nigerian Society for Biological Conservation, the Development Exchange Centre, Edom Development Group, the Farmers' Development Union, the Forum for Environmental Protection, Green-Crop-Foundation, the Nigerian Concerned Group for Environment, Population and Development, the Population, Environment and Development Agency, the Women Farmers Association of Nigeria, the Women's Health, Environment and Development Agency, Savannah Watch, and Friends of the Environment, amongst others (Federal Ministry of Environment 2006: 24).

5 These include the Movement for the Survival of Ogoni People (MOSOP), the Niger Delta Wetlands Center (NDWC), the Niger Delta Human and Environmental Rescue Organization (ND-HERO), the Ijaw Council for Human Rights (ICHR), the Niger Delta Focus (NDF), the Women Initiatives Network (WINET), the African Environmental Action Network (EANET-Africa), the Niger Delta Oil Producing Communities Organization (NIDOPCODO), the Anpez Center for Environment and Development (ACFED), the Niger Delta Peace Coalition (NDPC), and others. Recently another two organizations have come about: the Ijaw Youth Council (IYC) and the Movement for the Emancipation of Niger Delta (MEND).

References

Adedeji, O.H. 2002. "Political Economy and Ethics of Ecological Problems in Nigeria: Challenges to Environmental Management and Sustainability in the New Millennium." Unpublished paper. Olabisi Onabanjo University, Nigeria.

Adegoke, A. 1996. "The Challenges of Environmental Enforcement in Africa: The Nigerian Experience." Paper presented at the Third International Conference on Environmental Enforcement, Oaxaca, México.

Agbola, Tunde, and T.A. Olurin. 1998. "Sustainable Cities Programme: The Case of Sustainable Ibadan Project (SIP) and Lessons from Abroad." In Kunle Adenjiji and V.I. Ogu (eds.), *Sustainable Physical Development in Nigeria: A Book of Readings*. Ibadan: NISER. 271-89.

Brown, L.R., and H. Kane. 1995. *Full House: Reassessing the Earth's Population Carrying Capacity*. London: Earthscan.

CILSS (Permanent Inter-State Committee for Drought Control in the Sahel). 1991. "Analyse des stratégies et plans de lutte contre la désertification: gestion des ressources naturelles dans les pays membres de CILSS." Paper presented at Symposium international sur l'integration et l'évaluation des actions de lutte contre la désertification/gestion des resources naturelles, Oct. 14-18, Niamey, Niger.

Debailleul, G., É. Grenon, M.-M. Kalala, and A. Vuillet. 1997. *Regional Dimension of Environmental Management*. Ottawa: IDRC Publications.

Efole, M.A. 2004. "Environmental Degradation and the Impact of Oil Exploration in Isoko Island." Paper presented at the Convention of the Isoko Association of North America, New York, July 30-Aug. 1.

Egboka, B.C. 1993. *The Raging War*. Awka: God's Time Publishers.

Ezzati, M., and D.M. Kammen. 2002. "Household Energy, Indoor Air Pollution and Health in Developing Countries: Knowledge Base for Effective Interventions." *Annual Reviews Energy and Environment* 27: 233-70.

Federal Environmental Protection Agency. 1990. *Nigeria: Country Profile—Implementation of Agenda 21. Review Process Made since the United Nations Conference on Environment and Development.*

Federal Government of Nigeria. 1996. *Vision 2010 Report*. http://www.vision 2010.org. Accessed Sept. 15, 2006.

Federal Ministry of Environment. 2001. *National Action Plan*. http://www.unccd.int/actionprogrammes/africa/national/2001/nigeria-eng.pdf. Accessed Aug. 15, 2006.

—. 2006. *National Action Programme to Combat Desertification*. http://www.unccd.int/actionprogrammes/africa/national/2001/nigeria-eng.pdf. Accessed Aug. 25, 2006.

Federal Office for Statistics. 2001. *Annual Abstract Statistics. Facts and Figures about Nigeria*. Ibadan: FOS.

Global Gas Flaring Reduction Initiative. 2002. *Report on Consultations with Stakeholders*. World Bank Group in collaboration with the Government of Norway. http://www.worldbank.org/ogmc/files/global_gas_flaring_initiative.pdf. Accessed Feb. 11, 2004.

Hiernaux, P. 1999. *Crisis of Sahelian Pastoralism: Ecological or Economic*? Overseas Development Institute, Pastoral Development Network Series. No. 39a. http://odi.org.com. Accessed Sept. 15, 2006.

Iyayi, F. 2005. *Transparency and the Oil Dilemma*. Second-Year Anniversary of the Nigeria Extractive Industries Transparency Initiative. Abudja, Feb. 17.

Kindzierski, W.D. 2000. "Importance of Human Environmental Exposure to Hazardous Air Pollutants from Gas Flares." *Environmental Reviews* 8: 41-62.

Okebukola, P. 2001. "Our Environment, Our Diversity," Lecture delivered at the second distinguished lecture series of the Adeniran Ogunsanya College of Education. Oto, Ijanikin, Lagos State. March 1.

Onibokun, G. 1999. *Managing the Monster: Urban Waste and Management in West Africa*. Ottawa: International Development Research Centre, Books Online. http://www.idrc.ca. Accessed April 12, 2006.

Onibokun, G., and A.J. Kumuyi. 1991. *Urbanization Process in Nigeria*. http://www.idrc.ca/en/ev-42972-201-1-DO_TOPIC.html. Accessed Dec. 10, 2006.

Onwioduokit, E.A. 1998. "An Alternative Approach to Efficient Pollution Control in Nigeria." In Akinjide Oshuntokun (ed.), *Current Issues in Nigeria's Environment*. Ibadan: Davidson Press.

Osuocha, P.C. 1999. "Improving Refuse Management in Urban Nigeria." Paper presented at the 25th WEDC Conference, Addis Ababa.

Oyebande, L.E., S.O. Sagua, and I. Ekpenuona. 1980. "The Effect of Kainji on the Hydrological Regime Water Balance and Water Quality of the River Niger." *Proceedings of the Helsinki Symposium*, June 1980, IAHS-IASH, No. 13: 221-28.

Smith, K.R. 1999. "Fuel Emission, Health and Global Warming." Paper presented at the United Nations Food and Agriculture Organization's Regional Wood Energy Development Program meeting on Wood Energy, Climate, and Health, Phuket, Thailand. October.

Stebbing, E.P. 1935. "The Encroaching Sahara: The Treat to West African Colonies." *The Geographical Journal* 85: 506-24.

Stockholm Environmental Institute. 2003. *New Fact Sheets on Air Pollution in Africa*. http://www.sei.se. Accessed April 30, 2006.

Tadegnon, N.K. 2005. *Environment: Motorcycles Emit Air Pollution in West Africa*. Inter Press Service. http://www.ips.org/regionalcenters/africa.shtml. Accessed Dec. 7, 2006.

Toluope, A.O. 2004. "Oil Exploration and Environmental Degradation: The Nigerian Experience." *Environmental Informatics Archives* 2: 387-93.

UNEP (UNITED NATIONS ENVIRONMENTAL PROGRAMME). 2006. *Nigeria: The State of the Environment*. http://countryprofiles.unep.org/profiles/NG/profile/state-of-the-environment/referencemanual-all-pages. Accessed Dec. 7, 2006.

World Bank. 1996. *Restoring Urban Infrastructure in Nigeria*. Findings. African Region. No 62. May.

—. 1997. *Rural Development, Poverty Reduction and Environmental Growth in Sub-Saharan Africa*. Findings. African Region. No. 92. August. www.worldbank.org/afr/findings. Accessed Jan. 25, 2006.

Zeba, S. 1998. *Community Wildlife Management in West Africa: A Regional Review*. Evaluating Eden Series. Working Paper No. 9. http://www.poptel.org.uk/iied/docs/blg/eden. Accessed July 25, 2006.

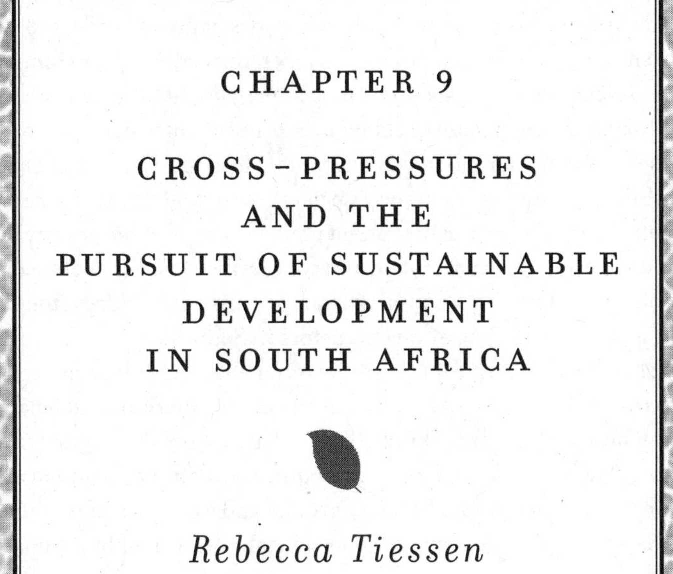

CHAPTER 9

CROSS-PRESSURES AND THE PURSUIT OF SUSTAINABLE DEVELOPMENT IN SOUTH AFRICA

Rebecca Tiessen

INTRODUCTION

In 2002, South Africa hosted one of the largest international events ever held in Africa: the World Summit on Sustainable Development (WSSD). The event lasted nine days and received tens of thousands of participants, including heads of state and government, national delegates, as well as leaders from non-governmental organizations (NGOS), businesses, and other major groups. The WSSD focused the world's attention on meeting the challenges of improving people's lives and conserving natural resources.

For South Africa, this was an important event for several reasons. First, South Africa's recent shift from an Apartheid system to democracy in 1994 has opened up new opportunities for the country to participate in key international fora and decision-making arenas.[1] Second, South Africa has, since 1994, adopted numerous new international and national policies as a demonstration of its commitment to sustainable development. The adoption of these policies, however, has been met with a range of financial and personnel challenges as the country struggles to deliver on the promises made in international agreements. Third, the changes since Apartheid include a shift in the focus of many civil society groups and NGOS who are encouraging the South African government to adopt policies that promote poverty reduction, sound natural-resource management, and conservation. These civil society groups put additional pressure on the South African government to address the range and, at times, competing interests of various actors in South Africa.

In this chapter, I elaborate on these and other challenges experienced in South Africa as a result of ten years of new commitments to environmental management and sustainable development. The challenges are examined in light of the numerous pressures experienced in the country both within the government and across government departments. Civil society groups and NGOS also add to these pressures, with their advocacy campaigns and special-interest lobby groups that have the objective of prioritizing some issues over others. All these cross-pressures are a factor on the international level as South Africa struggles to meet the international commitments it has made and to attract the requisite donor funds to turn policy into practice. Finally, the corporate sector contributes a number of cross-pressures that challenge the South African government to meet both its economic and environmental obligations. The pressures experienced in South Africa can also be attributed to the country's move to a more cooperative governance model that enables new partnerships both within government and between government and the private sector.

The cross-pressures examined here must be understood in the context of globalization and changing global trends, as well as in light of the emergence of new political and environmental actors both within South Africa and internationally. The international environmental agenda has had a significant effect on the politics and policy of South Africa as it has struggled to build an international reputation for sustainable development initiatives and meet the standards and policies to which it has agreed through its national and international policy commitments. Moreover, the country faces a challenge at the political level since the demands for services and resources to reduce poverty, address high rates of HIV/AIDS, and mitigate and prevent environmental degradation are at an all-time high. Supply, on the other hand, is at an all-time low, given the limited personnel available to implement programs (partly as a result of high HIV rates in the country), as well as the overall lack of financial resources to deliver on the government promises. Civil society actors and the corporate sector place additional pressure on the government to respond to varying demands, ranging from land redistribution to environmental conservation. South Africa therefore provides a unique example within the African context because of its political history and the salient and changing role of civil society groups in the country.

In the sections that follow I provide background information on South Africa's political context, including a summary of Apartheid and its legacy as well as some of the other contemporary political and social challenges facing the country. I also examine the numerous environmental opportunities and challenges experienced in South Africa. I then turn to the four key cross-pressures witnessed in South Africa: national, civil society, international, and corporate pressures.

SOUTH AFRICA'S POLITICAL CONTEXT

The system of Apartheid (an Afrikaans term meaning "apartness") was instituted in South Africa in 1948 as a way of segregating groups along racial lines. Through Apartheid, people were legally classified into racial groups as Whites, Blacks, Indians or Coloureds and forcibly separated into geographical areas. South Africa's Black majority were physically segregated into "homelands," faced forced removals, and had restricted political rights and sub-standard social services. Segregation also existed in public spaces such as beaches, churches, health services, and public transportation. Non-White South Africans were treated as second-class citizens under this system, which was condemned internationally as unjust and racist. In 1973, the General Assembly of the United Nations put forth the *International Convention on the Suppression and Punishment of the Crime of Apartheid*. The Convention provided a legal framework within which member states could apply sanctions to South Africa to force the country to change its policies. This convention was ratified in 1976.[2] Between 1973 and 1994, acts of resistance and sanctions combined to force the South African government to change its constitution and extend political rights to all groups. Since then, however, the legacy of the Apartheid system remains a core challenge for South Africa as it struggles to fight poverty, social inequality, and widespread HIV infection. Political activities in South Africa have been largely preoccupied with anti-Apartheid struggles over the past several decades, shifting attention away from environmental problems in the country.

Nevertheless, South Africa is today one of the more progressive African countries in terms of its policy efforts to promote sustainable development; the range of national and international environmental policies that have been adopted in South Africa is extensive, and the country's constitution has a provision granting every citizen the right to a healthy and protected environment. The country therefore has made some preliminary efforts to address environmental concerns. However, South Africa's commitments to sustainable development are much more evident in policy statements and discourse than in actual practice. In the next section I highlight some of the environmental opportunities and challenges facing South Africa.

BACKGROUND TO SOUTH AFRICA'S ENVIRONMENTAL OPPORTUNITIES AND CHALLENGES

South Africa is a country rich in natural resources, from gold, diamonds, and platinum to coal, iron ore, and uranium. Throughout the country there are numerous conservation areas and some of the world's most popular and attractive national wildlife parks. South Africa is also home to mountain parks, beautiful beaches, and coastal wilderness such as Table Mountain, Kruger National Park, and the Wilderness Lakes Areas. The rich and diverse natural environment of the country is a result of a long history of nature conservation strategies and government commitments that have been instituted to protect the natural environment. However, growing pressures on the government to redistribute land, as well as conflicts emerging between conservationists and communities living on the borders of national parks, have made wildlife habitat protection an increasingly contentious issue.

South Africa therefore faces a range of old and new environmental challenges that range from recurring drought, soil erosion, desertification, and limited clean water to polluted rivers, mounting urban waste, and air pollution that results in acid rain. These environmental challenges are compounded by political and economic realities of a country in which water privatization is on the increase[3] while so too are rates of poverty, unemployment, and the number of people living in areas where there is no potable water and no waste-management facilities. South Africa is also struggling to develop alternative sources of energy to replace the use of coal for electricity production, while at the same it is anxious to develop its industrial and manufacturing sectors to improve export earnings. Many of these environmental challenges are the result of global integration stemming from, among several factors, the introduction of new technologies designed in one part of the world, built in other parts, and used in another location. Other factors contributing to these environmental challenges are the result of changes in trade policies that have taken place in recent decades. Most significant among these have been the introduction of trade liberalization, lower tariff barriers, the reduction of the role and functions of the state, and the privatization of state-owned enterprises. For South Africa, these forces of globalization have translated into growing economic inequality and poverty in the country. Low-income neighbourhoods in South Africa are now faced with the realities of limited access to water as a result of water privatization schemes. The reduced role of the government in providing basic services means that service provision becomes a matter of profit rather than meeting the needs of people. Neo-liberal economic reforms therefore exacerbate social and environmental challenges by reducing access for low-income households to basic necessities such as water and wood for cooking. Furthermore, environmental policies designed in the context of neo-liberal economic reforms are done with economic growth—rather than conservation—in mind.

South Africa's unique political trajectory has also had an effect on environmental management practices. As we have seen, South African politics were dominated by a minority group until 1994. The concentration of power in the hands of a minority meant that policies did not represent or benefit all South Africans. The changes that took place in 1994 included widespread efforts to address the minority control over policy-making to become more responsive to the broader demands of society at large. The end of Apartheid also witnessed a burgeoning of civil society actors. While South Africa had a strong civil society prior to 1994, it became reinvigorated with the advent of democracy. The post-1994 civil society has diversified its political platform to include programs to address the impacts of HIV/AIDS on the country's social and economic development as well as the ongoing, and rapid, process of environmental degradation.

The Government of South Africa remains committed to increasing public participation and civil society activities and refocusing national attention on the promotion of policies aimed at achieving sustainable development rather than serving narrow group and sectoral interests (DEA & T 1999). The South African government is thus faced with a number of cross-pressures, including pressures to promote economic growth and create jobs in the industrial sector while also trying to balance land redistribution and a sustainable development approach in line with an increasingly vocal environmental movement in the country. However, public demands for government accountability on environmental issues have forced the South African government to adopt a range of policies that have resulted in a stretched capacity of the government to enforce the environmental policy framework; by 1997, South Africa had adopted at least 21 environmental policy development projects, placing tremendous pressure on the government's capacity, especially within provincial environmental departments (DEA & T 1999).

Environmental legislative changes have also been widely felt across South Africa and have altered the landscape of corporate sustainability. The introduction of 60 new or revised statutes on corporate management between 1994 and 2004 was geared at promoting safety, health, environment, socio-economic development, labour, governance, and ethics issues. These statutes were driven by national and international pressures. Issues around employment equity and Black economic empowerment reflected the pre-election aspirations of the ANC's (African National Congress) Reconstruction and Development Programme (RDP) agenda (KPMG 2004). The RDP was implemented in 1994 as a socio-economic policy framework that seeks to mobilize people and resources with the goal of eradicating Apartheid for the purpose of building a democratic, non-racial, and non-sexist country.

Despite widespread efforts to protect natural habitats and to promote policies and laws for environmental management and resource use, South Africa is one of the world's most polluted countries (Bond 2002). For example, South Africa has one of the world's worst records for carbon dioxide (CO_2) emissions, worse than those of the US (accounting for income and population size). Estimates suggest that CO_2 emissions in South Africa are 70 per cent above unsustainable levels. The country made policy commitments to address this problem but took no action to reduce emissions in the period between 1990 and 1998, allowing emissions to increase during this time (Bond 2002).

South Africa's high rate of HIV/AIDS infection is also relevant to environmental management. In a 2005 National Household Survey, the researchers estimate that 10.8 per cent of all South Africans over the age of two live with HIV, and those between 15 and 49 had a prevalence rate of 16.2 per cent (AVERT 2005). These high rates of infection mean that numerous trained experts in the field of environmental management have been unable to continue working or have died as a result of HIV/AIDS. The impact on the workforce is only now beginning to be felt, and the problem is likely to worsen as more people become ill.

NATIONAL PRESSURES AND POLICY COMMITMENTS

Within the South African government, the Department of Environmental Affairs and Tourism (DEA&T) is responsible for the vast majority of environmental initiatives. DEA&T evolved from a junior department prior to 1994 to a key one during the first ten years after the country's democratization. Relative to other departments, however, DEA&T remains financially and politically weak. Expenditure on environmental issues has increased in recent decades, but some of the funding has been allocated to the pursuit of two new national priorities: tourism, which became essential to South Africa's economic development model in 1993-94; and pollution control, which became a priority in 1996-97. Nonetheless, relative to other government programs, the environmental budget remains low, at about one to two per cent of total government expenditure (DEA&T 1999). Projections for the DEA&T budget show a marginal increase in spending, from 0.2 per cent of the national budget to approximately 0.3 per cent in 2003-04 (Seeliger et al. 2003).[4] Even with this increase, however, there are insufficient resources to manage environmental and resource issues in the country.

The financial and political weakness of the DEA&T translates into weak enforcement mechanisms and a lack of appointed officials to carry out these responsibilities (DEA&T 1999). This institutional weakness reflects broader regional trends where governments in southern African countries have limited capacity (in terms of human and financial resources) to implement existing environmental policies and regulations. Uday Desai (1998) extends this analysis to developing countries more generally by highlighting these weaknesses in terms of weak management capabilities of government agencies entrusted with environmental policy implementation as well as a serious lack of resources, personnel, and expertise. Relative to other government ministries in less developed countries, especially those in charge of the economic and industry portfolios, environmental ministries are generally financially and resource weak, and lack coordination capacity. Ultimately, economic interests are usually privileged over environmental sustainability measures in all countries (Desai 1998: ch. 1). For less developed countries, and African countries in particular, the privileging of economic interests over environmental sustainability has serious impacts on low-income households who rely more on their natural environments for day-to-day survival. Let us now turn to a discussion of the prioritization of economic interests.

Since 1994, South Africa's environmental focus has shifted from one of conservation and the protection of natural resources to a broader notion of sustainable development. Within this overall change in focus, DEA&T underwent a significant reorientation through the implementation of institutional changes, with the objective of providing the department with "user friendly environmental information services to the full spectrum of stakeholders, provincial and local, agencies, NGOs, CBOs and relevant environmental institutions" (DEA&T 1999). The restructuring of the DEA&T included an increased focus on service delivery, better coordination between programs such as tourism, conservation, and heritage, linking marine and coastal management, increased emphasis on information systems, and greater collaboration with research organizations, universities, and private companies (DEA&T 1999). Such restructuring reflects a South African trend toward consolidation of activities and an effort to minimize overlap in resources and activities as cost-saving measures. It also signals a move toward a more collaborative process of shared information and research. While DEA&T is weak in its policy implementation capacity, it has been responsible for taking on a number of new policy initiatives in recent years, initiatives that will be reviewed below.

Policies are important instruments for governments to promote and bring about change. They act as guidelines to encourage some actions and prevent or discourage others. The implementation of policy objectives enables governments to set priorities and determine the allocation of resources to various activities. Without clear policies, there can be considerable confusion over rights and responsibilities. Governments are not the only actors responsible for policy creation. In South Africa, civil society organizations and environmental experts have contributed immensely to the process of policy formation. National environmental policy in South Africa has also been influenced by international policies and agreements. For example, the *White Paper on the Conservation and Sustainable Use of South Africa's Biological Diversity* (Government of South Africa 2006) was influenced, in part, by the Convention on Biological Diversity, which South Africa signed at the 1992 United Nations Conference on Environment and Development, the Rio Summit, and ratified in 1995. Environmental policy can only have real impact, however, if it is translated into law and enforcement mechanisms are put in place to ensure compliance.

The 1990s saw a rapid growth in the number of environmental policies adopted around the world. The process of national environmental policy-making was spurred, in part, by the international agreements made by South Africa at the Rio Summit. Such policies include the Reconstruction and Development Programme (RDP) (1994), the Lorimer Report on the Council for the Environment (1995), the Integrated Pollution Control (1996), and the Consultative Environmental Policy Process (CEPP), which resulted in the White Paper on National Environmental Management Policy. The White Paper set out several objectives, including a more democratic and participatory process for policy creation, a move from a narrower understanding of the environment as a conservation approach to the broader concept of sustainable development, a broader understanding of environmental degradation in relation to poverty and inequality, and a focus on the linkages between social, economic, and environmental factors for a more holistic, integrated, and coordinated sustainable development.

In 1998, Cabinet approved the White Paper and, later in that year, DEA&T translated it into the National Environmental Management Act (NEMA) (Act 107 of 1998). In more recent years, DEA&T has made a number of amendments to the NEMA (e.g., NEMA Act No. 46 of 2003 and the 2004 Amendment Act). These amendments give greater power and authority to the NEMA to ensure enforcement and compliance. Some of the strengths of the NEMA include its emphasis on environmental management, the development of procedures and institutions to promote intergovernmental relations, facilitating public participation in environmental governance, and enforcement of environmental laws (Government of South Africa 2006).

DEA&T has also instituted a number of reforms to various laws, including the Environmental Law, the Tourism Law, and the Marine and Coastal Law. Other law-related priorities include the Atmospheric Pollution Prevention Act, No. 45 of 1965 (the APPA) and, more recently, the APPA's replacement, the National Environmental Management: Air Quality Act No. 39 of 2004 (the Air Quality Act), which was signed by the President in 2005. Another significant act is the National Environmental Management: Protected Areas Act, No. 57 of 2003 (the Protected Areas Act), which ensures the continued existence of South African National Parks and the declaration and management of protected areas in South Africa (Government of South Africa 2006).

Environmental policy formulation and implementation have been thwarted, however, by a number of key challenges. These include a lack of coordination and any clear set of responsibilities for various government departments responsible for environmental management, weak enforcement measures, and limited resources to put policies into practice. The enforcement of environmental laws, therefore, has been uncoordinated and problematic.

DEA&T is not the only department responsible for designing environmental programs, implementing environmental policy, and enforcing compliance. Other departments—namely the Department of Water Affairs and Forestry, Mineral and Energy Affairs, Transport, Health, Land Affairs, and Agriculture—have a budget and responsibilities for environmental management. Provincial authorities are also responsible for the administration of environmental laws and all have important roles in the areas of policy implementation and enforcement. There are additional institutional coordination challenges that result from the fact that government responsibilities are shared among the nine provinces and approximately 800 local authorities. Attempts to improve coordination among these numerous institutions have been made, however, through efforts to elaborate environmental management plans as set out in the National Environmental Management Act (1998).

At the national level, then, we can see a number of cross-pressures that have resulted in an uncoordinated and financially weak DEA&T. Such cross-pressures at the national level reflect the tensions and challenges of effective policy implementation. The government faces additional problems in its efforts to satisfy a range of political actors and interests represented in NGOs and civil society groups in the country.

NGOS AND CIVIL SOCIETY PRESSURES IN SOUTH AFRICA

Since 1994, civil society groups and locally based NGOs have applied increasing pressure on the government of South Africa to address environmental issues and their relationship to health concerns, poverty, inequality, (un)healthy workplaces, and community conflicts. The power of these organizations is strengthened by international coalitions and support networks between civil society groups and national and international NGO alliances. I examine the role of international civil society activities later in this section, after a brief discussion of the historical significance and dynamics of the civil society movement specific to South Africa.

The history of civil society mobilization in South Africa prior to 1994 was characterized by a concentration of efforts to promote political freedom, anti-racism, and the end of the system of Apartheid. The dismantling of Apartheid signalled a shift in NGO efforts from political freedom to concerns more central to poverty alleviation and development, and especially a shift to the promotion of sustainable development (Govender 2001). Civil society actors now respond to crises related to an increase in poverty, social exclusion based on gender and sexuality, weak citizenship, and the failure of the post-Apartheid government to meet the basic needs of the poor.

In very general terms, South African civil society consists of the following groups: trade unions, civic movements, the women's movement, the youth and students' movement, Black business and small business, the community of NGOs and community based organizations (CBOs), communities of faith, political organizations, and liberation movements (Ballard et al. 2005). There are numerous NGOs involved in environmental policy research, advocacy, and implementation. The largest NGO in South Africa is the Environmental Justice Networking Forum, which has a membership of some 500 institutions. The interests of NGOs involved in environmental issues is broad, ranging from pure conservation issues to dealing with toxic waste and other environmental hazards. The predominant sources of funding for the NGO sector are foreign donor assistance, the corporate sector, private foundations, government grants, and membership fees.

NGOs have also made important progress in fostering networks and fora through which they have been able to make their demands heard by government and various United Nations (UN) bodies, enabling them in turn to engage critically with governments and raise issues that might otherwise be ignored. Civil society groups and NGOs therefore play an important role in the promotion of alternative perspectives and solutions to environmental problems, while at the same time adding to the policy debates and holding government and international bodies accountable to their policy commitments. Furthermore, the civil society sector and key development NGOs in South Africa played a pivotal role in pushing for broad consultations with governments and United Nations agencies during the WSSD, so that the diverse views of the majority of South Africans were represented at this meeting.

While a diverse and remarkable number of civil society organizations participated in the broader WSSD events, these groups were nonetheless marginalized from the key debates and discussions held with government officials and heads of state, as their participation was largely limited to the parallel summits held throughout Johannesburg. The presence of the parallel civil society summit, however, demonstrated an upsurge of support for sustainable development among a range of political actors (Vina et al. 2003).

Civil society groups, therefore, played a prominent role in the WSSD and other important sustainable development fora. However, indigenous groups, environmentalists, and other civil society organizations do not have adequate resources and frequently play a marginal role in decision-making and discussions. The civil society community in South Africa, for example, does not often have sufficient financial resources, logistical support, interpreters, and other types of necessary assistance. Despite these weaknesses, the civil society movement has captured the media's attention and created an additional cross-pressure for the South African government; NGOs (both local and international) have played an important role in placing some issues on the national agenda and fighting for the rights of communities to have access to land as well as hunting rights. Locally-based resource management and control is an issue of particular relevance to communities located in regions that border wildlife areas, as communities face the day-to-day challenges posed by wildlife, such as the destruction of crops and homes. South Africa's national government, recognizing the increasing international and national pressure to preserve and protect wildlife, must also contend with communities increasingly at risk of losing their livelihoods. Conservation efforts for the preservation of wild animals and game parks are therefore frequently in direct contrast to the economic and environmental needs of local communities. Some of the issues raised by people living in rural communities near national parks highlight the problems associated with wild elephants trampling crops, wild animals eating farmers' chickens and other livestock, and the day-to-day hazards of those who are at risk of attack from wild animals.

Local communities gained leverage through commitments made by the government at the WSSD to protect the rights of communities and indigenous peoples. The WSSD's "Plan of Implementation" recognizes the centrality of community-based natural-resource management and the vital role indigenous peoples play in sustainable development (Vina et al. 2003). The South African government has therefore made public participation in rural communities a high priority in the development of policies.

Civil society groups and NGOs in South Africa have achieved some significant political gains in their efforts to participate in decision-making processes, as is evident in their increased presence at national and international events. While the whole of civil society is divided on a number of issues and priority areas, Patrick Bond suggests that there is a great deal of potential for unity between a range of actors from labour groups, environmentalists, gender equality organizations, and health activists (2002). Through a number of grassroots campaigns, the solidarity of a number of non-governmental groups is becoming more deeply rooted in South Africa. A collective of small groups can, in itself, be an impressive political force (Dwivedi et al. 2002) for which the South African government must develop better negotiation and communication strategies.

INTERNATIONAL PRESSURES

South Africa has faced a new set of international demands and opportunities since the country was welcomed back to the international community in 1994. Donors and international agencies began to place a great deal more emphasis on South Africa's environmental management initiatives. South Africa, eager to contribute to international decision-making, participated in a number of international meetings and signed several new international agreements and conventions. These new activities required South Africa to establish updated systems, personnel, and structures. One of the biggest challenges that South Africa has faced in terms of its commitment of personnel and organization was its agreement to host the WSSD, which was held in 2002 in Johannesburg.

The WSSD marked the tenth anniversary of the UN's Rio Summit, held in 1992. The main document that was produced at the conference, *Agenda 21*, marked an important watershed moment as the international community began to take environmental issues more seriously. It also increased the accountability of signatory countries to the United Nations in terms of the targets and activities upon which they agreed. The Plan of Implementation is the internationally agreed-upon document that emerged out of the 2002 WSSD and is meant to guide development, financial, and investment decisions by governments, international organizations, and other stakeholders. The signing of international policy documents has put South Africa, as well as all other signatories, in a binding position to adhere to the obligations these policies require. The South African government's limited revenues for environmental programs have meant, however, that the government has increasingly relied on donor funds (international donor agencies and government bilateral and multilateral initiatives) to implement environmental programs. These funds, furthermore, come with compliance issues around reporting and monitoring funds provided by donors. Thus, South African ministries experience external pressures from international bodies and foreign governments to comply with external demands.

Official Development Assistance (ODA) to South Africa has been essential for the country to implement programs that are required to fulfill the commitments the country made to implement *Agenda 21*. ODA from international donors has enabled South Africa to implement a range of new policies and programs, such as a National Waste Management Strategy, the Coastal Management Policy Programme, and the development of the White Paper on Biological Diversity, referred to above.

The majority of funds (56 per cent) channelled through the South African government from ODA sources has been directed to the African National Parks Fund and the National Botanical Institute. Approximately 40 per cent of this type of financing has been directed to a variety of projects related to pollution, waste management, and environmental management. Between 1994 and 1999, the amount of ODA that South Africa received totalled R286,629,201 (approximately US $49,000,000) (DEA&T 1999). Additional donor funds were made available to other departments to address environmental issues. In addition to international commitments, South Africa's international pressures stem from its efforts to increase foreign corporate accumulation of capital, to promote its local industries and manufacturers in an increasingly competitive global market, and to attract foreign direct investment, to which we now turn.

CORPORATE AND PRIVATE-SECTOR CROSS-PRESSURES

The role that corporations play in South Africa's political and economic life is reflective of the impact globalization has had on the country, as well as its global reach. In many countries throughout Southern Africa globalization has been a process that has generally been imposed from the outside. In South Africa, however, globalization has been generated both within the country and through external pressures. Internally, South Africa has spearheaded globalization initiatives primarily through its major business groups and international companies based in the country (Carmody 2002).

South Africa's return to the international stage in 1994 placed it in a new environment of international accountability, which forced companies to address sustainability and meet environmental standards in line with global market expectations. Environmental certification, at this time, became central to corporate activities. South African companies have also been actively pursuing new standards and guidelines for voluntary self-regulation in areas of social, ethical, and environmental issues. Regulatory measures include the implementation of codes such as ISO 14001 (the International Standard Organization's Environmental Management Standard) and "sustainability reporting" known as the Global Reporting Initiative.[5]

Since 1994, South African companies have increasingly identified the environment as a strategic priority (KPMG 2004). The implementation of various environmental objectives in the private sector is influenced by the Industrial Environmental Forum (IEF). The IEF was established in 1991 and represents the interests of 30 leading corporations (DEA&T 1999). This institution in many ways serves as an important barometer of environmental trends and practices within industry and the business sector. While national policy has had an influence on the manner in which the private sector behaves, international forces and agreements, such as changing consumer preferences, trade agreements, and environmental conventions, have also had an impact. The types of influences vary, but they tend to be focused mainly on the introduction of Environmental Management Systems (EMSs) such as the ISO 14001 standards and cleaner production technologies. Overall, the IEF report indicates an increased environmental awareness in the corporate sector.

The WSSD made explicit efforts to increase corporate responsibility and accountability, and the environmental movement played a central role in the process. The international NGO called Friends of the Earth International, as well as other civil society groups from the development and environmental sectors and labour groups, were at the fore of discussions around increased corporate accountability measures. Civil society groups pushed for more than voluntary initiatives on the part of corporations to guarantee the respect of citizen and community rights, to reduce ecological debt,[6] and to force companies to disclose environmental and social practices and impacts. Stricter accountability mechanisms were not included in the final WSSD Political Declaration, and statements were limited to the encouragement of corporate responsibility and accountability. However, important advances were made to develop a forum for the exchange of information regarding best practices within the context of sustainable development. There is potential, then, for the references to corporate responsibility and accountability that were included in both the Plan and the Political Declaration to promote intergovernmental processes that would in turn foster civil society input on an international regulatory framework for corporations (Vina et al. 2003).

However, it remains uncertain how far companies have in fact implemented these measures. A 1999 study conducted in South Africa reviewed 83 local and foreign organizations operating in the country. The study showed that South African companies still lag approximately five years behind their counterparts in other countries in the implementation of an environmental management plan (DEA&T 1999). Furthermore, South African companies are shifting away from the language of corporate social responsibility in favour of a new concept: "corporate social investment."[7] The move has been an effort to divert attention from the historical links that existed between corporations and Apartheid (Fig 2005). However, the ANC remains biased toward corporate accumulation of capital. The legacy of Apartheid-capitalism is thus deeply entrenched in the current activities of the ruling party (Bond 2002).

International trade has also had an impact on decisions made by corporations on environmental issues. The establishment of the World Trade Organization (WTO) in 1995 has increased trade liberalization globally. As a result, more goods, technology, and capital are flowing from developed to developing countries. The increased interaction of trade activities has a direct bearing on all countries and, as Patrick Bond (2002) suggests, this process has a tendency to commodify everything and to reduce the state to an accomplice in unsustainable economic growth and corporate expansion.

CONCLUSION

South Africa's environmental initiatives over the past several years have reflected significant advances in the country's overall sustainable development approach. The changes witnessed in South Africa need to be understood, however, in the context of national, civil society, international, and private-sector influences. The multiple cross-pressures suggest that the South African government faces competition for priorities, resources, and funding, thereby limiting what the country can realistically manage. These pressures must also be measured in relation to broader challenges South Africa is facing, namely globalization, poverty, and high rates of HIV/AIDS. In terms of the national pressures, the challenges can be summarized as limited resources and lack of coordination between government departments as funds for environmental programs are spread thin throughout a number of government units. The international pressures reflect South Africa's commitments to international policy-making, as well as the role played by donors in topping up environmental financing.

Civil society's shifting role since 1994 reflects a trend toward greater demands for environmental accountability, as anti-Apartheid associations have changed their focus to sustainable development and poverty concerns. South Africa's new constitution carved out a larger role for the public in the governance, ownership, and utilization of resources. It also calls for greater interaction and consultation between government and civil society to ensure that the public contributes more broadly to decision-making on sustainability issues. Some of the positive outcomes of the shift to greater participation from civil society actors in government decision-making include the opportunity for broader public awareness, participation, and better environmental management. As the public begins to understand more and more about environmental problems, it becomes better equipped to make demands on politicians to pay attention to harmful environmental practices.

The demands of the public (both nationally and internationally) have resulted in increased national and international participation in policy formulation on the part of the South African government. The creation of new policies and legislation has the potential to transform management systems and enforcement mechanisms. The environmental legislation that has been introduced and the enforcement strategies that have been developed have meant that environmental considerations are taken into account in development projects, thereby reducing or preventing detrimental impacts to biophysical and social environments.

Some of the challenges posed by the new cross-pressures on environmental management include those stemming from the numerous commitments that have been made to achieve sustainable development, which have not been accompanied by required resources to see that these policies be translated into measurable outputs. There also exist challenges of coordination: responsibility for action on policy commitments is spread across several government departments, authorities, NGOs, and the private sector. In the case of pollution, for example, there is some confusion over who is responsible for which actions. At the national level, the line departments include the Department of Water Affairs and Forestry (DWAF), Department of Minerals and Energy Affairs (DMEA), Department of Trade and Industry (DTI), Department of Land Affairs (DLA), and the National Department of Agriculture (NDA), all of which have functional responsibilities over certain sectors of the environment. Other problems experienced at the national level involve the neglecting of responsibilities or duplication of efforts and therefore the waste of limited resources.

Even DEA&T experiences internal challenges as it struggles to cope with limited financial and human resources with an ever-expanding policy portfolio. Personnel working on environmental issues in South Africa are frequently stretched beyond their work limits as they juggle a range of activities as part of the increased workload associated with the signing of international conventions. The workload attached to international policy commitments and environmental diplomacy includes a range of responsibilities, from high-level international summit activities to the day-to-day coordination that takes place between states and through international organizations (Dwivedi et al. 2002). Reporting to secretariats of international bodies alone is a time-consuming process that leaves little time for travel and attending meetings. Financial resources, which are limited to begin with, are further strained by the introduction of membership fees to international organizations.

South Africa is also faced with challenges stemming from high staff turnover rates, the loss of experienced staff, and a budget that has changed little in recent years despite the increase in environmental programs. Some of the challenges South Africa faces as a result of multiple cross-pressures from international bodies, civil society groups, and the private/corporate sector reflect similar challenges witnessed during the World Summit on Sustainable Development in 2002. The WSSD had some successes in terms of garnering renewed support for environment and sustainable development initiatives around the world. However, the summit ended on a low note as governments failed to agree on or adopt effective means of implementation (including financing issues) and institutional mechanisms. Furthermore, the divisions that became apparent at the WSSD—especially between governments and civil society, but also within civil society groups—reveal the extent of the challenges to achieve the goals outlined by the WSSD's Implementation Plan. Some have deemed the WSSD to be a failure in that it was unable to develop targets, timetables or enforcement measures for environmental accountability. The WSSD will therefore be remembered as a site for talk without action. As the WTO gains power and control over multilateral and national environmental regulations, the WSSD will also be remembered as a missed opportunity to reduce corporate control over nature (Bond 2002).

The success of the WSSD cannot be attributed to any specific action or challenge experienced by South Africa alone, but to a broader global process that favours economic growth and profit over environmental protection. South Africa's role in hosting the summit, however, reflects its increasing presence and the contributions it has made to international sustainability commitments. Its numerous commitments to environmental management, pollution control, and conservation, both nationally and internationally, demonstrate a level of commitment to sustainable development.

South Africa's success with environmental management will depend on a number of factors, including a sustained and increased investment from donors to support environmental initiatives, a commitment on the part of the corporate sector to regulate themselves in accordance with national and international environmental standards, and the extent to which the civil society groups contribute to or control the decision-making process.

Notes

1 The system of Apartheid will be elaborated upon below.

2 Canada applied partial import sanctions and voluntary export sanctions to South Africa in December 1977.

3 See MacDonald and Ruiters 2005. Privatization of water has been correlated with increased water prices. Low-income households are therefore less likely to access a sufficient amount of water as water prices rise.

4 Some environmental spending is allocated to various sectoral departments. However, it is unclear how much money is spent across the departments as it is not possible to separate out the budget expenses from the rest of the departmental budgets.

5 The Global Reporting Initiative (GRI) is a framework for reporting on economic, environmental and social performance by all organizations for the purpose of standardizing practices between companies in a similar way as financial reporting is routinized and standardized. More information about GRI is available at: http://www.globalreporting.org/.

6 The idea of "ecological debt" is based on the idea that the North (industrialized countries) are in debt to the South (less developed countries) because of the long history of exploitation by the North of the Southern countries' resources.

7 "Corporate social investment" or "corporate social responsibility" are terms used to refer to a company's obligation to be sensitive to the needs of all stakeholders, as well as to citizens on whom the business's operations may have an impact. A corporate social investment is then an investment in the communities and the people to ensure that profits from the business transaction are experienced by everyone and not exclusively by the company.

References

AVERT (AVERTING HIV AND AIDS). 2005. "South Africa HIV/AIDS Statistics." World Wide Web: http://www.avert.org/safricastats.htm. Accessed April 4, 2006.

Ballard, Richard, Adam Habib, Imraan Valodia, and Elke Zuern. 2005. "Globalization, Marginalization and Contemporary Social Movements in South Africa." *African Affairs London* 104(417): 615-34.

Bond, Patrick. 2002. *Unsustainable South Africa: Environment, Development and Social Protest*. Scottsville and London: University of Natal Press and Merlin Press.

Carmody, Padraig. 2002. "Between Globalisation and (Post) Apartheid: The Political Economy of Restructuring in South Africa." *Journal of Southern African Studies* 28: 255-75.

DEA&T (Department of Environmental Affairs and Tourism). 1999. "National State of the Environment Report—South Africa: Political Environment." World Wide Web: http://www.environment.gov.za/soer/nsoer/issues/politic/index.htm. Accessed March 15, 2006.

Desai, Uday. 1998. *Ecological Policy and Politics in Developing Countries*. Economic Growth, Democracy, and Environment. Albany: State University of New York Press.

Dwivedi, O.P., Patrick Kyba, Peter Stoett, and Rebecca Tiessen. 2002. *Sustainable Development and Canada: National and International Perspectives*. Peterborough, ON: Broadview Press.

Fig, David. 2005. "Manufacturing Amnesia: Corporate Social Responsibility in South Africa." *International Affairs* 81(3): 599-617.

Govender, Charm. 2001 (Dec.). "Trends in Civil Society in South Africa Today." *Umrabulu* 13. World Wide Web: http://www.anc.org.za/ancdocs/pubs/umrabulo/umrabulo13.html. Accessed April 20, 2006.

Government of South Africa, Department of Environment and Tourism. 2006. World Wide Web: http://www.environment.gov.za/. Accessed April 20, 2006.

KPMG. 2004. "Corporate Sustainability in South Africa: A Ten Year Review." A 2004 KPMG Survey on Integrated Sustainability Reporting in South Africa. December. World Wide Web: http://www.waynevisser.com/cc_in_sa_short.htm. Accessed April 20, 2006.

MacDonald, David, and Greg G. Ruiters, eds. 2005. *The Age of Commodity: Water Privatization in Southern Africa*. London: Earthscan Press.

Seeliger, Leanne, Carlene van der Westhuizen, and Albert van Zyl. 2003. "Budgeting for Long-term Sustainability of the Economy: Building Capacity in the Department of Environmental Affairs and Tourism (DEAT)." World Wide Web: http://www.essa.org.za/. Accessed March 3, 2006.

Vina, Antonio, Gretchen M. Hoff, and Ann Mare DeRose. 2003. "The Outcomes of Johannesburg: Assessing the World Summit on Sustainable Development." SAIS *Review* 23(1): 53-70.

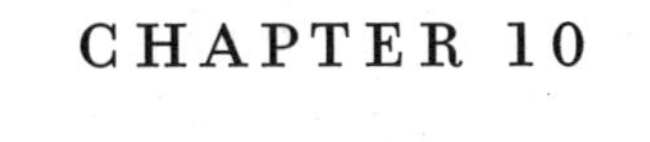

CHAPTER 10

GLOBALIZATION AND ENVIRONMENTAL POLITICS IN MEXICO[1]

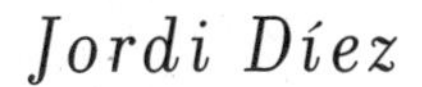

Jordi Díez

INTRODUCTION

Mexico is endowed with vast natural resources. It produces three million barrels of oil a day, and its proven and recoverable reserves are estimated at 50 billion barrels. It is the foremost producer of silver in the world and a leading exporter of copper, gold, lead, and zinc. It also counts for 12 per cent of the world's biodiversity and is among the 12 countries considered to be "mega-diverse." Mexico's natural wealth has been intertwined with the country's history, as the extraction of natural resources has been the basis for the country's economic development for centuries. During colonial times, Mexico, then called New Spain, was the largest supplier of silver and gold to Europe, making it the most important colony of the Spanish Crown. After the country gained independence in 1821, and following a fairly long period of social and political instability, Mexico's elites made concerted efforts to integrate the country into the international economic system through exports. By the end of the nineteenth century, it had become a primary exporter of minerals, grain, and timber as its economy became fully integrated with global trading systems. The extraction of resources continued apace during the twentieth century, even after the 1930s when the country changed economic strategies, adopting an economic model based on the substitution of imports, and began a process of industrialization.

The continued extraction of natural resources, rapid post-war industrialization, and a sustained increase in the country's population have all placed enormous pressure on its natural environment. The effects have been extremely severe: the groundwater quality in most of Mexico's 24 hydrological regions has been qualified as polluted or as strongly polluted; the air quality in the major urban centres is one of the worst among OECD (Organization for Economic Cooperation and Development) countries; over 90 per cent of its forest cover has been depleted; and land use practices have reduced soil fertility in 80 per cent of its territory (OECD 1998).

The severity of environmental damage in Mexico became apparent in the 1980s, and citizens began to mobilize to demand better environmental protection. Mexico thus witnessed the emergence of a Green Movement, which gained strength and national visibility during the 1990s and became an important actor in national politics. The growing visibility of the movement, combined with international commitments undertaken in the early 1990s, especially at the Rio Summit of 1992, brought about the establishment of national environmental institutions and a legal framework intended to manage the natural environment in a more sustainable manner. By the end of the twentieth century, Mexico could boast numerous environmental institutions and a range of environmental protection laws.

Mexico's experience in the management of its natural environment exemplifies many of the challenges faced by countries in the Global South. Over the last two decades, as environmental degradation has become increasingly evident, citizens have started to demand better protection and governments have responded by establishing institutions and enacting numerous pieces of legislation intended to protect the environment. This process has unfolded within a context of economic crises, growing social mobilization, the strengthening of numerous international environmental regimes, and the acceleration of the process of globalization. In this chapter I explore the country's experience with environmental management over the last two decades within the context of globalization. I argue that the manner in which environmental protection has evolved in Mexico has been largely influenced by global forces that must be examined when analyzing the politics of environmental management. The process of globalization has brought critical, and in many cases new, challenges to Mexico's natural environment. But it has also brought numerous opportunities. I argue that any analysis of Mexico's experience with environmental management must take into account the international context in order to provide fuller, richer explanations. In this chapter I therefore look at the impact that global forces have had on the politics of environmental management in Mexico.

The chapter is divided into four main sections. The first section provides a brief overview of Mexico's post-war economic and political development and looks at the various environmental challenges the country has had to face over the last three decades. The second section analyzes the rise of Mexico's green movement. It looks at the factors that led to its rise and at the movement's impact on environmental policy. The third section is an overview of the manner in which successive governments have responded to the institutionalization of environmental management and examines the various institutions and legislative frameworks that have been established. The final section studies the causes of the recent decline of environmentalism in the country and assesses the successes and limits of Mexico's efforts to protect the environment.

MEXICO'S POST-WAR DEVELOPMENT

In contrast to most other Latin American countries, Mexico's post-war development was characterized by notable social and political stability. While many countries in the region underwent successive democratic breakdowns and military coups, the Mexican political system proved remarkably stable. Such stability was largely due to the very distinct political system that emerged after the revolution during the first part of the twentieth century,[2] a system once described by Peruvian novelist Mario Vargas Llosa as the "perfect dictatorship" (1991).

The military leaders that emerged victorious from the revolution, also known as the "revolutionary family," reached a political pact with various sectoral interests and drafted a new constitution in 1917, setting about to fulfil the three ideological goals of the revolution: the adherence of the liberal ideal enshrined in the constitution, the achievement of social justice through improving living standards, and economic development. The revolutionary family responded to the country's desire for the restoration of political order—after the revolutionary's bloody experience during which a tenth of the population died—through the establishment of a centralized political system that vested considerable powers in the president. During the 1930s, the national party, which had been established in 1929, was reorganized and renamed the Institutional Revolutionary Party (PRI). Such reorganization involved the integration of the labour movement, the peasantry, and popular organizations vertically into the party's leadership. By this means, leaders representing the various groups (e.g., labour unions and the peasantry) exchanged party loyalty for material benefits. The new political pact guaranteed that the official party could rely on the various groups to win elections in exchange for the provision of economic and political benefits, all while controlling popular mobilization.

The system gradually evolved into a well-developed mechanism of control and redistribution which depended on a delicate balance among co-optation, selective repression, and limited political freedom. The regime's adherents were rewarded with material benefits, and opponents were either co-opted or ultimately eliminated. While the president assumed significant powers, the introduction of the non-re-election clause into the constitution guaranteed elite turnover by allowing those who had been in power the potential for a lucrative career and the various political groups within the party an opportunity to reach the top. From approximately 1940 until the mid 1980s, the PRI dominated all aspects of national life. Its authoritarian-corporatist structure allowed for the resolution of conflict within the party, thereby maintaining political stability.

The stability of Mexico's political system was also attributable to the very high and sustained levels of economic growth that Import Substitution Industrialization (ISI) generated. The adoption of inward-oriented measures following the Great Depression—in an attempt to protect internal markets from foreign competition, such as high tariff barriers—resulted in the expansion of internal production and consumption. The implementation of ISI policies began a process of state-led industrialization that produced the "Mexican miracle"; from 1940 until the late 1960s, Mexico's economy grew at an average of 6.5 per cent of Gross Domestic Product (GDP) per annum. Such growth gave the regime the resources necessary to distribute within its corporatist structure and along a very well-developed system of distribution based on clientelistic networks, guaranteeing stability.

The so-called Mexican miracle, however, came with serious environmental consequences. The establishment of industries in the urban centres resulted in very high levels of industrial emissions of pollutants. It has been estimated that from 1950 until 1970, industrial emissions in Mexico grew by 50 per cent, and then by an additional 25 per cent between 1970 and 1989 (INE 1994). The result was that the air quality in the country's largest cities deteriorated significantly. This was especially the case in Mexico City, one of the world's "megacities," in which, by the mid 1980s, the officially acceptable standard of air pollution was exceeded nine days out of ten.[3] The Ministry of Health estimated that 25 per cent of the city's residents developed respiratory problems (OECD 1998: 84). Industrial activity and the growth of urban areas, in large part due to migration from rural areas in search for job opportunities in cities, also had an effect on the water quality. Because only a very small percentage of municipal sewage was properly treated, waste-water discharges polluted many of the country's main aquifers, creating a host of problems, such as eutrophication,[4] and serious public health consequences, such as gastrointestinal diseases.

In rural areas, the adoption of imported technology to establish industrial agriculture that emphasized the production of high yielding varieties of crops during the 1950s and 1960s, the so-called "Green Revolution," required the introduction of high levels of fertilizers and pesticides.[5] Since 1960, the use of pesticides has increased by five per cent a year. The increase in the use of these chemicals resulted in high levels of reported cases of poisoning of rural workers and in soil erosion; soil fertility has been reduced in 80 per cent of Mexico's soil, and approximately 15 per cent of agricultural land faces problems with salinization and chemicals. Moreover, the use of nitrate as fertilizer has been one of the main causes of groundwater pollution; most of Mexico's hydrological regions are polluted or highly polluted.

The Green Revolution, along with the growth of more traditional forms of agriculture that relied on slash-and-burn practices, increased the land used for agriculture by 40 per cent from 1970 to 1990. The effects of such rapid agricultural expansion have been quite dramatic on the country's forests and biodiversity: from 1970 to 1990, the country's entire forest cover decreased by 15 per cent; Mexico has the third largest deforestation level in the world; and 790 of its species are threatened while 350 are in serious risk of extinction. Mexico's post-war development was therefore not that miraculous, as it came with serious environmental consequences. The World Bank calculated that, by 1996, overall environmental degradation in Mexico equalled 9.43 per cent of its GDP (Gilbreath 2003: 11).

Neither the perfect dictatorship nor the economic miracle proved permanent. By the beginning of the 1970s, the system began to crack. The economic model began to show signs of exhaustion as firms became highly uncompetitive, subsidies did not follow economic rationale, agricultural production declined, and macroeconomic policy became overly politicized. On the political front, the regime's authoritarian structure came under strain as the PRI lost a great deal of legitimacy after the 1968 Tlatelolco Massacre, in which more than 200 students were killed by government forces. The country experienced a period of economic boom during the late 1980s, primarily due to the discovery of oil reserves.[6] But this boom soon became a bust. When international petroleum prices fell in 1981 and the price of borrowing increased, Mexico was unable to service its foreign debt or secure the foreign exchange necessary to pay for essential imports, thus forcing a steep devaluation of the peso. The Mexican economy consequently crashed, and what remained of the old economic model crumbled. In 1982 alone, GDP shrank by 1.5 per cent, inflation reached 100 per cent, unemployment doubled, and the public deficit soared to 18 per cent of GDP (Lustig 1998). Internationally, the meltdown of the Mexican economy provoked the Debt Crisis of the Global South and, nationally, it unleashed a series of very profound economic and political reforms that fundamentally changed the country. During the 1980s and early 1990s, the administrations of Miguel de la Madrid (1982-88) and Carlos Salinas de Gortari (1988-92) implemented a fairly sweeping series of market reforms based on neo-liberal tenets that culminated with the adoption of the North American Free Trade Agreement (NAFTA) in 1994. By as early as 1987, tariffs had been reduced from levels of 50-100 per cent to 20 per cent, all import licences had been eliminated, and, in the same year, Mexico joined the General Agreement on Tariffs and Trade (GATT). By the early 1990s, the Mexican economy had been liberalized and, with the implementation of NAFTA as well as numerous other bilateral free-trade agreements,[7] fully integrated into the global economy.

Concurrent with this profound economic transformation, Mexico also witnessed important political changes during the 1980s. The crash of 1982 and the ensuing economic deterioration, along with the economic reforms, together had a severe impact on social conditions: unemployment increased, real wages dropped, and standards of living generally declined. This had serious repercussions for the regime. Within the PRI, the economic meltdown strained the party's heterogeneous coalition, as it could no longer afford to provide resources to its various allies—the peasants, organized labour, the federal bureaucracy, and the employees of state-owned enterprises. This provoked tensions and infighting within the party as factions began to fight over national economic policy. The internal struggles culminated in 1987, when Cuauhtémoc Cárdenas Solórzano defected from the PRI and launched an independent presidential bid in 1988—the first real challenge to the PRI since its coming to power. The elections of 1988 were marred by widespread allegations of fraud, further eroding the PRI's legitimacy. By the end of the 1980s, the PRI had effectively lost legitimacy with the Mexican population and the old regime was in crisis. During the 1990s, support for the regime eroded even further, and the PRI gradually lost seats in Congress as well as an increasing number of elections at state and municipal levels. The old regime came to an end with the election in 2000 of Vicente Fox Quesada (2000-06) as president of Mexico.

By the late 1990s, as the process of democratization gained full force, Mexico's economy had been liberalized and fully integrated into the global economy. Whereas during the ISI period the state was seen as the main promoter of economic development, since the introduction of economic reform successive administrations have placed their faith in the ability of market forces to bring about growth and prosperity and have, as a result, been strong supporters of global economic integration. Similar to what has occurred in many other Latin American countries, the effects that economic liberalization and integration into the world economy have had on Mexico have been mixed: while modest economic growth has been restored, poverty and social inequality have not been reduced in any significant way (Middlebrook and Zepeda 2003). Supporters of globalization argue that time is needed for the process to bring about the promised benefits. However, there is no doubt that the integration of Mexico's economy into the global economy has so far brought about significant social and environmental challenges.

One of these challenges has been a considerable increase in regional and geographic inequalities within the country. Disparities between rich and poor have grown significantly between urban areas, which have become a lot more connected with global trade and have benefited from investment, and rural areas, where the rural population has been excluded from economic growth and has not seen any significant improvement in their livelihoods (Garza 2003; Grammont 2003). Socio-economic disparities have also grown between regions as the southern states have lagged behind the central and northern ones, which have become more integrated with the North American economic market. This growing regional differentiation has had several subsequent environmental effects. Because sources of employment have not increased in rural areas, and in many cases they have in fact decreased, rural populations have been forced to migrate. While many rural workers (millions, in fact) decide to seek employment in the US, mostly illegally, many others migrate to urban centres in search for jobs. However, the bulk of the employment created by NAFTA has been primarily for skilled labour (Wood 2006: 16-18). Consequently, the majority of these migrants are unable to find formal employment. They have therefore not been able to raise their living standards and have settled in the numerous shanty towns that surround the main urban centres. This has applied great pressure on local governments to provide basic services. Since the majority of local governments do not have the capacity to manage waste, for example, the consequent pressures on the environment are enormous; it has been estimated that more than half of the waste that is released in these areas is either not managed or thrown into illegal landfill sites and that only 27 per cent of the residual water is treated (OECD 2003: 30, 80). From 1993 until 2001, municipal waste increased by 28 million tons, 84 per cent of which is estimated to come from households (OECD 2003: 99).

Given that poverty has increased in rural areas, pressures on natural resources have also augmented, especially on forests. There are currently 12 million people living in forested areas, the poorest and most marginalized in the country. Because they rely on subsistence agriculture for survival, they embark on slash-and-burn practices which have greatly expanded agricultural lands. This process has largely contributed to continued and very high levels of deforestation; as mentioned, Mexico has the third highest deforestation level in the world. As a result, the number of endangered species has increased. Even though numerous Natural Protected Areas have been established over the last decade, they cover only 10 per cent of the nation's territory.

Mexico's integration with global market forces and the North American economic area, through NAFTA, has brought multiple other environmental effects. The increase in its manufacturing sector (an annual increase of 3.1 per cent from 1995 to 2005),[8] which was been primarily fuelled by demand from the US, has resulted in turn in an increase in overall emissions to very dramatic levels. Indeed, according to the first data released on Mexico's emissions, the country's industrial sector releases *three times* more pollutants (including highly poisonous ones such as arsenic) than

Canada and the US *combined*![9] Moreover, the very significant increase in trade flows within the NAFTA region has increased dramatically the number of transportation trucks, increasing in turn the emissions of pollutants into the air: from 1990 until 2001, the traffic of transportation trucks on Mexican roads increased by 78 per cent (OECD 2003: 34). Emissions generated by these trucks are in turn the main source of urban pollution in the country's largest cities.[10] A direct result has been a steady increase in CO_2 emissions; during the 1990s these emissions increased by 23 per cent (OECD 2003: 54). Free trade has also encouraged the importation of hazardous and toxic waste, especially from the US, and Mexico has become a net importer. Even though it has signed several international conventions stipulating that these materials can only be processed and recycled and must be returned to the country of origin, the level of compliance of these requirements is low, and the mechanisms to track their movement are weak (OECD 2003: 107).[11]

The environmental effects of Mexico's integration into the North American economic region have also affected its rural areas. Based on the logic of comparative advantages, and in an effort to benefit from the new markets opened through the establishment of NAFTA, the administrations of Ernesto Zedillo (1994-2000) and Vicente Fox (2000-06) developed a rural policy that supports the production of certain agricultural products for exports. This has primarily involved the establishment of an important subsidy program for commercial producers (PROCAMPO). The program reaches a third of the total number of agricultural producers in Mexico and, in 2002, the total amount disbursed was US $1.2 billion a year. The program has been responsible for a significant increase in exports of some types of produce as well as for the expansion of agricultural land. For example, Mexico is today the number one producer of avocados in the world and, despite opposition from US producers, Mexican avocados now reach 31 states in the US. However, the increase of this type of production has had several negative environmental effects, as the subsidy program does not take into account environmental concerns (OECD 2003: 207-08). The program emphasizes monocultures, which contribute to soil erosion; commercial agribusinesses tend to overuse water, and in the case of Mexico, its use has steadily increased over the last ten years and it is being overexploited; it is believed that current agricultural practices are not sustainable (OECD 2003: 37). There has also been an increase in the use of some fungicide chemicals, such as methyl bromides, which deplete the ozone layer (OECD 2003: 38).

As we can see, the integration of Mexico's economy with global trade and commerce has brought about numerous environmental challenges. Indeed, in a recent report, the Mexican statistical bureau (*Institutito Nacional de Estadística e Informática*) has calculated that, during the 1999-2004 period, environmental degradation cost the Mexican economy 10.2 per cent of its GDP (INEGI 2006). While the economic and social benefits of free trade can be debated, there is no question that, in the case of Mexico, the results of globalization have had an impact on the country's natural environment.

THE RISE OF MEXICAN ENVIRONMENTALISM

Both economic reform and the crises of the 1980s had important social repercussions. During the latter part of the decade, the country witnessed the emergence of large-scale social mobilization as citizens began to withdraw from the PRI to channel their demands directly to the government. Unlike previous economic crises, the deterioration of socio-economic conditions of the 1980s affected severely various sectors of society, from the urban and rural poor to the middle classes. This unleashed general social discontent that contributed to the emergence of significant social mobilization, as new social groups began to bypass the corporatist structure in an attempt to place demands directly upon the state. This process accelerated with the 1985 earthquake. This powerful earthquake (7.6 on the Richter scale) hit Mexico City on September 13, 1985, and claimed the lives of approximately 20,000 residents.[12] The Mexican government proved highly inadequate in providing relief and assistance to the hundreds of thousands of victims and homeless people. Due to delayed government action and sheer incompetence, residents of Mexico City began to organize swiftly and in large numbers to provide food, water, shelter, and medical supplies to the victims. Such social mobilization witnessed the formation of a significant number of social organizations, a phenomenon that is regarded as a catalyst in the crystallization of large-scale social movements in contemporary Mexico (Foweraker 1990).[13] Vikram Chand has referred to this "strengthening" of Mexican civil society as the country's contemporary "political awakening" (2001).

It is against this backdrop of increased social mobilization during the 1980s that several catalytic events impelled the formation of the Mexican environmental movement. In November 1984, an extremely potent explosion at a gas plant run by the state-owned corporation Mexican Petroleum (PEMEX) in San Juan Ixhuatepec, outside Mexico City, killed over 500 people. The explosion not only caused outrage, but it heightened environmental sensibilities as the environmental damage it caused became evident through widespread television coverage. The 1985 earthquake also contributed to environmental mobilization; along with various kinds of NGOs that surged following the disaster, ENGOs were created as "green brigades" to support people who, as a result of the earthquake, were living in squatter communities around the ruined homes and in the suburbs of Mexico City (González Martínez 1992).

Two months after the earthquake, and in an attempt to coordinate efforts and share information, fourteen civil associations called for the first National Meeting of Ecologists in Mexico City. The meeting was attended by representatives of more than 300 regional groups, civil associations, and scout groups, which discussed a wide variety of themes. At this meeting participants created Mexico's first network of ENGOs, the Pact of Ecologist Groups (PGE). The PGE brought together 50 organizations and established ten working commissions that dealt with issues ranging from pollution in the Valley of Mexico to deforestation. The foundation of the PGE was a rather important development as it was the first time ENGOs had become organized under a formal structure within a larger network. It also played a pivotal role in organizing and amalgamating opposition to the government's nuclear-energy program shortly after the network was formed. The PGE's anti-nuclear campaign contributed further to the strengthening of the environmental movement in Mexico.[14]

Two events at the beginning of the 1990s added further momentum to the movement: the preparatory discussions for the 1992 United Nations Conference on Environment and Development (the Rio Summit); and the joint declaration by Presidents Carlos Salinas and George Bush Sr. (1988-92), in June 1990, that their respective administrations planned to undertake discussions to draft a free-trade agreement. Environmental activism before the Rio Summit was mostly spurred by the fact that Mexican NGOs did not believe they had sufficient discussion space during the official preparatory meetings (Umlas 1996: 97-99). Twelve NGOs and networks therefore called a meeting—entitled the First National Forum of Civil and Social Associations of Environment and Development—seeking to open discussions on alternative development models and to promote interest in participating at a parallel summit, the Global Forum. The meeting resulted in the formation of the Mexican Civil Society Forum for Rio 92 (FOROMEX), which, at one point, incorporated 103 organizations.[15]

The prospects of signing a free-trade agreement with the United States, and eventually with Canada, also strengthened the environmental movement and increased ENGO activity. First, there was a galvanization of public opinion in Mexico with regard to the benefits of free trade. Media coverage of the national debate increased considerably, with some sectors of society, such as PRI supporters and business, strongly supporting the agreement. Opposition to the agreement came mostly from Mexican environmentalists, who were opposed mainly because it ignored sustainable development and environmental protection. There was a concern that free trade would further degrade Mexico's national resources and increase pollution levels (Peña 1993: 124). Environmentalists thus saw the need to organize and collaborate in order to oppose the agreement, and various networks, working groups, and associations were created, such as the Mexican Action Network on Free Trade (RMALC). These networks encouraged the creation and registration of NGOs (Hogenboom 1998; Ávila 1997).

Second, because the Bush-Salinas declaration was unprecedented, there was little information on the effects that free trade would have on the environment. Consequently, national and international collaboration among environmentalists increased due to the necessity to share information.[16] In effect, prior to NAFTA, US NGOs had hardly dealt with Mexico, had little knowledge of Mexico, and had few contacts with Mexican NGOs (Fox 2003: 363). Barbara Hogenboom points out that "within three years (from the summer of 1990 to the summer of 1993) many contacts were established [between US and Mexican NGOs], information shared and experience gained" (1998: 151). All these factors contributed significantly to the strengthening of the environmental movement in Mexico. By the mid-1990s, environmental issues had gained national attention, in great part the result of increased environmental activism.

Mexico's environmental movement also benefited hugely from the increased interaction between Mexican environmentalists and their international counterparts. Indeed, the internationalization of the movement has been a contributor to its institutionalization, which has resulted in the rather marked proliferation of ENGOs.[17] During the NAFTA negotiations, there was an unprecedented increase in funding for Mexican ENGOs from international organizations. Organizations such as the National Audubon Society, the Natural Defence Council, the National Wildlife Federation, the Word Wildlife Fund, and the Action Canada Network made funds available to Mexican ENGOs (Hogenboom 1998).

Moreover, several US conservation organizations, such as the World Wildlife Fund and Conservation International, received substantial financial aid from the US government to promote the park approach to biodiversity conservation in Mexico, and they collaborated with their Mexican counterparts to channel donations from the Global Environment Facility to manage Natural Protected Areas (ANPs) (Fox 2003: 363).[18] Access to financial resources from international organizations greatly contributed to the formation of Mexican ENGOs as well as to the establishment of offices in Mexico of some of these international organizations (Hogenboom 1998; Gallardo 1997; 1999; Torres 1997; Gilbreath 2003). It was during the first half of the 1990s that a rapid increase in the number of ENGOs took place; whereas in 1985 there were no more than 30 registered ENGOs, their number had increased to approximately 500 by 1997, and by the mid-1990s, approximately 5 per cent of Mexicans belonged to an ENGO (Díez 2006: 33).

The increased and sustained interaction that members of the movement have had with their international counterparts has resulted in the institutionalization of Mexican environmentalism. The integration of Mexico into the North American economic market has been central to this phenomenon. The debate over the effects of NAFTA created an opportunity to encourage the interaction between national and international ENGOs, but that interaction has been sustained and has contributed to the strengthening of Mexican ENGOs. Most of the ENGOs in Mexico have received most, and in certain cases all, of their funding from international NGOs, especially from the US, and they have benefited from the transfer of knowledge and expertise. Such transfer has greatly contributed to the professionalization of ENGO members in Mexico and to the institutionalization of the activities, which has helped them tremendously in their activities and interaction with the government. In effect, the institutionalization of the movement accounts for the strengthening of Mexican ENGOs during the 1990s, as evidenced by several successful campaigns to stop a number of projects that would have had important environmental impacts.

In 1995 a coalition of local activists and national ENGOs mobilized to halt the construction of a $478-million project to build a development complex consisting of a golf course, a hotel, and 880 houses in the city of Tepoztlán, south of Mexico

City (Stolle-McAllister 2005: 143-54; Díez 2006: 83-84). In the same year, another coalition formed by ENGOs, local residents, and municipal councillors successfully stopped the establishment of a toxic-waste treatment plant in the northern city of Guadalcázar by the California-based Metalclad Corporation. The project was personally supported by the President and federal environmental authorities, but strong environmental mobilization was successful in convincing the municipal government to deny the issuance of the permit to allow construction (Ugalde Saldaña 2001; Borja Tamayo 2001). A third, and perhaps most notable, successful campaign relates to the cancellation of a project to build the world's largest salt mine in the state of Baja California Sur. On March 2, 2000, President Zedillo made the unexpected announcement that his government had decided to cancel the project to expand the operations of a company in the San Juan Lagoon, which serves as a sanctuary for whales that migrate from Alaska and British Columbia in the winter. The cancellation of the project represented the culmination of a very successful five-year campaign waged by a coalition of Mexican and international ENGOs and was a definite triumph for Mexico's environmental movement.[19] By the late 1990s, then, the environmental movement had not only become highly visible in Mexico, but it could claim several important victories. As we will see in the next section, they were also successful in impressing the government to establish the legislative and institutional tools necessary to better protect the environment.

THE INSTITUTIONALIZATION OF ENVIRONMENTAL MANAGEMENT

The 1990s witnessed Mexico's most serious efforts to address environmental degradation through the establishment of environmental institutions and the enactment of numerous important pieces of legislation. This is not to say that prior to 1990 these did not exist. In effect, Mexico's first environmental law, the Law for Environmental Pollution and Prevention Control, dates back to 1971. In 1972 the Under-Ministry for Environmental Protection was also established. However, this environmental law dealt mostly with air and water pollution and was generally not implemented, since the under-ministry established was given inadequate technical and financial resources. A renewed effort was made during the 1980s to establish more effective mechanisms for environmental protection. In 1982, the Ministry for Urban Development and Ecology was established, and in 1981 the environmental law was reformed. However, although environmental concerns had been elevated to national attention, the economic crisis of the 1980s effectively restricted environmental policy to symbolic and educational activities (Mumme 1988: 187).[20] A new Environmental Protection Law (General Law for Environmental Protection, LGEEPA) was enacted in 1988 and, in 1992, two institutions were established: the National Institute of Ecology (INE), as an environmental research institute; and the Environmental Protection Office (Attorney General for Environmental Protection, PROFEPA), as the country's national environmental inspectorate.[21] Although it was important, observers agree that Salinas's contribution to environmental policy was mostly symbolic and lacked the political will to be implemented (Umlas 1996; Hogenboom 1998; Mumme 1988). Indeed, Salinas introduced environmental reform at a time when public debate over the effects that NAFTA could have on the environment was intense. It has even been suggested that his environmental reform was driven mostly by his interest in conveying a positive international image prior to the signing of the agreement, without the intention to implement policy (Hogenboom 1998).

The arrival of Ernesto Zedillo to the presidency of Mexico in 1994 signalled a new era in the government's attention to environmental problems. Zedillo took a number of steps soon after assuming power that elevated environmental issues to a more important place in the national agenda. Within a few hours of assuming office, he established Mexico's first Ministry of the Environment (Ministry of the Environment, Natural Resources and Fisheries, SEMARNAP). The new ministry assumed the responsibility of overseeing the areas of the environment, natural resources, water management, and fisheries, areas previously overseen by other ministries. Giving the environment full cabinet-level status, also for the first time, allowed for the formulation of a more comprehensive and cohesive environmental policy with increased institutional and financial resources.

Second, upon taking office in 1994, Zedillo appointed an environmentalist, Julia Carabias Lillo, as his new environment minister. The appointment of Carabias was an important development because she was, unlike most cabinet ministers under the PRI who were career politicians, a natural biologist by training as well as a widely respected environmentalist within Mexico's green movement. The environmental community welcomed her appointment because of her proven experience and knowledge, which she had established through a series of important publications that she had authored and co-authored while she was a researcher at Mexico City's National University (UNAM).[22] Third, and certainly most important, Zedillo gave his new environment minister a significant degree of autonomy over the setting of environmental priorities and the formulation of environmental policy.

The Environment Minister used that autonomy to launch an ambitious environmental reform program. In 1995, she established Mexico's first National Consultative Council for Sustainable Development, as well as several regional councils. These councils act as consultative and advisory bodies, and their role is to advance proposal and advice to the minister, as well as to coordinate the sharing of information, ideas, and experiences with social, business, and environmental organizations. They are also in charge of disseminating information, education training, and the supervision of National Protected Areas (ANPs). The National Council is composed of 76 representatives, of which 12 are government officials, with those remaining coming from the academic world, the business community, and NGOs.

The establishment of the council was followed by a series of important reforms of existing environmental laws and the enactment of new ones. In 1996 and 1997, the minister launched a reform of the Environmental Protection Law and the Forestry Law. These reforms were significant as they introduced numerous legal mechanisms intended to reduce environmental degradation. The reform of the Environmental Protection Law, for example, increased the number of activities for which Environmental Impact Assessments are required, decentralized environmental responsibilities to sub-national levels of governments, increased environmental penalties, and enhanced the notion of "environmental responsibility" whereby every party that contaminates is legally liable and must repair the damage. The reform of the Forestry Law was also important. During the early 1990s, and within the overall context of economic liberalization, the forestry sector had been liberalized through a dismantling of the regulatory system established in the 1980s. The 1997 reform of the law introduced a new regulatory framework intended to reduce deforestation levels. Central to this effort was the introduction of the requirement to prove that timber that is transported or stored is accompanied with documentation establishing that it comes from areas in which logging has been allowed, making it a crime not to comply.

In 2000, the Minister also enacted Mexico's first Law on Wildlife. The new legislation established a Council of Wildlife (National Technical Council on Wildlife) with the responsibility to develop and manage the National List of Endangered Species and oversee the various policies implemented for their protection. Moreover, it instituted the National Commission for Protected Areas with the mandate to administer the country's National Protected Areas (ANPs), whose number increased dramatically during Carabias's administration: by 2000, Zedillo had established 30 new ANPs, bringing the total number of hectares from over 10 million to close to 16 million (INE 2000). By the end of Zedillo's administration, environmental management had thus been institutionalized as Mexico took many steps to increase environmental protection. In an assessment of Mexico's environmental management, the OECD concluded that Mexico's environmental and legislative institutions are "complete" and "solid" (OECD 2003).

What accounts for the establishment of these wide-ranging tools for environmental management in Mexico during this time? There is no doubt that it was in part due to the government's decision, under President Zedillo, to give environmental issues more attention during the 1990s, as well as to the appointment of a reformist as environment minister. But two additional contributing factors played important roles. First, the push for environmental reform was in accordance with international obligations Mexico had undertaken at the June 1992 United Nations Conference on the Environment held in Rio. According to Agenda 21 of *The Rio Declaration*, produced at the end of the conference and agreed upon by all signatory countries, including Mexico, governments agreed to take steps to foster the promotion of sustainable development through the establishment of appropriate institutions and the elaboration of effective environmental policies. In effect, according to the former environment minister herself, as well as other senior environmental officials, their environmental reform program was partly in response to the President's willingness to conform to these international obligations.[23]

Second, and as we have seen, the environmental movement had gained strength by the 1990s and had begun to apply pressure on the government to take action. Environmental issues had been brought to the national agenda by the increasing awareness of the severity of environmental degradation in the country. In effect, opinion poll data showed that, by late 1993, 83 per cent of Mexicans expressed "a great deal" or a "fair amount" of personal concern about the environment, and 80 per cent expressed a willingness to pay higher taxes for environmental protection (Brechin and Kempton 1994: 250, 259). But more importantly, ENGOs applied strong pressure on the Environment Minister to undertake environmental reform and were very active participants in the reform process, having in fact had significant input. Indeed, one of the most distinctive characteristics of environmental reform during the Zedillo administration was the significant and rather unprecedented influence ENGOs had on environmental policy-making and the participatory nature of the process (Díez 2006). The 1996 reform of the Environmental Protection Law, for example, was a very open process that took 19 months to complete and in which representatives of more than 108 ENGOs participated. Most of these ENGOs declared, at the end of the reform process, that they were highly satisfied with the final bill, which was unanimously passed through Congress.

Such a participatory process was to a great extent due to the opening created by a reformist environment minister who believed strongly in the inclusion of civil society groups in the formulation of environmental policy. But it was also the result of international factors; environmental reform in Latin America is considered part of what has been termed "second-generation" reforms. These reforms followed the structural adjustment programs of the 1980s and were more inclusive than previous economic reform programs introduced in the region. International organizations, such as the World Bank and the Inter-American Development Bank, began to call for the inclusion of civil society actors in the formulation, implementation, and delivery of government policies and services by the mid-1990s.

There is no doubt that ENGO influence was also largely due to the level of organization and expertise that many of these organizations possessed. The transfer of resources, both technical and financial, and expertise from international actors allowed them to present well-crafted proposals during the reform process. This is especially the case with those ENGOs that pursue issues relating to conservation and bio-diversity. These organizations have the most resources and the most highly trained staff, and all of their funding comes from international donors.

By 2000, Mexico had thus established a host of institutional and legislative tools for the management of the environment. Some additional reforms were undertaken during the administration of Vicente Fox (2000-06). Most notable among these are a reform of the Environmental Protection Law in 2002, through which the Pollution Release and Transfer Registry (RETC) was established. The RETC, similar to the US

Toxic Release Inventory, requires firms to submit information to a registry on the type, location, and quantity of pollutants released on site and transferred off site by industrial facilities. The establishment of the registry had been long overdue, given the international commitments Mexico had made.[24] However, fierce opposition from industry during previous reforms resulted in a limited version whereby industry agreed to release information about pollutants on a voluntary basis. The RETC makes this obligatory, and it is accessible by the public. The Environmental Protection Law was again reformed in 2003 to require that tenders for public works include environmental provisions that ensure the "efficient and responsible use" of water and energy. The Penal Code was also reformed in 2002, increasing penalties for environmental crimes, making it a grave crime to stockpile more than four cubic metres of illegal lumber, and expanding environmental crimes to include "genetic contamination." The Law on Wildlife was also reformed to prohibit the "extraction" of marine mammals from their habitats, and, finally, the Law on Toxic Waste was enacted in 2003. The Fox administration also developed the Program of Payments for Environmental Services. Launched in 2003, the program provides individuals with monetary compensation for conservation activities in areas with high forest density that are categorized as being critical. Individuals that qualify receive 400 pesos a year per hectare. However, the program is small.[25]

Despite these reforms and the establishment of these programs, the importance of environmental issues under Fox dropped precipitously. Fox also appointed a renowned environmentalist as environment minister: Victor Lichtinger Waisman. Lichtinger had been the Secretary General with the North American Commission for Environmental Cooperation (CEC)[26] from 1994 until 2000 and had won the respect of the environmental community in North America during his tenure. However, unlike the previous administration, Fox's environment minister was neither given autonomy to formulate environmental policy nor supported in many of his decisions. According to the Minister, Fox's interest in environmental issues was minimal. He stated that Fox made it clear that economic growth was a priority for his government and that environmental protection was not compatible with this priority: "Fox became upset when I spoke about the environment at cabinet meetings. He argued that economic growth and environmental protection were not compatible. At first I thought that I could educate him, well, 'de-educate' him, but it soon became apparent that it was not the case. He had a personal prejudice against the environment."[27] In 2003, the Environment Minister and his team were discharged from their posts because of his decision to publish the results of studies that indicated high levels of pollution in several Mexican beach resorts. The Minister intended to launch a "clean beaches" program to reduce pollution levels. However, because of the potential negative effects the program would have on tourism, Fox, having been heavily influenced by the private sector (especially Mexican hoteliers), decided to stop the program and fire the minister (Díez 2006: 159–60, 166). This event perhaps best illustrates the scant importance given to environmental issues under Fox. Indeed, with the exception of one interviewee, all other people I interviewed stated that the environment dropped in level of priority from the previous administration, and they believed that there had been a *retroceso* (a step backward) in environmental policy.

CHALLENGES TO ENVIRONMENTAL MANAGEMENT IN MEXICO

A question that naturally arises is this: how effective have these environmental management tools been in addressing environmental degradation in Mexico? It appears that important strides toward greater environmental protection have been made in some areas. In very general terms, air pollution has decreased in Mexico's largest cities during the last ten years. The concentrations of some pollutants, such as CO, CO_2, and lead, have decreased in many urban areas, and there is evidence pointing to a reduction of respiratory afflictions among children (OECD 2003: 26). This has been in part due to various anti-pollution programs launched by local governments, as well as to the elimination of lead in some kinds of gasoline sold by the state-owned oil company, Mexican Petroleum (PEMEX). In terms of water management, many of the objectives set by the National Water Programme, launched in 1995, have been met. Although some rural areas have not yet been covered by the program, over 95 per cent of the water that is administered has been disinfected, causing a reduction in gastrointestinal illnesses. In regard to biodiversity, despite the fact that deforestation has continued apace, the establishment of the numerous ANPs over the last ten years has reduced this tendency dramatically within these parks; it is estimated that deforestation rates within ANPs is ten times lower than in unprotected areas (OECD 2003: 114-15).

Nevertheless, as seen earlier, environmental degradation continues in many other areas. Even with a strong legislative and institutional framework, environmental damage has not been significantly redressed. It appears that there are several factors behind this problem. The first, and certainly most important, relates to the very significant gap that exists between the formulation and implementation of policy. While the environmental legislative framework in Mexico is strong on paper, its implementation is not. A very clear case in point is the implementation of forestry policy in some parts of the country. For example, the regulatory framework introduced with the 1996 reform of the Forestry Law has not been fully implemented in some of the country's rainforests, such as the Selva Lacandona of Chiapas (Díez 2006: 127-49). The lack of implementation seems to be widespread, as it is found in many other parts of the country and across policy areas (Gilbreath 2003: 32; CESPEDES and CEMDA 2002: 15; PROFEPA 2000; Molnar et al. 2001; OECD 1998; 2003).

The causes of weak implementation are multiple. Similar to many other Latin American countries, there exists in Mexico a strong dualism between legal formalism and compliance with the law. It has been argued that weak compliance is mostly the result of an Iberian legacy that permeates Latin American political culture, where there are those who make the law, interpret it, and are above it, and those who have no option but to comply with it (Anglade 1986; O'Donnell 1999). Although there is undoubtedly an element of truth to this, it is also true that weak compliance is primarily the result of inadequate institutional and financial resources. In Mexico, total government expenditure on environmental protection is very low compared with OECD countries (OECD 2003) as well as several other Latin American countries (Barcena and de Miguel 2002). There have been variations on the funds assigned to the Environment Ministry since its creation. But most environmental officials admit that the funds are not enough to guarantee the implementation of the legal system. Limited funds result in weak institutional capacity and a very limited number of environmental inspectors: in 2006 there were a total of 700 inspectors in charge of enforcing all environmental policy areas (water, oceans, forests, and industrial facilities) in the entire country. In the forestry sector, for example, a study concluded that the limited capacity of the ministry's regional officers "makes the new regulatory framework largely unenforceable" (Molnar et al. 2001: 672). The same study affirms that there exists "weak institutional capacity in both budget and manpower to apply and enforce norms and standards, and has resulted in a continued incentive for illegal extraction [of timber]" (675). The diminished importance of environmental issues since 2000 has also been a factor, since there has not been the political will to increase funding. As mentioned before, the administration of Vicente Fox did not make the environment a priority, and this has translated into a decrease in funding.

An additional factor has been the weakening of Mexico's environmental movement since 2000. Whereas the movement experienced a remarkable strengthening during the 1990s that produced several policy triumphs, the movement lost strength during the Fox administration. In part, such weakening was caused by the recruitment by Fox's first environment minister of several of its leaders into senior positions within the Environment Ministry in 2000, thus creating a leadership vacuum. One interviewee even referred to the "beheading" of the movement. Lichtinger now believes that this shortfall was a mistake, as it reduced the vibrancy of ENGOs.[28] A weaker movement has not been as successful in exerting pressure on the government to strengthen environmental compliance.

Finally, the strong influence the private sector exerts on the policy-making process frequently results in the blocking of policy and, at times, in its thwarting. A clear example is the blocking of the launch of the Clean Beaches program, referred to above, which ultimately resulted in the environment minister's dismissal. But examples abound. According to the Mexican political system, federal laws require the enactment of secondary pieces of legislation, called regulations (*reglamentos*). Whereas in other presidential systems, such the US, regulations are generally drafted by independent institutions established by Congress, such as the Interstate Commerce Commission, in Mexico they are drafted by the President's Office. Because of the inordinate influence the private sector has on the policy-making process in Mexico, lobbying results in the blockage of the drafting of these regulations, making it impossible to implement the law. For example, while Mexico's Law on Toxic Waste was enacted in 2003, its Regulation had not been published by the end of Fox's administration in December 2006, even though a provision in the law states that it be published within six months of the promulgation of the law (which would have been on April 8, 2004). The Senate's Standing Committee on the Environment has passed several resolutions demanding the Executive to elaborate the regulation—but to no avail.

CONCLUSION

Mexico's post-war development was characterized by political stability and economic prosperity. The political system that was established after the revolution proved very successful at mitigating conflict within the regime through a complex balance among co-optation, repression, and limited political freedom. Through its authoritarian-corporatist structures, the regime managed to control opposition while rewarding its supporters. The material goods distributed to the regime's adherents were in turn greatly facilitated by the high and sustained levels of economic growth that the ISI model generated, the so-called Mexican miracle. However, as this chapter has shown, Mexico's development did not prove all that miraculous as, by the late 1980s, the environmental costs of such a development model had become more than evident; the degradation of the country's natural environment had reached unprecedented levels.

The 1980s brought about profound changes to Mexico. The exhaustion of the ISI model and the onset of the Debt Crisis in 1982 resulted in the worst economic downturn since the Great Depression, a downturn that affected various sectors of society as socio-economic conditions deteriorated rapidly. Given the country's precarious economic situation, and encouraged by international financial institutions, Mexico's elite embarked upon a process of economic reform, grounded on neo-liberal tenets. The deterioration of socio-economic conditions created by both the Debt Crisis and the process of economic reform unleashed general social discontent that contributed to the emergence of significant social mobilization, as new social groups began to organize outside the regime to place demands directly on the state. It is within this context of profound economic change that Mexico witnessed the emergence of its green movement, which began to lobby the government for better environmental protection.

Economic reform in Mexico saw the dismantling of the ISI model and a gradual but continued liberalization of the economy, and the implementation of a market-friendly model that culminated in the adoption of NAFTA in 1994. By the late 1990s, Mexico had been fully integrated into the global economy. As I have attempted to show in this chapter, while the economic and social results of the adoption of this model can be debated, the integration of Mexico's economy into the global economy, Mexico's economic globalization, has brought about numerous environmental challenges. The increased extraction of natural resources and the expansion of agricultural land that is fostered by a model based on exports, significant rural migration to the cities, and increased transportation have all applied increased pressure on Mexico's natural environment.

Nevertheless, Mexico's globalization has also brought about opportunities. The opening of its economy allowed for increased interaction between Mexican environmentalists and their international counterparts, especially from the US. Largely due to the debate that surrounded the adoption of NAFTA, Mexican environmentalists began to interact with foreign environmental organizations in the early 1990s. Such interaction has been sustained and has resulted in the transfer of resources and expertise that contributed to the institutionalization, or "NGOization," of the Mexican environmental movement. The number of ENGOs in Mexico therefore grew dramatically during the 1990s. But the strength of these organizations was not simply numerical. Because of their increased professionalization and institutionalization, Mexican environmentalists were able to mobilize and apply stronger pressure on the government to increase environmental protection. As a result, they were able to reverse a number of projects that would have affected the environment, and they have been very successful in influencing the government's environmental policy. Indirectly, then, the opening up of Mexico's economy has created opportunities for Mexican environmentalists to demand better environmental management.

The green movement's increased visibility and influence, the opening of the political system, and the various international environmental commitments that Mexico has undertaken have resulted in the introduction of a series of environmental legal statutes and several environmental institutions. By the turn of the twenty-first century, Mexico had a strong institutional and legislative framework of environmental management. As we have seen, this new framework has been responsible for the reversal of environmental degradation in some areas and a reduction in the total emission of certain pollutants. Nevertheless, as I have shown in this chapter, the implementation of Mexico's strong environmental laws has been very poor. Largely because of the limited funds given to environmental institutions, the state's capacity to implement the environmental legal framework is very weak. Consequently, the extent to which environmental degradation will be significantly reduced is likely to be very limited and will be largely dependent upon the political will forthcoming from the government in turn. If the Fox administration is any indication, it appears that any significant improvement on this front is unlikely to transpire any time soon.

Notes

1 This chapter is based on data obtained through more that 30 interviews with government officials, members of NGOs, and representatives of business chambers in the summers of 2004 and 2005.

2 The Mexican revolution lasted from 1910 until 1917, but social and political turmoil continued until 1929, when national order was restored.

3 A study conducted by the National Autonomous University of Mexico City found that from 1976 until 1980, air pollution had increased by 90 per cent in Mexico City (Gilbreath 2003: 19).

4 Eutrophication refers to the excessive release of chemicals, such as nitrogen or phosphorous, which, as nutrients, enrich ecosystems and promote excessive plant growth. When the phenomenon occurs in aquatic environments, it fosters the growth of certain vegetation, such as algae, which in turn choke other forms of vegetation, thereby disrupting the ecosystem.

5 After Brazil, Mexico was the largest importer of fertilizers and pesticides in Latin America during this time (Maltby 1980).

6 The economy grew at an average of 8.5 per cent of GDP per year from 1976 until 1981.

7 By 2005, Mexico was the country with the most trade agreements in the world, including those signed with Chile, Bolivia, Colombia, Venezuela, Costa Rica, Nicaragua, the European Union, and Japan.

8 Data obtained from the World Bank http://www.worldbank.org/. Accessed Sept. 15, 2006.

9 North American Commission for Environmental Protection http://www.cec.org/news/details/index.cfm?varlan=english&ID=2728. Accessed Nov. 10, 2006.

10 A study conducted between 1995 and 1998 found that emissions from transportation trucks accounted for the following percentages of total emissions: 90 per cent in the border city of Ciudad Juárez; 85 per cent in the Greater Mexico City Area; 75 per cent in Guadalajara; 70 per cent in Mexicali and the border city of Tijuana; and 50 per cent in Monterrey (OECD 2003: 53).

11 These treaties refer primarily to the Basel Convention, which Mexico ratified in 1991, and the Convention on the Traffic of Hazardous Waste adopted by the Organization for Economic Cooperation and Development (OECD) in 2002. Mexico joined the OECD in 1994.

12 Mexican officials placed the number at 10,000, but this figure is widely believed to be an underestimate.

13 According to Daniel Moreno, by 1995 the number of NGOs stood at 1,300 (according to the Directory from the Ministry of the Interior); 2,600 (according to the Mexican Philanthropic Centre, Centro Mexicano de la Filantropía); 3,500 (according to the Ministry of Social Development, SEDESOL); 5,000 (according to the network Front for Mutual Support, Frente de Apoyo Mutuo or FAM); and 10,000 (according to the NGO Grupo San Ángel) (1995: 2).

14 The PGE's opposition began to brew in September of 1986, eight months after the nuclear disaster of Chernobyl, when President de la Madrid announced that the project to build a nuclear power plant in Laguna Verde, in the Gulf state of Veracruz, was to go ahead. Although he eventually decided to build the plant in 1988, the anti-nuclear campaign was successful in bringing together a large number of environmental groups, in raising awareness further and, ultimately, in opposing and defying the government through actions such as highway blockades. In effect, the Laguna-Verde mobilization is considered to be one of the watershed events of Mexico's environmental movement (Berlin 1988; Payá Porres 1994; García-Gorena 1999).

15 The network disappeared fifteen months after its establishment, but it was very influential in the dissemination and exchange of information among ENGOs, and in the encouragement of debate outside formal fora, thus invigorating social participation in environmental issues (Umlas 1996).

16 For example, in October of 1990, 35 Mexican and 30 Canadian ENGOs held a two-day meeting in Mexico City to exchange information. Then, in January of 1991, a tri-national forum on agricultural, environmental, and labour issues was held on Capitol Hill attracting more than 400 ENGOs from the three countries. The purpose of the meeting was to stimulate debate and share information on social and environmental issues.

17 As Sonia Álvarez has argued, the level of institutionalization of a movement can be seen through its "NGOization" (1998).

18 The donations have been administered by a newly created organization, the Mexican National Conservation Fund.

19 For a more in-depth analysis of these mobilizations, see Díez 2006: 83-89.

20 The World Bank (1992) reported that, between 1986 and 1989, Mexico's operating budget declined by 69 per cent in real terms.

21 Both of these institutions were created as "semi-autonomous" institutions within the Ministry of Social Development (SEDESOL).

22 Carabias was a full-time professor and researcher at the Laboratory of Ecology at UNAM's Programme for the Management of Natural Resources (PAIR).

23 Interviews with Martín Díaz y Díaz, Director General of Legal Unit, Ministry of the Environment, Mexico City, Sept. 18, 2000; Antonio Azuela de la Cueva, Attorney General for Environmental Protection, Mexico City, Oct. 24, 2000; Julia Carabias Lillo, Minister of the Environment (1994-2000), Mexico City, July 18, 2001.

24 As a signatory of the North American Agreement on Environmental Cooperation, Mexico agreed to resolution 97-04, which encourages the three countries toward the adoption of comparable registries. Also, as a member of the OECD, Mexico agreed to harmonize its registry with all member states. The establishment of the registry is also in line with commitments made to Agenda 21, whose principle 10 stipulates that states should facilitate and encourage the dissemination of information.

25 In 2003, 198 million pesos were distributed to 272 individuals; in 2004, 288 million pesos were distributed to 350 individuals.

26 The CEC was created by the NAFTA's environmental side agreement (the North American Agreement of Environmental Cooperation) in 1994. It is headquartered in Montreal.

27 Interviews with Victor Lichtinger, Environment Minister (2000-03), Mexico City, June 24, 2005, and Adolfo Aguilar Zinser, National Security Adviser (2000-02), Mexico City, June 29, 2004.

28 Interview with Victor Lichtinger, Environment Minister (2000-03), Mexico City, June 24, 2005. Miriam Alife Cohen argues that the weakness of the Mexican environmental movement is due to the inability of its members to unite forces (2005: 223). However, as we saw earlier, ENGOs presented a rather strong, united front in the various campaigns they waged during the 1990s, questioning such analysis.

References

Alife Cohen, Miriam. 2005. *Democracia y desafío medioambiental en México: Riesgos, retos y opciones en la nueva era de la globalizacion*. Barcelona: Ediciones Pomares.

Álvarez, Sonia. 1998. "NGOization of Latin American Feminisms." In Sonia Álvarez, Evelina Dagnino, and Arturo Escobar (eds.), *Culture of Politics, Politics of Culture*. Boulder, CO: Westview. 306-24.

Anglade, Christian. 1986. "Sources of Legitimacy in Latin America: The Mechanisms of Consensus in Exclusionary Societies." Essex Papers in Politics and Government, University of Essex, Colchester, UK, 38: 1-48.

Ávila, Patricia. 1997. "Política ambiental y organizaciones no gubernamentales en México." In José Luis Méndez (ed.), *Organizaciones Civiles y Políticas Públicas en México*. México: Miguel Ángel Porrúa.

Barcena, Alicia, and Carlos de Miguel. 2002. *El financiamiento para el desarrollo sostenible en América Latina y el Caribe*. Santiago: CEPAL y PNUD.

Berlin, Thomas. 1988. *Laguna Verde: ¿El próximo desastre?* México: Planeta.

Borja Tamayo, Arturo. 2001. "The New Federalism in Mexico and Foreign Economic Policy: An Alternative Two-Level Game Analysis of the Metalclad Case." *Latin American Politics and Society* 43(4): 67-90.

Brechin, Steven R., and Willett Kempton. 1994. "Global Environmentalism: A Challenge to the Postmaterialism Thesis?" *Social Science Quarterly* 75(2): 243-69.

CESPEDES and CEMDA. 2002. *Deforestación en México: Causas Económicas e Incidencias del Comercio Internacional*. México: Centro de Estudios para el Desarrollo Sustentable y Centro Mexicano de Derecho Ambiental.

Chand, Vikram K. 2001. *Mexico's Political Awakening*. Notre Dame, IN: University of Notre Dame Press.

Díez, Jordi. 2006. *Political Change and Environmental Policy-making in Mexico*. New York and London: Routledge.

Foweraker, Joe. 1990. "Popular Movements and Political Change in Mexico." In Joe Foweraker and Ann L. Craig (eds.), *Popular Movements and Political Change in Mexico*. Boulder, CO, and London: Lynne Rienner. 3-20.

Fox, Jonathan. 2003. "Lessons from Mexico-U.S. Civil Society Coalitions." In David Brooks and Jonathan Fox (eds.), *Cross-Border Dialogues: Mexico-U.S. Social Movement Networking*. La Jolla: University of California, San Diego, Center for U.S.-Mexican Studies. 341-418.

Gallardo, Sofía C. 1997. "Participación ciudadana frente a los riesgos ambientales de la globalización y el TLCAN." In María Teresa Gutiérrez and Daniel Hiernaux Nicolas (eds.), *En busca de nuevos vínculos: Globalización y reestructuración territorial en la Américas*. México: Universidad Autónoma Metropolitana, Unidad Xochimilco e Instituto.

—. 1999. *Acción Colectiva y Diplomacia Social: Movimientos Ambientalistas frente al Tratado de Libre Comercio de América del Norte*. PhD Thesis, El Colegio de México.

García-Gorena, Velma. 1999. *Mothers and the Mexican Anti-Nuclear Power Movement*. Tucson: University of Arizona Press.

Garza, Gustavo. 2003. "The Dialectics of Urban and Regional Disparities in Mexico." In Kevin J. Middlebrook and Eduardo Zepeda (eds.), *Confronting Development: Assessing Mexico's Economic and Policy Challenges*. Stanford, CA: Stanford University Press. 487-521.

Gilbreath, Jan. 2003. *Environment and Development in Mexico: Recommendations for Reconciliation*. Washington, DC: Center for Strategic and International Studies.

González Martínez, Alfonso. 1992. "Socio-Ecological Struggles in Mexico: The Prospects." *International Journal of Sociology and Social Policy* 12(4-7): 113-28.

Grammont, Hubert C. 2003. "The Agricultural Sector and Rural Development in Mexico: Consequences of Economic Liberalization." In Kevin J. Middlebrook and Eduardo Zepeda (eds.), *Confronting Development: Assessing Mexico's Economic and Policy Challenges*. Stanford, CA: Stanford University Press. 350-81.

Hogenboom, Barbara. 1998. *Mexico and the* NAFTA *Environment Debate: The Transnational Politics of Economic Integration*. Amsterdam: International Books.

INE. 1994. *Informe de la situación general en material de equilibrio ecológico y protección al ambiente*. México: Instituto Nacional de Ecología.

—. 2000. *Áreas Naturales protegidas de México con decretos federales*. México: Instituto Nacional de Ecología.

INEGI. 1994. *Estadísticas Históricas de México*. México, DF: Instituto Nacional de Estadística, Geografía e Informática.

—. 2006. *Sistema de cuentas económicas y ecológicas de México 1994-2004*. Mexico City: Instituto Nacional de Estadística e Informática.

Lustig, Nora. 1998. *Mexico: The Remaking of an Economy*. Washington, DC: Brookings Institution.

Maltby, C. 1980. *Report on the Use of Pesticides in Latin America*. Washington, DC: Inter-American Development Bank.

Middlebrook, Kevin J., and Eduardo Zepeda. 2003. "On the Political Economy of Mexican Development Policy." In Kevin J. Middlebrook and Eduardo Zepeda (eds.), *Confronting Development: Assessing Mexico's Economic and Policy Challenges*. Stanford, CA: Stanford University Press. 3-52.

Molnar, Augusta, Thomas A. White, Luis F. Constantino, and Adolfo Brizzi. 2001. "Forestry and Land Management." In Marcelo M. Giugale, Olivier Lafourcade and Vinh H. Nguyen (eds.), *Mexico, A Comprehensive Development Agenda for the New Era*. Washington, DC: World Bank. 671-81.

Moreno, Daniel. 1995. "ONG: Los nuevos protagonistas." *Reforma* (México): 2.

Mumme, Stephen P. 1988. "Political Development and Environmental Policy in Mexico." *Latin American Research Review* 23(1): 7-34.

O'Donnell, Guillermo. 1999. "Polyarchies and the (Un)rule of Law in Latin America: A Partial Conclusion." In Juan E. Méndez, Guillermo O'Donnell and Paulo Sérgio Pinheiro (eds.), *The (Un)Rule of Law and the Underprivileged in Latin America*. Notre Dame, IN: University of Notre Dame Press: 303-38.

OECD (Organization for Economic Cooperation and Development). 1998. *Environmental Performance Reviews: Mexico*. Paris: Organization for Economic Cooperation and Development.

—. 2003. *Evaluación del Desempeño Ambiental en México*. Organización para la Cooperación y el Desarrollo Económico.

Payá Porres, Víctor Alejandro. 1994. *Laguna Verde: la violencia de la modernización. Actores y movimiento social*. México: Instituto Mora.

Peña, Devon. 1993. "Letter from Mexico: Mexico's Struggle against NAFTA." In *Capitalism, Nature Socialism* 4(4): 123-28.

PROFEPA (PROCURADURÍA FEDERAL DE PROTECCIÓN AL AMBIENTE). 2000. *Informe 1995–2000*. México: Procuraduría Federal de Protección al Ambiente.

Stolle-McAllister, John. 2005. *Mexican Social Movements and the Transition to Democracy*. Jefferson, NC, and London: McFarland.

Torres, Blanca. 1997. "Transnational Environmental NGOs: Linkages and Impact on Policy." In Gordon J. MacDonald, Daniel L. Nielson and Marc A. Stern (eds.), *Latin America's Environmental Policy in International Perspective*. Boulder, CO: Westview.

Ugalde Saldaña, Vicente. 2001. "Las relaciones intergubernamentales en el problema de los residuos peligrosos: el caso de Guadalcázar." *Estudios Demográficos y Urbanos* 44(2): 77-105.

Umlas, Elizabeth. 1996. *Environmental Non-Government Networks: The Mexican Experience in Theory and Practice*. Ph.D. Dissertation, Yale University.

Vargas Llosa, Mario. 1991. "Mexico: The Perfect Dictatorship." *New Perspectives Quarterly* 8(1): 23-25.

Wood, Duncan. 2006. "Sharing the Wealth? Economic Development, Competing Visions, and the Future of NAFTA." In Jordi Díez (ed.), *Canadian and Mexican Security in the New North America: Challenges and Prospects*. School of Policy Studies, Queen's University. Kingston and Montreal: McGill-Queen's University Press: 11-24.

World Bank. 1992 (March 9). *Staff Appraisal Report: Mexico Environmental Project*, Report no. 1005-ME. Washington, DC. 7.

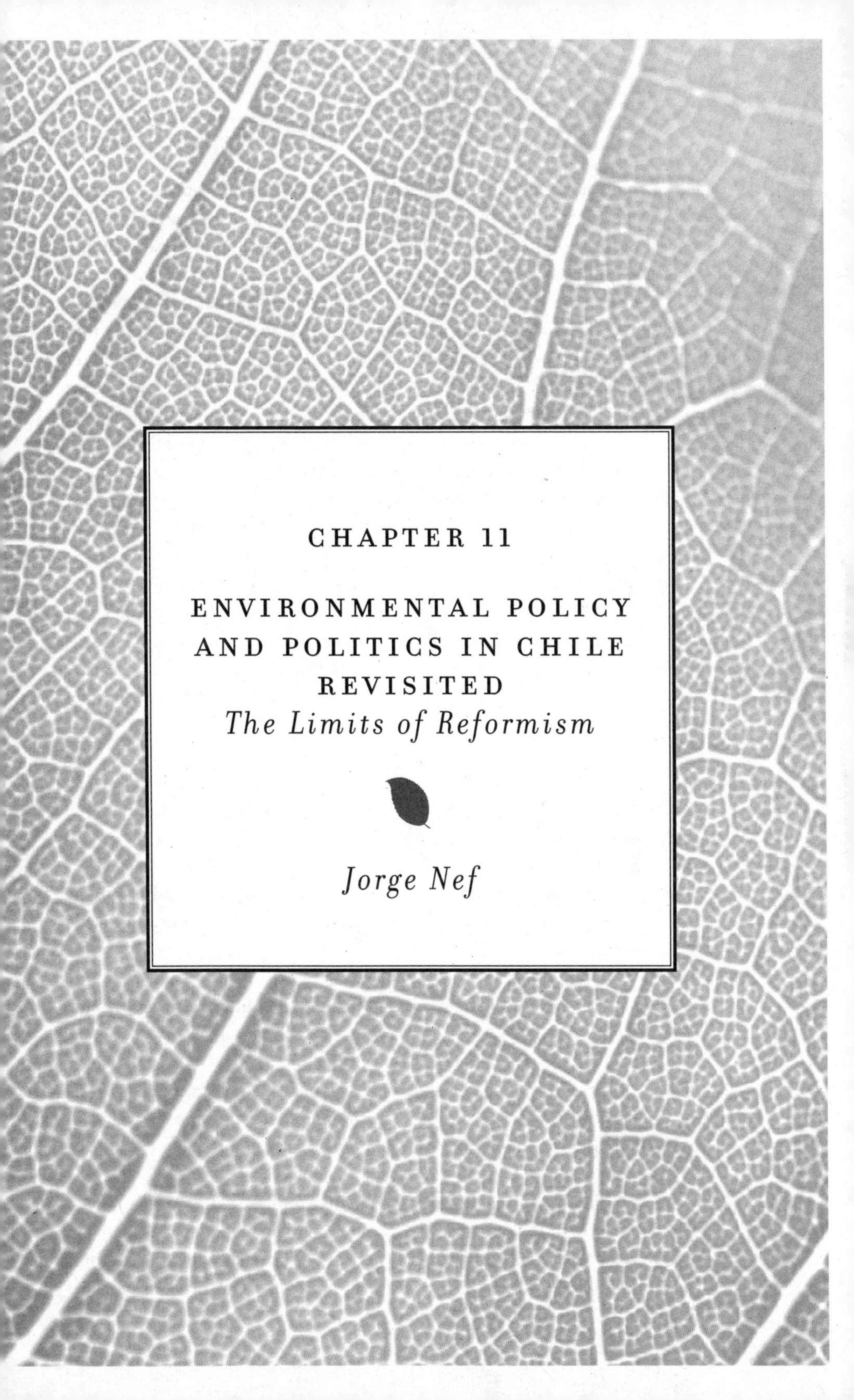

CHAPTER 11

ENVIRONMENTAL POLICY AND POLITICS IN CHILE REVISITED

The Limits of Reformism

Jorge Nef

INTRODUCTION

The study of environmental policy and administration offers a convenient analytical window through which to study the interface between politics, social forces, and the broader issue of human security (Nef 1999a). In this chapter I examine this relationship in one Latin American country that has become emblematic in the application of neo-liberal recipes for Third World development. It builds upon a previous study conducted a decade ago (Nef 1995). This juxtaposition allows for a significant insight into two different diachronic contexts and the continuities and changes taking place between these two "moments." One such context is that of the early transition to democracy under the presidency of Patricio Aylwin (1990-94) and the beginning of the administration of Eduardo Frei (1994-2000). The other historical context is the subsequent and slow institutional consolidation that took place between the Frei government and that of Ricardo Lagos (2000-06). The perspective taken to study both historical moments is broadly systemic, as well as interpretative (Laswell 1950; Easton 1957).[1] It uses a political economy framework (Staniland 1985), attempting to link environmental policies with the broader development model, the constellation of socio-economic forces affecting these policies, the institutional constraints and mechanisms to policy formulation and implementation, and the overall outcomes of the policy process (Stallings 1978: 5-10, 21-23, 51-52; Nef 1991: 11). The present study, as well as its 1995 predecessor, explores a number of operational questions:

1. What are the basic environmental problems that have characterized Chilean development strategies, and how have these problems been articulated by diverse social actors?

2. What are the main alliances and coalitions supporting and opposing various environmental perspectives, and how are they organized in the Chilean political arena?

3. What are the main ideological and programmatic underpinnings of such policy options?

4. What are the principal channels of influence and mechanisms through which these environmental perspectives and policy options became both public policy objectives and official policy instruments?

5. What has been the explicit and implicit environmental policy menu? That menu refers to the definition of problems, the articulated alternatives, and the courses of action offered as options within a given power interplay among political brokers, either as part of one or many broader political

agendas. More specifically, how is the environmental agenda connected with the larger developmental model, and in particular with the issue of sustainable development?

6. What are the consequences of these policies upon the economic development strategies, the different sociopolitical actors, and ultimately upon the existing environmental problems and development process?

7. Finally, as a conclusion, what are the nomothetic "lessons" one could generalize from an idiographic case study of Chile for the larger Latin American region and beyond? More specifically, how does the country's environmental predicament of the 1990s compare with that of the first decade of the twenty-first century: what has changed; what remains the same?

To analyze Chilean public policy and administration, it is essential to come to grips with a number of distinctive and persistent traits that permeate the country's administrative culture and institutions (Cleaves 1974; Heady 1984). The first trait is the high degree of bureaucratic salience and continuity conditioned by a highly legalistic and formalistic administrative culture (Nef 1989). Despite a pervasive neo-liberal orientation grafted on the state apparatus since the counter-revolution of 1973, the Chilean public sector is still by far more decisive in national life than that of any other country in Latin America, with the sole exception of socialist Cuba. The second trait is that in one generation Chile has undergone a dramatic reshaping of public policies and priorities. These various phases have included Keynesian import-substitution industrialization (1938-70), a short-lived experiment with democratic socialism (1970-73), the forceful establishment of unbridled neo-liberalism under military rule (1973-89), and finally redemocratization with economic orthodoxy under the aegis of the Washington Consensus. The third trait is the obvious fact that, despite much hoopla, Chile is still an underdeveloped (albeit outwardly modernized) and dependent country, subject to the "boom-and-bust" oscillations of commodity export cycles.

The above observations are particularly relevant to understanding the place of "environment" as a relatively new policy area in the Chilean agenda. Chile's environmental problems are not new. What is new is the codification of a number of otherwise symptomatic dysfunctions (land degradation, air pollution, sewage and chemical contamination of streams and water resources, fauna and flora depletion, and the consequent health hazards and decline of quality of life associated with these) under the term environmental *problems*. The very presence of environment in the political agenda is contingent upon the administrative, political, and legal semantics that label and structure the problem as a goal to be tackled by government intervention. Brokerage and agency are essential here, for labelling and constructive (as well as deconstructive) problematization also shape public views and attitudes. Until recently, there was no significant unifying concept in the country to deal with environmental issues. The latter were traditionally handled by diverse levels and jurisdictions. They were attended by various territorial (national or municipal) and functional legislation—health and sanitation, agriculture, forestry, public works, industry, transportation, fisheries, and the like.

HISTORICAL OVERVIEW

Since the middle of the nineteenth century, Chilean élites have shared a positivist view of progress: the environment-as-resources view has been dominant. It has been shared by right-wingers, centrists, centre-leftists (radicals, Christian democrats, moderate socialists), and Marxist-Leninists, irrespective of their specific utopias, and, of course, the political methods to reach those utopias. As Stutzin has noted, "To a certain extent all these models share the basic fallacies of modern industrialism and the limitations of short-term, non-ecological thinking. Moreover, each model is supported by, and supports in turn, political and economic interests the aims of which are rarely in accordance with the interests of the natural environment" (1975: 38).

Import substitution policies, born out of the necessity to arrest the 1930s economic recession, often failed to examine environmental impacts. In the 1960s, it became apparent that industries were having severe effects on air quality in large industrial centres. This was especially the case in Santiago, the capital, but it was also a problem in the area near the southern city of Concepción, the port of Talcahuano, and the Bay of San Vicente where smelters were located. But the damming of rivers for electrification, the wholesale exploitation of forests, strip-mining, and other practices hardly raised an eyebrow. Chile's present environmental crisis was shaped during these years. It combined, as in most underdeveloped nations, the worst of both worlds: "modern" contamination of high chemical toxicity with "traditional" bacterial pollution; underutilized resources side by side with indiscriminate resource depletion.

With increasing industrialization, urbanization, and the expansion of government and business activities into industry, environmental damage also increased. This lack of environmental concern and absence of regulation cut across political barriers. It was shared by such dissimilar political coalitions as those of the populist and corporatist administration of Carlos Ibáñez (1952-58), the rightist government of Jorge Alessandri (1958-64), the reformism of Eduardo Frei's Christian Democrats (1964-70), the brief Socialist experiment of Salvador Allende (1970-73), and the military dictatorship of General Augusto Pinochet (1973-90). Until then, outside the pioneering work undertaken at the UN Economic Commission for Latin America and the Caribbean (ECLAC), environment had not evolved into a distinctively political issue, nor had environmental problems been put in a larger developmental context (Sunkel 1980; Gligo 1980, 1981; Bifani 1982).

The need to present a clear democratic alternative to the policies of the Pinochet dictatorship politicized environmental concerns. Paradoxically, however, it was the military regime that elevated the rhetoric of environmentalism to unprecedented heights. It included in Article 19 (No. 24) and especially Article 20 of its own tailor-made authoritarian constitution of 1980 a provision guaranteeing citizens the "right to live in an environment free of contamination" (Nef 1983: 314-16). Obviously, this statement had little to do with the real public policies pursued. That section of the constitution meant first and foremost an entrenchment of the absolute right to private property. It constituted, however, good public relations and asserted that Chile was "in tune with the times."

By the mid-1980s, what had been pyrotechnics and political opportunism had taken the form of an important component of the political agenda, especially when opposition to the dictatorship seized the environmental cause. The timing was right, since the country had already plunged into a real environmental crisis that defied rhetoric. The discourse had finally caught up with reality. Moreover, the growing opposition to the regime had become both more unified and more internationalized. On the one hand, environmental organizations were receiving significant external support in financing, technical cooperation, and ideas. On the other, the very phenomenon of exile had meant that numerous Chilean intellectuals had been exposed to global issues to an extent hard to imagine scarcely a decade before, and had become "cosmopolitanized."

When the prohibition for dissidents to come back was gradually lifted, many foreign-educated Chilean academics and professionals returned home. Among them there were *ambientalistas*—physicians, engineers, economists, geographers, anthropologists, sociologists, and physical scientists—with a new professional agenda and a more holistic outlook. They had been influenced by "environmental studies" in Europe and North America, by the UN efforts,[2] especially by the aforementioned studies by ECLAC (Gligo 1980), and by the Brundtland Report (WCED 1987). Departing from the established Chilean tradition, democratic politics came to converge with a "clean" environmental posture, if not in substance, then at least as a politically correct stand. In fact, it could be argued that "environment"—and in particular the relationship between environment and development—had become a paradigm or surrogate for ECLAC-style structuralism (Sheahan 1987: 103-10) with respect to the onslaught of the neo-liberal monetarism pursued by the authoritarian regime.[3] However, this new-found environmental stance did not have the centrality and symbolic resonance in Chile that it had, for instance, in Costa Rica, where environmentalism has become a discursive device for national identity. Chile has not been marketed as a "green" country, but as one of economic efficiency and institutional probity.

CHILE'S ENVIRONMENTAL CRISIS

By the late 1980s, it was increasingly obvious that most of Chile was affected by a host of interrelated environmental dysfunctions, many of which have continued, or become worse, two decades later. Even a sketchy list provides a dramatic picture of negative synergies at work:

1. There has been a dangerous and ever-present air pollution problem in metropolitan areas. The culprits are particle and carbon monoxide, as well as dust resulting from soil erosion and the destruction of vegetation. In Santiago, poor air quality has approached and at times surpassed the levels of Mexico City (Prickett 1992). In other cities, such as Rancagua or Concepción, south of Santiago, the source of the problem of atmospheric contamination has been industrial smokestacks rather than vehicle emissions as in the metropolitan region.

2. There has also been a water pollution problem: untreated sewage combined with industrial waste is dumped into rivers, whose waters are used for irrigation. Both the Mapocho River, which divides Santiago, and the Zanjón de la Aguada, built in colonial times, for decades essentially became open sewers where fecal matter, industrial refuse, and garbage ended up. These waters were mainly used in Santiago's green belt, producing a persistent situation of bacterial contamination. Given the leakage of bacteria into sources, drinking water has been a problem, though in recent years standards have improved significantly in major urban areas. In the northern city of Antofagasta and in numerous mining centres in the north, mining by-products (such as arsenic) have been finding their way into the limited water resources, as with some CODELCO (Copper Corporation) and ENAMI operations. In Concepción, Arauco, Bio-Bio, and Valdivia in the south, wood-processing and paper mills contaminated lakes and streams in varying but significant degrees (Tomic and Trumper 1990).

3. The aforementioned poisonous liquid mixtures finally make their way to the coast, creating an ocean pollution problem affecting flora, fauna, resort areas, and the marine food chain. The most contaminated area has been the central zone (the San Antonio-Algarrobo resort area and the long coastal strip from Valparaiso to Quintero). But in the Eighth Region, some 500 kilometres to the south (Concepción, Talcahuano, and Lota, with its steel mills, refineries, wood-processing plants, and canneries), the level of contamination has reached disastrous proportions. In recent years, massive intensive fish farming in the southern provinces has produced severe water contamination and affected native natural fish stocks. On top

of this, oil extraction, transportation, and refining have compounded the problem both in the southernmost Magellanic region and in the south-central and central ports, where petroleum is transported, stored, and refined.

4. Industrial and vehicular air pollution has brought about an acid-rain problem, concentrated especially in Santiago but not exclusively limited to it. The main effects have been on trees, but overall water acidity levels have also increased.

5. The rapid expansion of the forestry industry in recent years has created a severe and growing deforestation problem for native species. This currently affects mostly the central-south regions, and in particular the *araucaria*, *alerce*, *raulí*, and other native-tree forests further south. It should be remembered, however, that the near-northern semi-arid regions of Atacama, Coquimbo, and Aconcagua are currently suffering drought and desertification resulting mainly from deforestation started in colonial times. Attempts at reforestation with competitive species, such as eucalyptus, have compounded the problem.

6. The increasing demand for electricity for both industrial and domestic use has brought about the development of hydroelectric mega-projects such as the Rapel, Laja, and Pangue dams. The damming of rivers has affected flora, fauna, and the lives of peasant cultivators, many of whom are native peoples, whose livelihoods have been severely and arbitrarily disrupted.[4]

7. In recent years, with export-led policies concentrating upon non-traditional agricultural exports—mostly fruits—the use of toxic substances in agriculture, particularly pesticides, fertilizers, and other chemicals to preserve or improve the appearance, texture, and longevity of the products, has increased. Many of these substances—such as Alar or DDT, or its local equivalent, TANAX—have been widely employed in the industry. Some of these have been banned in other countries. Recently there have been reports of fertilizers reaching the water table in the fruit-growing regions. A related problem is posed here by the expansion of cold-storage facilities and the wide use of fluorocarbon-based refrigeration.

8. A more recent and threatening environmental problem only became known in 1989: the thinning, or "hole," in the protective ozone layer. Its effects, so far concentrated in the far southern regions of Aysén and Magallanes and the Argentinean and Chilean Patagonia, have been

visible in sheep and cattle (causing blindness) and in wild animals. Direct effects on humans have not so far been evaluated, but they are potentially alarming. Unlike the other seven categories, the ozone problem is one whose causes are mostly "external": the massive use of fluorocarbons by industrial countries.

There are even more potential environmental dysfunctions looming on the horizon. Two of these future environmental threats are worth mentioning. One is the eventual processing of rare metals, such as uranium, molybdenum, and especially lithium, of which Chile has a large supply. The processing of these rare metals releases highly toxic substances. The other threat is perhaps less defined but also alarming. Since the 1973 coup, the military has had a free hand in controlling what is today a significant "military-industrial complex." Firms such as the Army's Factory of Material and Equipment (FAMAE), the National Explosives Enterprise (ENAEX), and the privately owned Cardoen Enterprises are reportedly engaged in the development of lethal substances. In 1991, there were reports of an industrial accident with loss of life in the town of Melipilla, in which nerve gas was allegedly released from an army factory.

The manner in which some of these multiple dysfunctions interact is quite complex. The mere physical chain of events (e.g., the complex interconnection between soil erosion, dust, transportation, air pollution, and acid rain, and the impact of the latter upon flora and water acidity) is, however, far less complicated than the economic, social, and political linkages underpinning Chile's environmental crisis. These are issues of culture and perception, of "economic rationality" and interests, and, above all, of power—or, more to the point, lack of power. While Chile's environmental problems have a long historical lineage, the present crisis is distinctively connected with the acceleration of the contradictions of the present method of development. As Osvaldo Sunkel observed more than two decades ago, there is a definite relationship between present development styles and environmental problems (1980).

THE POLITICS OF THE ENVIRONMENT

Most orthodox economists and their associates in the international business community describe Chile as an "economic miracle" (Kandell 1991) to be imitated by other countries (Graham 1992; Santiso 2006). Yet there has been a persistent flip-side to this view. As a 1992 report by the US-based Natural Resources Defense Council stated,

> Two decades of rapid economic growth have had significant impacts on the quality of Chile's environment and natural resources. Santiago, for example, suffers from the second worst air quality in the Western hemisphere, following Mexico City. Copper smelting operations in

> northern Chile have produced dangerous blood levels of arsenic among local populations. Commercial logging of southern Chile's native forests is rapidly eliminating some of the world's oldest temperate rain forest ecosystems. Over-fishing has left all of Chile's important commercial fisheries in critical condition ... (Prickett 1992: 2)

Many of these observations are rather old, yet the substance of the assertion still holds true. What clearly emerges from an examination of Chile's environmental predicament is that the country's environmental crisis has been related to its development strategy—one characterized by a policy of drift on environmental matters. Contrary to a commonly held belief, the pro-business military regime did not create the necessary conditions for environmental catastrophe. What it did was to accelerate the environmental dysfunctions present in the existing mode of development by imposing an extreme view of "cowboy" capitalism and by eradicating the possibilities for a meaningful democratic debate on issues of human concern. The democratically elected government of the *Concertación*,[5] under President Patricio Aylwin, "declared a commitment to improve the environmental situation [but] his government's overriding concern for continued economic growth ... delayed progress in developing an effective environmental regulatory program" (Prickett 1992: 2). Fifteen years later, and under much more favourable political conditions, this commitment has failed to materialize in the realm of policy and administration. A brief examination of the policy process helps in explaining such immobility.

Actors, Alliances, and Agendas

The constellation of actors participating in the environmental policy process is relatively large and complex. It is also characterized by a high level of transnationalization; that is, the boundary between the "domestic" and the "international-regional" dimensions of environmental issues and alliances is unclear. For the sake of simplicity, we can identify three major blocs of agents in the process, all possessing transnational linkages. The first is what could be called, for lack of a better term, the "anti-regulation" or "pro-development" bloc, formed by public, parastatals, and large business concerns and their associates at home and abroad. Second, there is the equally transnationalized "proregulatory" or "environmentalist" alliance of numerous research centres, non-governmental organizations (NGOs), and grassroots groups and constituencies. Third, there is what can be generally referred to as "the government." It is constituted by a number of policy-making institutions and public bodies where conflicting interests eventually become regulatory norms and where rule-making, rule-implementing, rule-adjudication (arbitration), and rule-enforcement take place. The military establishment, even under the present democratic administration, and the attempts at re-civilianization during the Lagos administration have to be considered quite outside "the government." They are extremely powerful and autonomous, maintaining close personal, institutional, and ideological ties with Chile's aristocracy and business classes and with the dominant regional superpower. The Pinochet regime forged an enduring "reactionary coalition" (North 1978: 79) between the officer corps, local capitalists, and US military and business constituencies. After the decline of Pinochet's grip on power, the military effectively did not play a direct role in the environmental policy debate. However, in specific matters such as the controversial acquisition of Pumalin Park by American environmentalist Douglas Tompkins (Escobar 1996; Rubio 2003; Rother 2005), the officers have expressed concern about matters of "sovereignty."

THE "ENVIRONMENTALISTS"

Over ten years ago, Glenn Prickett identified a vast web of environment-related organizations as participants in the policy process. Although his "mapping" was originally intended to describe the main official and NGO participants with regard to hydroelectric projects, this "environmental network" offers an insight into the structure and modus operandi of environmental groups in Chile from the 1990s to the present. Although from afar environmentalist groups and their allies may appear to be a highly cohesive, even powerful, coalition, they in fact suffer many of the same problems present in NGOs worldwide. The Chilean setting adds some specific complexities. They experience the chronic underfinancing that results from operating under limited resources; their supports come mostly from outside sources. In addition, while there is a degree of cooperation and solidarity among the groups, there are also profound cleavages stemming from ideology, as well as professional or personal distrust, compounded by fierce competition for scarce resources.

The largest, oldest, best-organized, and most influential environmental interest group in Chile has been the National Committee for the Defence of Fauna and Flora (CODEFF). It was formed in 1968 during the Christian Democratic administration of Eduardo Frei (Sr.). Because of its "technical" conservationist (and seemingly "apolitical") orientation, the organization has been able to function under dramatically different development projects (Stutzin 1975). Today it constitutes a sort of nodal organization, able to network with more functionally or geographically specific groups. Among its successes, CODEFF coordinated an effective research and advocacy campaign to win legal protection for native forests of *araucaria* and *alerce* trees. They also took an interest in the legal challenges to mining companies using desert waters. In the 1990s and in subsequent years, the committee participated in the study of the impact of hydro projects on the Bio-Bio River in southern Chile. While toward the end of the century it devoted less direct attention to the issue of hydro power, it chose to endorse the action of a regional group, the Bio-Bio Action Group (GABB), to which it provided assistance.

Another environmental group is the Centre for Environmental Research and Planning (CIPMA), established in 1979 as a think-tank to study the relationships between economic development and the environment. It is a policy-oriented group that seeks economically efficient and socially equitable solutions to environmental problems. It acts as host to annual symposia on environmental policy and publishes a journal, *Ambiente y Desarrollo* (Environment and Development). In addition, there has also been a National Network for Ecological Action (RENACE) affiliated with the Institute of Political Ecology (IEP), an advocacy group that publishes a bulletin, *Ecoprensa*, and also holds seminars and workshops. RENACE has been a liaison and information exchange among ecologically oriented NGOs, independent research centres, and the public. Other environmental organizations have included the Centre for Ecological Extension, geared to public education, and the Ecological Centre Canelo de Nos, an experimental program that implements alternative pilot projects for both urban and rural grassroots organizations.

These organizations interact with external linkage groups[6] such as the Natural Resources Defense Council (NRDC) in the United States, Probe International in Canada, and Greenpeace International. The latter has a Chilean chapter, while representatives of the Council and of Probe have visited Chile on a regular basis. International NGOs have been instrumental in developing an environmental partnership (Pickett 1992: 10-11). The latter coordinates and assists in lobbying foreign governments, international organizations (such as the World Bank or the Inter-American Development Bank), foreign business, and transnational corporations.

But the environmental coalition in Chile involves much more than environmental interest groups. It comprises a myriad of other organizations that either have environmental components in their agendas or share some of the environmentalists' interests and concerns. These concerns range from specific technical or social issues to broader considerations such as "sustainable development." Central in the environmental front have been native organizations, like the Council for all the Lands and numerous other groups. While environment and First Nations issues intersect, indigenous environmental demands have been linked with questions of land ownership, identity, and especially political autonomy (Mariman 1994). The environment has been a point of convergence with ecological and human rights issues articulated by non-native progressive groups, namely the question of sustainable development and livelihoods. Those organizations whose mandates encompass "sustainable development" include the Corporation for Economic Research in Latin America (CIEPLAN) and the Centre for Alternative Research (CEPAUR). The largest and most influential of these think-tanks is CIEPLAN, associated with the Christian Democratic Party.[7] It was created in 1976 as a public research institute, with significant external support, including a large and sustained contribution from the International Development Research Center (IDRC) in Ottawa. The corporation's mandate is to identify and promote equitable, democratic, and sustainable economic development policies. Smaller but equally important, CEPAUR is a policy research group whose efforts have also been concentrated upon the formulation of sustainable development strategies. Manfred Max-Neef, the winner of the "alternative Nobel Prize," was active within the organization. CEPAUR has enjoyed high international visibility and has many contacts at home and abroad, especially with European research centres and organizations such as the Dag Hammarskjöld Foundation.

Several independent technology-research centres are also part of the environmental coalition. These include the Centre for the Study of Alternative Technologies for Latin America (CETAL), the Energy Research Program (PRIEN), and a US group with offices in Santiago—the International Institute for Energy Conservation (IIEC). The CETAL was established in the main port city of Valparaíso as a technology-research group promoting the development of alternative energy technologies, including energy-efficient and nonconventional renewable sources. It has provided support to other groups on the issue of energy and environment. PRIEN was created by a group of engineers from the University of Chile conducting research on energy production, use, and conservation. It has done studies at the local (municipal) level in Santiago and in other cities and provided technical advice to authorities and electric utilities in energy-saving measures, especially in integrating end-use energy-efficient programs into investment planning.

An effective functional linkage has evolved between environmental interest groups and the Chilean Human Rights Commission (CCHDH) created in 1976. The commission, which has a Programme for Indigenous Peoples, has acted as a catalyst for environmental rights, especially with regard to native claims. In this respect, CCHDH has been a bridge-builder between legal, social, and technical concerns on environment. Likewise, through the active membership of José Aylwin, then a young human-rights attorney—and also a member of the Bio-Bio Action Group and a son of President Aylwin—international linkages were established between native Pehuenche representatives and Cree Indian leaders in Canada struggling against the James Bay II hydroelectric development.

This list included here is not exhaustive, since there are numerous regional and more issue-specific groups. Most significant among these is the aforementioned GABB, which carried the environmental banner in the proposed ENDESA hydroelectric Pangue project in the Bio-Bio River. Other groups such as Vida Magallánica (Magellanic Life) in the southernmost region and the Institute for Andean Development in the Andean ranges of the far north are associations dealing with sustainable local development. The environmental network, despite the "elite," upper-middle-class nature of most environmental interest groups, has been able to connect successfully with civic action groups (e.g., Citizens' Action Assembly), NGOs, and grassroots movements. In the specific cases of forests, land, and rivers, an important though not always easy collaboration has developed between environmentalists, human rights activists, and native peoples' associations. These associations represent aboriginal ethnic minorities within Chile: *Aymara* in the North Andean region, *Rapa-Nui* in Easter Island, *Mapuche* (*Pehuenche* and *Huilliche*) in the south, and *Kawashkar* in the far south. This new role for groups that had been historically relegated, discriminated against, and often brutalized, entails a drastic change for a society marred by deep-seated racism. Native and pro-native organizations include the Aymara Studies Workshop (TEA), *Pacha Aru*, the Regional Institute for Aymara Development and Promotion (IRPA), the Coordinating Body of *Mapuche* Institutions (CIM), the *Pehuen Mapu* Guild Association, the *Mapuche* Council for All the Lands, the umbrella organization National Native Commission (1989), and the government-appointed National Commission for Native Peoples (CEPI, established in 1990) to mention just a few.[8]

To these national and international groups, especially NGOs, that make up the environmentalist field, one should add the highly influential presence of the aforementioned American entrepreneur-ecologist Douglas Tompkins. A wealthy global businessman with a perspective of "deep ecology," since the 1990s Tompkins has always been a central figure in the Chilean environmental debate. He has been demonized and canonized at the same time, yet he has been able to reach decision-makers and create alliances—and some enemies too. Tompkins has acquired vast tracts of pristine or near-pristine lands in southern Chile and has offered to transform these into national parks, away from the reach of developers and the forestry industry.

THE "DEVELOPERS"

This category is perhaps even more complex than the environmental network. On the surface it comprises a multiplicity of firms of all sizes, both public and private, associational interest groups to which these firms are linked, foreign corporations and financial organizations with participation in the firms' operations, international organizations, and foreign governments' development agencies. It is by and large a loose and seemingly disjointed alliance constructed specifically to overcome opposition to projects and support deregulation. But this appearance is deceiving. Despite its looseness, the developers' bloc still possesses a hard-core hegemonic centre and is capable of mustering an enormous amount of power and resources, particularly money. The developers' bloc is extremely influential at the ideological level, too. Leaving aside the outer rims of the alliance constituted by associated small firms and support groups, the developers' main constituencies are linked among themselves by interrelated activities, common interests, interlocking directorships, associational bonds, access to official channels, control of the media, common ideology, and often party affiliation. In other words, this is the Chilean "establishment" (Ratcliff and Zeitlin 1988). Its interests are articulated in the private-sector-wide National Confederation of Production and Commerce (CNPC or COPROCO, established in 1932).[9] COPROCO, in turn, encompasses seven interest groups covering all of Chile's major productive and financial activities, all of them with strong transnational links and representation in numerous boards and commissions in the parastatal sector. These are the National Agricultural Society

(SNA, 1838), the National Mining Society (SONAMI, 1883), the Society for Industrial Development (SOFOFA, 1883), the Chilean Chamber of Construction (CCHC, 1951), the National Chamber of Commerce (CNC, 1857), the Association of Banks and Financial Institutions (ABIF, 1980), and the Santiago Stock Exchange (BCS) (Nef 1999b; Tomic and Trumper 1990).

The mining-sector lobby comprises all major extractive corporations. The largest is the autonomous parastatal CODELCO, established in 1965. CODELCO is one of Chile's "sacred cows," dealing with the country's single most important export commodity: copper. In the 1990s, it had over 30,000 employees, upwards of $1.5 billion in sales, net assets of over $500 million, and high profitability. CODELCO is charged with the exploration, exploitation, and international marketing of the largest share of Chile's copper market.

CODELCO and another parastatal ENAMI—charged with processing and smelting—are currently contributing in large scale to air and water contamination in the northern First and Second Regions as well as in the Ventana and Rancagua areas west and south of Santiago. Other mining parastatals include the National Coal Company (ENACAR), and the Aysén Mining Enterprise. Though they are functionally dependent on either the Ministry of Mining or the Ministry of Economics, there is very little public control and real accountability of these entities. In practice, each agency is directly responsible to a board of directors, many of whose members were appointed to tenured positions by the military regime. Privately owned medium- and small-sized mines throughout the territory equally share a rather spotty environmental record. Most privately owned mining companies are represented by SONAMI, one of the seven major business interest groups mentioned earlier.

In the industry and manufacturing sector, a constellation of semi-private and private corporations dominates the scene. Many of the largest are privatized detachments from the Production Development Corporation (CORFO) hub. Others are individually owned, varying in size from large-scale to small operations. The largest industrial concern is the Pacific Steel conglomerate, today an investment company centred in the production of steel at its Huachipato mills. As Kandell has noted, Conglomerado de Acero del Pacífico (CAP) "owns 19 firms—enterprises ranging from mines to real estate to forestry—with annual sales of $600 million. Exports account for half of these revenues. CAP's steel division claims to be among the most efficient producers in the world" (1991: 18). It is also a main contributor to the contamination of air, sea, and water in the Concepción region. In a similar vein, there is ENAP, the National Petroleum Company. With extraction operations in the southernmost island of Tierra del Fuego, off the Strait of Magellan, and with several refineries along the Chilean coast, ENAP'S activities are far from "clean." The same could be said about the expanding canneries, fishmeal processors, paper mills, and petrochemical plants that dot the South. Vertically integrated corporations, such as the Paper and Cardboard Manufacturing Company (CMPC), and highly diversified conglomerates, such as the Luskic Group, function by and large on the logic of production and profits with little or no concern for environmental or social impact. But the rather cavalier use of toxic substances does not stop here.

The extensive deregulation that characterized the post-1973 economic model has allowed both a lack of strict controls for formal industries and a total absence of controls for the mushrooming "informal sector." This sector—often a form of disguised unemployment—exists and even sub-contracts at the margins of the weakest labour, sanitary, taxation, and environmental regulations. It contributes in no small manner to air pollution and to the dumping of toxic waste into sewers and rivers. The deregulatory ideology enshrined in the legislation enhances the anti-environmental effects of industrial and manufacturing operations.

The principal association representing the manufacturing sector is SOFOFA, though there are numerous industry-specific lobbies. Forestry and forestry-related industries have been one of the fastest growing sectors in the Chilean export economy. Three mega projects—Celpac, Arauco II, and Santa Fe—were started in the early 1990s to increase the volume of exports. Foreign investments in the wood-producing area of Bio-Bio far exceeded $450 million a decade ago, most of it coming from the Ibn Mafuz, Fletcher Challenge, and Shell consortia (Kandell 1991: 18). The forest product industry has dramatically skyrocketed in recent years, as have tree depletion and conflicts with native settlers of the forests. The public sector, through CORFO's holding operations, has retained a small presence in the field, such as the *Panguipulli* Forestry and Wood Complex. This and other interests have been increasingly sold to private foreign investors by means of "debt-for-equity" swap mechanisms. In fact, a high degree of integration between domestic and foreign capital has taken place. The interests of forestry and related industries are represented by a powerful lobby and research entity, the Wood Corporation (CORMA), an affiliate of SOFOFA. It reportedly had direct access not only to government agencies such as the National Forestry and Renewable Resources Corporation (COMAF), but also to the Executive and members of the Legislature.

The expansion of electrification is a fundamental component of Chile's energy-intensive development strategy. For the 1979-89 period, the rate of increase in energy consumption (at 5.2 per cent per year) outpaced the average rate of economic growth (4.2 per cent annually), a trend that has continued at a faster pace since then. According to Prickett, "[t]he main users are transportation, industry and mining (copper, petrochemicals, paper and cellulose, steel, sugar, iron, cement and nitrates). Growth in electricity demand has been even greater, averaging more than 6 per cent a year" (1992: 3). ENDESA, established in 1943 as a COFO-(Committee on Forestry) affiliated parastatal, is a gigantic power monopoly. Its annual output is 7,550 GWh (equivalent to 45 per cent of the country's total needs), which it sells to the equally monopolistic retailing utility, CHILECTRA.

In 1990, ENDESA was completely privatized in a rather shady deal. At present, its stock—divided among some 50,000 stockholders—is heavily concentrated in a handful of private financial conglomerates related to insurance, medical services, and the like, known as *financieras*. The company owns several subsidiaries such as INGENDESA (engineering services), the *Pehuenche* Corporation of Maule and Curanilahue, and the *Pangue* Corporation in Bio-Bio. Zavaleta has pointed out that "ENDESA is the holding company for about 1600 MW of hydro capacity and 320 MW of thermal plant and remains the leading power producer in the country" (1991: 22). The company's mega-projects involve a comprehensive scheme for the expansion and interconnection of hydro power by damming Chile's major river systems. This development doubled the country's electrical capacity before the end of the twentieth century. Needless to say, since the ecological and social impact of these projects can be devastating, opposition to these plans has been growing. However, ENDESA's real power means more than electricity. It is a financial conglomerate connected with dominant economic and political groups and it has an enormous capacity to lobby, both inside and outside the government and through its linkages with other sectors of the business community. It enjoys a reputation of being "invulnerable" and "free from public oversight" (Prickett 1992: 1).

As with the environmentalist bloc, the structure of the "pro-development," anti-regulation alliance includes a myriad of other groups. One of the important sectors involved in both the problem and the controversy surrounding environmental effects is the transport sector. Also, as cities and private automobiles have expanded, particle contamination has dramatically increased. Santiago possesses a gigantic, unregulated, complex, and inefficient, privately owned public-transportation system. Private automobiles have compounded the problem. The city has probably twice as many buses and cars as are warranted for a city of its size—most of them obsolete, poorly maintained, and sources of high levels of emissions. Since the measures to control the buses and industry are seen as anathema, it has been extremely difficult to bring bus and truck owners into line (*La Nación* 1992a). *Transportistas* are a relatively powerful group, with a demonstrated ability to paralyze the country. At any rate, the bus owners' power to blackmail depends to a considerable extent on their ability to elicit support from their employees. All in all, bus owners have been far less cohesive and clear in their corporate positions than have truck owners. They have also been more politically and economically divided.

Other groups that can be included in the "developer" and anti-regulation side are large export-promotion associations, such as the *Fundación Chile* (Chile Foundation) and fruit producers and packers. Rural entrepreneurs use large amounts of fertilizer and pesticides and rely heavily upon fluorocarbon-based refrigeration. The land is being gradually poisoned and bird populations are reportedly declining. Farmers have become part of a highly "competitive world market in fresh produce; they thus are pressured to use ever-increasing amounts of pesticides as well as fertilizers" (Goodwin 1992: 67). Likewise, large-scale producers at the edge of technological innovation have largely depended upon hybridization and GMOs (genetically modified organisms), thus threatening the genetic variability of many local crops. For as long as there is a profitable market for exports, these environmental dangers are often dismissed by large and small agricultural producers alike. Commercial agriculture is represented by the established National Society of Agriculture (SNA), while some smaller and medium-sized producers articulate their interests in the Confederation of Guild Associations and Federation of Chilean Farmers, both of right-wing persuasion.

Last, but not least, three sectors are of crucial importance in the developer side. The first consists of the large Chilean financial institutions (banks and finance companies) that provide the internal capital for large-scale ventures and their corporate mouthpiece, the association of Banks and Financial Institutions (ABIF). Second, and in close collaboration with the local finance capital, there are the international private businesses, including foreign banks, joint-venture capitalists, and transnational corporations, with operations in Chile—names such as Citicorp, Aetna, Dole, The Bond Group, or the aforementioned Shell, Fletcher Challenge, and Ibn Mafuz. These groups are particularly influential with, and have direct access to, foreign governments and international organizations that participate in the financing of development projects. These include entities such as the World Bank, the Inter-American Development Bank (IDB), and the International Finance Corporation (IFC). The previously mentioned international sector acts as an external linkage group, or lynchpin for the "developer" alliance, giving its constituencies a distinctive transnational dimension.

THE GOVERNMENT

Government involvement in the environmental field is generally weak and fragmented, despite the fact that, as mentioned earlier, Chile as a whole still possesses a robust public sector. There is not a single agency dealing specifically with environmental issues, let alone one with the capacity to regulate and act. In the intricate body of administrative law and municipal law that prescribes the functioning of Chile's bureaucracy, there are very few and indirect references to the environment other than the vague constitutional provisions of Articles 19 and 20 of the charter.

A Comisión Nacional del Medio Ambiente (CONAMA) was established in the early 1990s and began preparing a comprehensive package of environmental legislation to be submitted to Congress. When it was submitted in 1997, it elicited a strong rejection by environmental and indigenous groups,[10] in the midst of allegations of pro-business conflicts of interests. Meanwhile, all governments have also faced a formidable legislative backlog in other critical areas of legislation. For instance, the Administration and Congress signalled, as early as 1990, their intent to rewrite the country's forestry law and water code, establish a national energy policy, and enact a new law to protect the rights of indigenous peoples (Prickett 1992: 2-3). Other pieces of legislation, such as the Fisheries Law, which were passed through Congress with drastic concessions to the anti-regulation forces of the right, contain some important environmental dispositions, such as the notion of common property management. Most of these initiatives, however, encountered the relentless opposition of the forces of the *ancien régime*—with some unexpected help by the new establishment—and became distorted or stalemated. But also the environmental policies had only weak support within the state, especially among technocrats and economists. The pro-growth policies of the governments of Presidents Frei and Lagos won the day, especially after the latter implemented a most successful anti-poverty program. Therefore, despite public opinion, the once promising environmental agenda has remained largely still a promise.

In retrospect, the "transitional" and timid Aylwin government appears by comparison to be a pro-environmentalist administration. The alternatives for the constitutional government then were few. Alternatively, it chose to pursue a painfully slow process of negotiating and watering down more "radical" proposals. What was a painful recognition of constraint for the Aylwin team became a comfortable situation for his two predecessors, including the highly rated, progressive, and effective administration of Ricardo Lagos. This gave the forces of deregulation and the status quo an overwhelming advantage from the start. Even when a negotiation began, the sheer power of the "developers," not to mention their legal ability to stalemate and veto initiatives or to sabotage the process, gave (and still gives) them once again the upper hand.

In this "toothless" state, the regulatory structure remains extremely weak and fragmented. A decade later, there is no such thing as an "environmental management system." Rather, there is a series of discrete bodies dealing with very partial aspects of environmental regulation. The most important of these is the aforementioned CONAMA, created in June 1990. The commission is attached to the minuscule Ministry of the National Patrimony, the least powerful and most problematic of all the cabinet portfolios. It still lacks personnel and funding, being staffed with a handful of qualified experts concentrated in its Santiago headquarters. In addition, CONAMA has small "field offices" spread among the country's thirteen territorial administrative zones. It is presided over by the minister of national patrimony and managed by a commission secretary. Its mission is still basically to frame the environmental legislation for environmental regulation and management.

So far, the most operational of Chile's environmental offices is the Santiago Special Commission on Decontamination (CDS). The office has a territorial jurisdiction confined to the metropolitan region, the largest urban concentration in Chile. Its prime directive is to deal with the dramatic and persistent smog problem. CDS has operated as both an inter-ministerial and an inter-governmental (municipal and central) entity and has a special working relationship with the Ministry of Transport. It possesses emergency powers to pursue an active plan to curtail carbon monoxide and particle emissions by establishing restrictions in coal and wood burning, vehicle circulation, pollution control, and dust-prevention measures. Private vehicular traffic and emissions are regularly checked, with some success, though it has been more difficult to deal with smokestacks and diesel contamination.

Early in 1992, the commission claimed a significant victory when one of Santiago's major bus associations started to fit new converters to its bus fleet, and electric-powered trolley buses went into operation (*La Nación* 1992a). Contamination levels were the lowest in many years (*El Mercurio* 1992a). However, this early optimism proved to be premature; by mid-year, pollution levels surpassed the 300 pre-emergency mark and approached the 500 emergency level. Acute respiratory infections continued to be the highest cause of death among children under five in poor urban areas (*La Nación* 1992b). In addition to these new agencies, there are a handful of established semi-regulatory bodies with a potential influence in implementing environmental programs. One of the most important is the National Energy Commission (CNE). The agency is preparing "a national energy strategy, with assistance from the US Department of Energy (DOE). As part of this strategy, CNE hopes to propose administrative and policy reforms, that would transform CNE into a true regulatory agency" (Prickett 1992: 8). Another is the Directorate-General of Waters (DGA) of the Ministry of Public Works. The DGA is given the authority to grant water rights to any prospective engineering work. The agency must verify that the use of the water will not affect the security of third parties or produce contamination. Yet in practice its power is very limited. The Superintendence of Electrical and Fuel Services, the Economic Regulatory Agency, and the National Irrigation Commission, all from the Ministry of Economics, have limited power in granting "concessions" to enterprises. Other agencies include more than 300 autonomous municipalities and the National Sewers and Sanitation Services of the Ministry of Public Works.

Two other mechanisms of an institutional nature deserve to be mentioned. One is, of course, the legislature. In theory all general normative matters fall within the jurisdiction of Parliament, and it is there that the aforementioned environmental legislation is being debated. However, in a presidential system, most of the initiative in legislation rests with the executive, and the political agenda is established by the president and the cabinet, with the approval of the ruling party coalition. As party configurations change, so does the nature of the cabinet and the government agenda. Most pressure-group activity is directed to executive agencies, with parliamentarians playing second fiddle.

The other is the judiciary. This is a relatively novel aspect in Chile, which is unlike the countries under British common law in that courts have played a minor role, except for strictly dealing with conflicts between contending parties or in the field of criminal law. In 1985, a farmer in the northern Valley of Azapa sued the Ministry of Public Works for having authorized the use of a natural water reservoir by the large parastatal SONAMI.[11] The challenge was supported by CODEFF, and a clever legal strategy was adopted based upon the precepts of the 1981 constitution. The Court of Appeals of Arica decided for the plaintiff. What is important in this case is that the challenge took place during the dictatorship years, under a supreme court that usually rubber-stamped all decisions from the executive and involved a major parastatal enterprise.[12] The confirmation was based upon constitutional grounds.[13]

While this case is quite remarkable in legal terms, its overall impact on Chilean practices regarding environmental protection can be overestimated. So far, there is no such thing as "environmental judicial review": the operation of the Chilean judicial system is not like that of its Anglo-American counterpart. The courts have limited jurisdiction, applicable only to the specific case before them. Moreover, jurisprudence in Chile is only a minor source of court decisions. There is no general and automatic application to analogous cases. Each instance requires a specific judgment.

THE ENVIRONMENTAL POLICY PROCESS AND ITS EFFECTS

In the context of Chilean bureaucratic politics, public-sector agencies often find themselves, by design or default, advocating one or the other side of the debate (and the two main alliances). In some cases, there are strong ties between agencies and interest groups. The policy process, though relatively fluid and not devoid of a degree of "pragmatism," does not occur on a level playing field. The possibilities for consensus-building, cooperation, and compromise between the sides exist, providing two conditions are met. One is that the specific issue at hand does not radically challenge the neo-liberal economic model. The other is that the issue discussed be either narrowly "technical" or of such dramatic magnitude (e.g., smog in Santiago, the cholera epidemic of 1991, or—more to the point—the ozone "hole" over the Antarctic region) that some declaratory agreement could occur.

Limited Bargaining and Intersectoral Cooperation

Given the heterogeneous nature of the blocs, it is also possible that some weaker actors (e.g., bus owners' associations) may agree to move away from a relentless opposition to regulation. The same could be said with regard to farmers. The opposite is also the case: "environmentalists" and their likely allies in the government become increasingly willing to accept the "pragmatic" and "realistic" fact that they cannot implement their agendas. The model that seems to characterize best the Chilean policy process regarding environmental issues is that of small changes in the correlations of forces (or power whirlpools), resulting from shifts in the outer rims of both the environmentalist and the developer alliances, ultimately affecting government action (or inaction). The different configurations of alliances and menus result from a combination and recombination pre-eminently among elites and around usually narrow issues. Popular groups, though formally represented in the alliance (e.g., natives in the case of hydro projects), have little to say in the final decisions. In this context, the process is exceedingly slow and frustrating. The most common policy outcome is either a non-decision or a decision resembling the status quo.

Intersectoral Conflicts

More common, however, are situations of profound disagreement. An example is the discussion of the Development and Recovery of Native Forests Law, also known as the Forestry Law. Although CODEFF played an important role in developing the proposals for the conservation of native forests, early in 1992 the environmental group asked the government to review substantial parts of the Executive's draft sent to Parliament for discussion in April of that year. CODEFF strongly criticized those parts of the legislation making possible the substitution of native forests for forestry plantations. The committee charged that the draft appeared "to be directly dictated by the wood chip industry."[14] It proposed instead a comprehensive and sustainable management system for native forests, recommending grants in aid for their "protection and development, an immediate end to the use of native woods in the wood chip industry, greater control by the state of exploitation permits and more resources to the organizations designed to enforce these measures." The committee's view was that industry should be able to develop the country without destroying its heritage.

This motivated a response by the socialist (PPD) president of the Chamber of Deputies, who had been among those most critical of the deregulatory policies of the former military regime. He called upon the parties "to find a happy medium to allow rational forest exploitation without destroying native species."[15] Heavy pressures on Parliament may have been coming from the executive, via the "development" ministries of finance and economics, as well as from those in the government coalition negotiating a series of delicate balances with the right-wing opposition. This conciliatory stand was apparently timed as well with the opening of the first International Conference on the Commercialization of Forestry Products in the Pacific Basin, organized by the Forestry Department of the *Fundación Chile*. More than 100 businessmen from all over the world were meeting to create new opportunities and improve communication among their industries. Environmental protection has been traded off for political stability and growth, even by subtle but legalized forms of corruption. The overall attempt seems to have been geared to prevent "rocking the boat," especially at a critical time in the process of democratic consolidation, requiring business support.

Another example of the dynamics of the environmental policy process is the case of one of ENDESA's affiliates, the Pangue Dam project. It is part of the enterprise's long-term energy development strategy, which contemplates a series of dams in the Bio-Bio River—Pangue, Ralco, and Huenquecura. The initial stage of the project (Pangue) entailed flooding some 1,250 acres of land along the river banks, with a peak generating capacity of 450 MW. According to environmental studies, the effects were devastating for an extremely fragile ecosystem, with destruction of forests and of land and river species, as well as soil erosion and the traumatic uprooting of three communities of Pehuenche Indians who see the dams as "the end of Pehuenche life" (Probe International 1992).

The cost of the operation was over $470 million, to be financed with foreign loans. One loan for $125 million was negotiated with the World Bank, and its feasibility was studied by the International Electro-Technical Commission (IEC). Another loan of $50 million came from the IDB. In addition, there was a tied third loan negotiated with the Spanish government. The approval of the World Bank loan was strategic to clear the way for an international finance package (International Rivers Network 1990). ENDESA obtained all the required authorizations from the pertinent agencies—the Ministry of Public Works' Directorate-General of Waters, the Ministry of Economics, and the National Energy Commission (Kunkar 1991: 16).

The abovementioned Bio-Bio Action Group (GABB) was formed in 1991, specifically to coordinate a campaign to protect the river and the populations living within its confines. The group was extremely successful in providing public education, articulating the public interest, and establishing a wide national and international network, along the lines discussed earlier, Despite ENDESA's powerful public relations machine, which dismissed criticisms as the actions of "wealthy white-water rafters," GABB was able to maintain the interest on the issue and to generate a good deal of external support. The Bio-Bio campaign became perhaps symbolic of a struggle to "save the planet." The tactics had changed from the domestic to the international arena. Since the ostensible weak link in the process was external finance, in particular the environmental assessment required from IFC prior to its recommendation to the World Bank, pressure by organized environmental groups in North America was brought to bear upon both IBRD and IFC. It should be remembered that such environmental assessments of international agencies involve heavy pressure by public interest groups in Europe and North America upon the "development establishment." Thus, while formal environmental reviews and assessments are, so far, inoperative in Chile, external assessments of project financing are not. It was, however, a case of too little too late, or worse, a case of too powerful a developmental coalition against a well-organized yet outgunned environmental campaign. Despite contradictory and optimistic indications for the environmentalists, the development side was to prevail in the long run (*El Diario* 1992).

CONCLUSION

Whatever their outcome, campaigns of this nature have a number of secondary and unintended effects. One of these is the institutionalization in Chile's governmental process of an environmental policy agenda, as well as mechanisms around which future political debates are likely to emerge. Here there have been some encouraging signs. For instance, in a speech at the International Seminar on Mining and the Environment organized by the Chilean Copper Commission, the Institute for International Studies of the University of Chile, and the International Law Institute, then-mining minister Juan Hamilton was quite blunt. He expressed the view that companies "that fail to respect environmental protection regulations" or those with poor environmental records in their own countries will not be tolerated (*La Época* 1992a). He indicated that his ministry was ready to implement "environmental protection measures in new mining enterprises and that in the near future CODELCO and ENAMI must present their decontamination plans for their Chuquicamata and Ventana plants, in accordance with the Mining Ministry's Decree 185" (*La Época* 1992b). He added that a number of foreign companies have already made significant investments for controlling their impact upon the environment. The minister also expressed optimism about the possibilities of getting through parliament a legislative package "to regulate substances harmful to the ozone layer, such as the use of combustibles with high sulphur content, related to mining and industrial production" (*La Época* 1992b). On the down side, that speech was probably at the very pinnacle of official environmental consciousness; in subsequent years globalization, productivity, and competitiveness displaced environmental concerns.

Another latent effect of environmental campaigns is the establishment of an organizational network of brokers, especially on the public-interest side, which allows for learning new strategies and for advancing environmental goals. This is particularly the case with both the incorporation of "grassroots" organizations, such as those advancing the natives' cause, and the external constituencies made up of environmental groups in industrialized countries. Globalization and transnationalization have benefited large business conglomerates, but generally not popular organizations and public interest associations. To some extent in the case of environmental politics, there has been some crucial transnational support for environmental causes. Groups like Greenpeace and Probe, the connections among indigenous movements throughout the Americas, and the presence of environmental entrepreneurs (like Douglas Tompkins) are a manifestation of a "Third System"-style globalization in the making (Nef 2006).

A less discussed, but equally important, consequence of bringing the environmental debate to the public is the increased salience of environmental issues among the population at large. Public attention generates an "environmental awareness" (or consciousness) in the unofficial public mind. My own assessment of the Chilean situation is that the culture is still dominated by a "predatory" attitude toward the natural environment. Conservation and environmentally friendly behaviour are not highly developed; instead, "modernism" and the engineering mode have been hegemonic in Chilean culture. Concern for environment at the mass level—perhaps with the exception of a minuscule and vanishing Amerindian subculture—was often confused with elite snobbism: a "keep-it-green, keep-it-clean" kind of discourse. Since 1989, this has begun to change. Environment has been incorporated, as in Europe and

North America, into the political menu and has become part of the normal, everyday discourse and practice. There is still a long way to go, but the qualitative nature of the change is significant. Prickett's optimistic assessment nearly two decades ago may not be too far off: "Public concern about the environment is widespread and rapidly growing. A large number of environmental organizations have formed to channel this public sentiment into action by the government and the private sector. These groups have had a number of notable victories" (Prickett 1992: 3).

There is, however, another less optimistic perspective. Chile's present predicament is clearly one where environment has been easily pushed aside for economic considerations, which is currently happening. As I indicated earlier, for as long as the social construction of environment does not appear threatening, and for as long as it does not involve the mobilization of popular actors who could upset the pact of elites, it will be permissible. Should it fail to remain within what the socio-economic elites define as "acceptable bounds," it could be disposed of by less-than-gentle means. After all, this menu-restructuring happened to the "old" discourse of democracy, social justice, import substitution, and the welfare state.

In addition, there are powerful regional forces that have a crucial impact upon environmental policies and, more broadly speaking, on human security.[16] In this sense, one needs to keep in mind the fact that Chile, like the rest of Latin America, is highly vulnerable to external pressures in the area of policy formulation and implementation. The free trade agreement between Chile and the United States (*El Mercurio* 1992a), modeled upon the United States-Canada Free Trade Agreement (FTA) and the North American Free Trade Agreement (NAFTA), had the effect of largely watering down rather than enhancing the environmental agenda. This has also been contingent upon who is in power in Washington. The Bush Sr. administration represented a clear anti-environmentalist position, which contributed to the ill fate of the Rio Summit because of "pressure from the United States" (*La Época* 1992a). Even *Agenda 21*, which committed "nations to carrying out environmental protection measures, [lacked] reference to any concrete programs in terms of energy conservation and ozone layer protection" (*La Época* 1992a). In fact, the outcome of the conference surpassed some of the worst predictions, signalling the corporate world that the status quo is acceptable. To add it up, subsequent US Supreme Court decisions have limited class-action suits and have virtually erased the possibilities of US environmental groups' launching legal challenges on behalf of foreign groups. This has been a clear victory for environmental conservatism, not conservation. Thus, the chances for effective interlinking and environmental partnership have been dealt a severe setback. The election of Bill Clinton in 1992, as well as his administration's efforts to give priority to environmental goals, signalled a significant departure from the previous anti-regulatory record. Environmental initiatives in Chile in the mid-1990s were reflective of this new international climate. However, the radically conservative swing in the US political pendulum in

2000 and 2004 has had deleterious repercussions in the already weakened Chilean environmental agenda.

In a way, the Chilean experience is relevant to other countries in the region to the extent that its "development model" is being hailed by the international business community as the "best case" for Third World development. As suggested earlier, its environmental record and policy are a reflection of this development style. This, in turn, is conditioned by the country's insertion into a global and regional political economy of development. The preceding exploration may have already shed some light on the relationships between environment and development that such "best case" really implies.

A 2003 article focusing specifically on the Bio-Bio hydroelectric projects noted, "Local communities, whether indigenous or not, have little power to exercise control over decisions with a profound impact upon them ... Center-Left government officials have become apologists for projects which compromise future social well-being, the quality and quantity of native natural resources, and discount the conservation of species and the ecosystems" (Haughney 2003: 13). The problem now is that after the long period of consolidation of a restricted democracy, the economic model imposed by force and its environmental corollaries have become entrenched (Bensabat and Nef 1992; Galleguillos and Nef 1992). However, the consolidation of an environmentally dysfunctional economic model is not limited to Chile, Latin America, or the two-thirds of humanity once referred to as the Global South. Developed and fully democratic countries without the traumatic political adjustments experienced by Chile appear to be caught in a similar "no-choice" predicament.

Notes

1 For an elaboration of Max Weber's notion of "interpretative" analysis, see his *Theory of Social and Economic Organizations* (1974).

2 The United Nations Environmental Programme (UNEP) was established in 1973 after the Stockholm meeting of 1972. It served as a centre for a series of official and NGO initiatives in Europe and the Americas. Several organizations in Europe, especially in Sweden (under the umbrella of the Dag Hammarskjöld Foundation) and Spain (CIFCA, Centro Internacional de Ciencias Ambientales, established in 1975), as well as in the Netherlands, France, and the United Kingdom, among others, served as pivotal points for a network of Chilean scholars dealing with environmental issues and constituting the vast intellectual diaspora under the military regime.

3 In a conversation at ECLAC in 1991, Osvaldo Sunkel explained that after the military coup of 1973 many left-of-centre academics shifted their focus to environment as a safer realm of permissible critical analysis, still not taken over forcefully by right-wing monetarists. In this sense, "environment" became the main venue for an alternative discourse to neo-liberalism.

4 "Fight for Bio-Bio Wages On," *International Rivers Network* Special Briefing, December 1991: 2; also September 1991: 10; December 1990: 6; March 1992: 11.

5 Chile's governing alliance since 1990 has included the centrist Christian Democrats (of former presidents Aylwin and Frei), the left-of-centre Socialists of various brands (currently, President Michelle Bachelet is the first Socialist to sit as Head of State), the "instrumental" Party for Democracy (PPD) of former

president Lagos, and the traditionally centrist and minuscule Radical party. In 1989, with the support of other mini-parties outside the coalition, such as the Humanists, the Communists, and the Ecologist (Green) party, the *Concertación* won the presidency by a wide margin, with over 56 per cent of the popular vote, and obtained a commanding majority in the Chamber of Deputies, though it failed to reach a majority in the Senate. From very early on, the coalition sought to distance itself from the Communists, Greens, and Humanists. As a result of the constitutional manipulation by the previous military regime, the coalition has generally lacked the capacity to implement legislation past the Senate and the other institutional safeguards built into the transition pacts. Likewise, in other arenas of decision-making, such as the Constitutional Tribunal, the Comptroller's Office, the Central Bank, the National Security Council, the Telecommunications Commission, the vast senior levels of the bureaucracy, the military, and the police establishment (until recently stacked with hand-picked appointees of the Pinochet regime), the elected government for over a decade did not have the necessary effective power to rule.

6 For a definition of "linkage groups" see Chalmers 1972.

7 CIEPLAN provided institutional support for important Christian Democratic researchers and intellectuals who became President Aylwin's close associates, cabinet ministers, or advisors (such as finance minister Alejandro Foxley). The corporation's openly critical stands against the neo-liberalism of the Pinochet regime have been quickly toned down as a neo-liberal "pragmatism" has increasingly permeated the government's "economic growth policies," with which CIEPLAN has become identified.

8 Presentation by GABB and other organizations to the International Finance Corporation, 27 Feb. 1992, p. 10. For an overview of native politics with special emphasis on the Aymara, see Arratia 1991.

9 Throughout Chilean history, the connection between the CNPC (COPROCO) and the right (conservatives, liberals, since 1965 fused into the National party) was axiomatic. Today, close contacts exist between the National Renovation party (RN), which recycled and modernized the old Liberal-Conservative and National "old guard" and the various components of the National Confederation of Production and Commerce. The same is the case between the CNPC and the more ideological and corporatist Democratic Independent Union (UDI). Most business sectors have been supportive of the military regime, while distancing themselves from its "excesses." *Pinochetismo* without Pinochet is still very strong in these quarters.

10 See "Declaración publica de mujeres indígenas y campesinas de Chile" (June 1997) http://www.mapuche.info/fakta/Domodec.htm.

11 Humberto Palza Carvacho subsequently ran as a Christian Democrat and was elected to Congress as a Deputy in 1989.

12 Supreme Court, trans. copy of verdict, Dec. 19, 1985: 252-57.

13 "Comments" to the verdict, pp. 252-57.

14 "Comments" to the verdict, pp. 252-57.

15 "Comments" to the verdict, pp. 252-57.

16 See Nef 1999a and Nef 2001 for a brief treatment of the subject.

References

Arratia, María-Inés. 1991. "Khistispxtansa? Who Are We? Rethinking Aymaraness and Chileanness in the Nineties." In Harry P. Díaz et al. (eds.), *Forging Identities and Patterns of Development in Latin America and the Caribbean.* Toronto: Canadian Scholars' Press. 253-62.

Bensabat, H., and J. Nef. 1992. "'Governability' and the Receiver State in Latin America: Analysis and Prospects." In Archibald Ritter, Maxwell Cameron, and David Pollock (eds.), *Latin America to the Year 2000: Reactivating Growth, Improving Equity, Sustaining Democracy.* New York: Praeger. 161-76.

Bifani, Pablo. 1982. *Desarrollo y medio ambiente*, vols. 1, 2, and 3, nos. 24, 25, and 26 of Cuadernos del centro interncional en ciencias ambientales (CIFCA). Madrid: Linoexpres S.A.

Chalmers, Douglas. 1972. "Developing in the Periphery: External Factors in Latin American Politics." In Yale Ferguson (ed.), *Contemporary Inter American Relations: A Reader in Theory and Issues*. Englewood Cliffs, NJ: Prentice-Hall. 11-16.

Cleaves, Peter. 1974. *Bureaucratic Politics and Administration in Chile.* Berkeley: University of California Press.

Easton, David. 1957. "An Approach to the Analysis of Political Systems." *World Politics* 9(3): 389-95.

El Diario. 1992 (April 25). "World Bank to Take Decision on Pangue in Next 60 Days."

El Mercurio. 1992a (April 25). "Buses Fitted with New Filters in Santiago," and "Changes in Make up of Santiago Pollution."

—. 1992b (April 25). "Possibility of Free Trade Agreement with the United States."

Escobar, Gabriel. 1996 (Nov. 29). "Rain Forest Gift Raises Suspicion." *Washington Post Foreign Service*. A01.

Galleguillos, N., and J. Nef. 1992. "Chile: Redemocratization or the Entrenchment of Counterrevolution." In Archibald Ritter, Maxwell Cameron, and David Pollock (eds.), *Latin America to the Year 2000: Reactivating Growth, Improving Equity, Sustaining Democracy*. New York: Praeger. 177-93.

Gligo, Nicolo. 1980. "The Environmental Dimension in Agricultural Development in Latin America." CEPAL *Review* 12: 129-44.

—. 1981. "Estilos de modernización y medio ambiente en la agricultura Latinoamericana." *Estudios e informes* CEPAL *4*. Santiago: EPAL.

Goodwin, Paul. 1992. *Global Studies: Latin America*. 5th ed. Guilford, CT: Sluice Dock, Dushkin Publishing Co.

Graham, Carol. 1992. "A Development Strategy for Latin America?" In Paul Goodwin (ed.), *Global Studies: Latin America*. 5th ed. Guilford, CT: Sluice Dock, Dushkin Publishing Co. 173-74.

Haughney, Diane. 2003. "The Protests Against the Bíobío Hydroelectric Projects: The Limits of Democratic Participation in Neoliberal Chile." SECOLAS *Annals* XXXV: 13.

Heady, Ferrell. 1984. *Public Administration: A Comparative Perspective*. 3rd ed. rev. New York: Marcel Dekker.

International Rivers Network. 1990. "Six Dams Threaten Chilean River." *International Rivers Network Special Report: The Bio-Bio: A River in Peril*. 6.

Kandell, Jonathan. 1991 (July 7). "Prosperity Born of Pain." *The New York Times Magazine*. 15-18.

Kunkar, Susana. 1991 (July 1). "Represas en Alto Bio-Bio. 'Si el río suena ...'" *Análisis*: 16.

La Época. 1992a (April 13). "Greenpeace Criticizes U.N. Earth Letter."

—. 1992b (April 25). "Chile Will Not Harbour Companies That Harm Environment."

La Nación. 1992a (April 20). "Chile Buys 50 Trolleys from Russia."

—. 1992b (June 18). "No End in Sight to Santiago's Pollution."

Lasswell, Harold. 1950. *Politics: Who Gets What When and How.* New York: McGraw-Hill.

Mariman, Jose. 1994. "Transición Democrática en Chile ¿Nuevo ciclo reivindicativo Mapuche?" http://www.xs4all.nl~rehue/art/jmar5a.html.

Nef, J. 1983. "Economic Liberalism and Political Repression in Chile." In Archibald Ritter and David Pollock (eds.), *Latin American Prospects for the 1980s: Equity, Democratization, and Development*. New York: Praeger. 314-16.

—. 1989. "Latin America: The Southern Cone." In V. Subramaniam (ed.), *Public Administration in the Third World*. Westport, CT: Greenwood Press. 352-84.

—. 1991. "Development Crisis and State Crisis: Lessons from the Latin American Experience." In O.P. Dwivedi and P. Pitil (eds.), *Development Administration in Papua New Guinea*. Boroko: Administrative College of New Guinea. 10-34.

—. 1995. "Environmental Policy and Politics in Latin America: The Chilean Case." In O.P. Dwivedi and D.K. Vajpeyi (eds.), *Environmental Policy and Developing Nations*. Boulder, CO: Greenwood Press. 145-74.

—. 1999a. *Human Security and Mutual Vulnerability. The Political Economy of Development and Underdevelopment.* Ottawa: IDRC Books.

—. 1999b. "Contradicciones en el modelo Chileno." In Amparo Menéndez-Carrión and Alfredo Joignat (eds.), *La Caja de Pandora. El Retorno de la Transición Chilena.* Santiago: Planeta/Ariel. 97-98.

—. 2001. "Human Security: Perspectives for Human Resources and Policy Management." In M.K. Tolba (ed.), *Our Fragile World: Challenges and Opportunities for Sustainable Development*, Vol. I. Oxford: EOLSS Publishing Co. 813-31.

—. 2006. "Third Systems, Human Security and Sustainable Development." In Rebecca Harris (ed.), *Globalization and Sustainable Development: Issues and Applications.* Tampa: The Kiram C. Patel Center for Global Solutions, University of South Florida. 43-58.

North, Lisa. 1978. "Development and Underdevelopment in Latin America." In J. Nef (ed.), *Canada and the Latin American Challenge*. Guelph, ON: OCPLACS. 79.

Prickett, Glenn. 1992 (Feb. 26). "Partnership Opportunities to Protect Chile's Bio-Bio River." Memo from Natural Resources Defense Council to Probe International.

Probe International. 1992. "World Bank Loan to Private Sector Spells Doom for Famed Chilean River Pehuenche People." *Probe International*: 1.

Ratcliff, Earl Richard, and Maurice Zeitlin. 1988. *Landlords and Capitalists: The Dominant Class of Chile.* Princeton, NJ: Princeton University Press.

Rother, Larry. 2005 (Aug. 7). "An American in Chile Finds Conservation a Hard Slog." *The New York Times*.

Rubio, Lorena. 2003 (Feb. 28). "70 mil hectareas en la mira del clan Tompkins." *La Tercera*. http://www.mapuche.info/facta/terc04229.html. Accessed April 2006.

Santiso, Javier. 2006. *Latin America's Political Economy of the Possible: Beyond Good Revolutionaries and Free-Marketeers*. Cambridge, MA: MIT Press.

Sheahan, John. 1987. *Patterns of Development in Latin America: Poverty, Repression, and Economic Strategy*. Princeton, NJ: Princeton University Press.

Stallings, Barbara. 1978. *Class Conflicts and Economic Development in Chile 1958-1973*. Stanford, CA: Stanford University Press.

Staniland, Martin. 1985. *What Is Political Economy? A Study of Social Theory and Underdevelopment.* New Haven, CT: Yale University Press.

Stutzin, Godofredo. 1975. "A Note on Conservation and Politics." *Environmental Policy and Law* 1: 38.

Sunkel, Osvaldo. 1980. "The Interaction between Styles of Development and the Environment in Latin America." CEPAL *Review* 12: 15-50.

Tomic, Patricia, and Ricardo Trumper. 1990. "The Contradictions of Neo-Liberalism in Chile, 1973-1989: The Case of the Concepción Region." *Canadian Journal of Latin American and Caribbean Studies* 15(30): 233-35.

WCED (World Commission on Environment and Development). 1987. *Our Common Future.* New York: Oxford University Press.

Zavaleta, H. 1991 (December). "Recent Hydro Development in Chile." *Water, Power & Dam Construction*. 2.

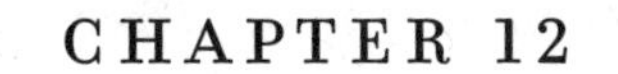

CHAPTER 12

IS A BETTER WORLD POSSIBLE?

The Experience of the Brazilian Environmental Movement and the "Construction of Citizenship"[1]

Angus Wright

INTRODUCTION

Any attempt to describe Brazil's environmental situation inevitably sounds a little like the introduction to Charles Dickens's *A Tale of Two Cities*, which begins, "It was the best of times, it was the worst of times," and continues with a set of stark contrasts. Brazil outpaces any other nation in its rates of deforestation and species loss, while it still has far more tropical forests and more species of plants and animals than any other country. One Brazilian city, Cubatão, became world-famous as "the valley of death" for its startlingly high rates of environmental pollution and a consequent public health nightmare, while not far away, and less than twenty years later, the city of Curitiba was applauded internationally as a model of good environmental planning. Brazil has an unusually thorough and sophisticated body of environmental legislation, based on strong environmental provisions in the national constitution of 1988. But it also continues to experience so many incidents of scandalous flouting of such laws by the rich and powerful that serious observers have concluded that impunity from the laws is built firmly into the legal culture. There are thousands of highly qualified environmental professionals in government, industry, and universities, and many Brazilian scientists do world-class environmental research. However, environmental agencies are chronically short of the people, money, and practical authority they need to do their job. There are hundreds of lively, diverse, and often effective environmental organizations, but issues and policies that are irrelevant or contradictory to environmental protection dominate politics and society at every level from national to local. Brazil hosted the 1992 United Nations Conference on Environment and Development, the Rio Summit, which unleashed the establishment of an unprecedented set of international agreements, but Brazil, along with most other nations, has failed to meet most of the promises of the 1992 Conference. Few countries have more engaged the interest and concerns of environmentalists around the world, and few countries have suffered more from the oversimplifications, distortions, and assorted stereotypes applied to it by foreigners and Brazilians alike (Hochstetler and Keck 2007).

The argument presented here is that most Brazilians care about the unparalleled natural treasures of the nation and are concerned about the continuing damage done to it. However, while people and organizations have been vigorous in attempting to address the problems, the environmental performance of the society and government remains disappointing. This is due to profound problems caused by deep social inequalities and a closely related chronic inability or unwillingness of public institutions to govern effectively.

However, any brief treatment of the way Brazilians deal with their environment inevitably risks reinforcing one or more of the common distortions. If for no other reason, this is because Brazil is so large and diverse. It is a nation with 186 million people whose territory is slightly smaller than the US, including Alaska. Its highly distinctive regions range from the world's largest rainforest, to a marshland the size of France with one paved road crossing it, to fertile sub-tropical farmlands, to semi-arid savannas. It has significant unique ecosystem types that require a special Brazilian lexicon to describe them properly.

Like its environment, Brazil's peoples are highly diverse, though they enjoy a certain unity from the fact that nearly all speak Portuguese. Approximately half the population is at least partially descended from slaves brought from Africa. There are significant numbers of Amerindians, some highly assimilated and intermarried with other ethnicities, and a much smaller number who still live in largely isolated settlements. There are people descended from immigrants from nearly every European nation. The people of São Paulo are predominantly Italian in origin, but there are about two million Japanese-Brazilians, most of them living in and around the city of São Paulo. Many Arab-Brazilians have achieved wealth and high political positions. German and Central European cultures strongly flavour the life of the southern states. There is a rich local vocabulary for describing the innumerable mixtures of ethnicity and race.

In the last half-century, Brazil has undergone extraordinarily rapid social and political transformation. Formal democracy was overthrown in a military coup in 1964. A military dictatorship ruled the nation until the mid-1980s, when the military surrendered power in a tumultuous but largely peaceful process of return to civilian rule. In the same period, the nation has gone from being about two-thirds rural and predominantly agricultural to being more than two-thirds urban and highly industrial. In the 1960s, the rate of population growth was one of the world's highest at around 3.4 per cent per year. It then fell precipitously to 1.1 per cent per year, well below the world average. Brazil has moved into the ranks of the top three to five countries in terms of agricultural production, and the top ten or so in industrial production (varying from year to year). It is difficult to exaggerate the importance of all of these changes for the lives of Brazilians, and for the fate of its natural environment (US State Department 2006).

UNDERSTANDING BRAZIL'S ENVIRONMENTAL SITUATION: FIVE MAJOR THEMES

There are five central themes that help to shape a coherent perspective of Brazil's environmental situation while recognizing its dynamic complexity.

The *first theme* is that Brazil's awareness of the value and unique character of its natural environment is not new, nor is it a foreign import. Throughout Brazilian history, many in Brazil have been struck by how kind nature has been to those fortunate enough to live in Brazil. One of Brazil's most prominent historians, Sergio Buarque de Hollanda, called this persistent and influential perspective "the vision of paradise," arguing that it is one of the keys to understanding Brazil. Some Brazilians and foreigners, of course, have seen more of hell than of heaven in Brazilian nature, especially in the Amazon and the drought-plagued Northeast.

But whatever the particular viewpoint, throughout Brazil's history, poets, novelists, essayists, artists, scientists, military men, and politicians shaped and reshaped a variety of perspectives on the nation, insistently incorporating the importance of the character of the Brazilian natural environment on its people's history, economy, and politics.

More specifically, there is a long Brazilian tradition of concern over the damage that humans have been doing to the land and forests of the nation. Understanding this piece of Brazilian cultural and intellectual history is important for several reasons, and it does much to combat the arrogance that many foreigners have brought to Brazil and to the analysis of Brazilian environmental problems.

A *second theme* essential to understanding Brazil's environmental governance and politics is the importance of the experience of living through twenty years of a repressive military dictatorship. The dictatorship killed thousands of its opponents, tortured more, and systematically dismantled political parties, trade unions, student organizations, and other groups that it saw as a threat to a conservative social order. The military regime practised rigorous press censorship, while at the same time it set up two political parties—an official loyalist party and an official and tightly circumscribed "opposition party"—in a kind of mockery of any real democratic process. It shoved through scores of economic development projects with little attention to the environmental costs. Indeed, at one point, it supported advertisements in *The Wall Street Journal* showing a belching smokestack with the invitation, "Come to Brazil, Your Pollution Haven." It is only by considering the dictatorship, the struggle against it, the emergence from it, and the creation of a new political and civil life that we can understand a great deal of the distinctive character of Brazilian contemporary culture. This very recent historical influence is comparable to the influence of the Civil War on American culture or the independence struggles for much of the post-colonial world. This is especially true for political culture and, in particular and important ways, for Brazilian environmentalism.

It is impossible to understand any significant feature of Brazilian politics and policy without reference to a *third theme*: Brazil's extraordinary levels of economic and social inequality. Brazil's "middle income" status among nations (with a per capita Gross Domestic Product of $8,400) masks one of the most unequal societies in the world. One good measure of inequality is the comparison between a fairly large section of those at the top with a similarly large section of people at the bottom. In Western Europe the income of the richest twenty percent is four times greater than the income of the bottom twenty percent. In the United States the ratio is eight to one. In Mexico it is approximately 17 to one. In Brazil, it is about 33 to one, the most severe inequality in any large nation on earth, and comparable with many of the smallest and poorest countries. The inequalities, which began with the patterns of land distribution and corrupt administration of land law established under the colonial regime, have been reinforced by the tendencies toward inequality in Brazil's pattern of dependent capitalist development.[2] The contrast between the fabulous wealth of Brazil's rich and the misery of its poor is reflected in every aspect of life: access to land and housing, health, education, proper diet, and social services. It is also unavoidably linked to the ability to gain fairness from the justice system—police, regulatory agencies, and the courts. Systematic inequality and injustice make it vastly easier for the rich to seize or otherwise acquire land and resources and to exploit the environment with a ready supply of cheap labour. The powerful often can ignore the damage done to air, water, fisheries, forests, and the land in ways that are disproportionately devastating to the poor. The rich can and do live in heavily guarded areas in which the "vision of paradise" is maintained, recreated, and enhanced. In such enclaves it is mostly (though not entirely) possible to pretend that air, water, land, and forests are cherished and protected. Throughout Brazil, there are those who attempt to resist both the inequity and the damage done by overuse and careless abuse of land and resources, but there is a scandalously high rate of such people being ignored, pushed aside, jailed, beaten or even murdered. General public awareness of the price to be paid for resistance discourages many from further resistance, and it is a powerful source of the cynicism that feeds a credibility crisis in government (Wright and Wolford 2003: xv-xix).

A *fourth theme*, and one that is more hopeful, is the way in which Brazilians are attempting to face the challenges of emerging from the dictatorship, building greater credibility in government, and reducing inequalities. It is their success in doing so that will greatly determine whether Brazil will be able to better protect its natural resources and environmental quality. In the wake of the deadening effect of the military government on Brazilian democracy and the exercise of citizenship, many Brazilians now discuss the broad challenges of creating a better society in terms of the building, or rebuilding, of civil society and the construction, or reconstruction, of Brazilians as citizens. Brazilians have come to use the term "the construction of citizenship" in many varied contexts. In the more specific context of environmental matters, this puts the emphasis on analyzing the quality and strength of the processes through which people can effect positive environmental changes. The changes themselves are clearly important, but not as important as the necessity for strengthening and universalizing the processes of democratic change. The emphasis on citizenship is necessary because without stronger and more universal citizenship, inequality, corruption, and cynicism put all accomplishments at peril.

It will clearly not be possible to construct a full sense of citizenship and a flourishing civil society without meeting the challenges created by *a fifth theme*, what I will call here the "credibility crisis" of the Brazilian state. This crisis is not so severe that it immediately threatens the stability or the continuity of the Brazilian government, but it is serious enough to undermine chronically the authority and effectiveness of Brazilian national, state, and local governments. The lack of confidence in government fosters a corrosive and sometimes pervasive cynicism. This cynicism itself operates as a feedback mechanism that further weakens government effectiveness, thereby creating stronger undercurrents of cynicism. At various points in the last twenty years—the end of military rule, the successful fashioning of a new constitution, the election of Tancredo Neves as the first civilian president after military rule, the election of Fernando Henrique Cardoso, and, most remarkably, the election of Lula (Luis Inacio da Silva)—it has seemed to many Brazilians that the vicious cycles that feed the problem might be broken. In each case, however, various disappointing events have been widely interpreted and understood in ways that have further weakened belief in the integrity and reliability of government. The government's constant inability or unwillingness to enforce the laws evenly and reliably has a devastating effect on the public's confidence in the government. The corrupt use of influence and money comes to seem pervasive. The lack of belief in government makes it difficult for even the best-intentioned and most capable people in government to apply environmental law and policy effectively, no matter how well it is designed. It is difficult for citizens to believe that their most dedicated efforts can be implemented and defended, and so citizenship and commitment to a civic culture can all too easily seem futile.

THEME ONE: THE TRADITION OF ENVIRONMENTAL AWARENESS

The first governor general of colonial Brazil, Tome de Souza, warned the King of Portugal that the granting of vast stretches of largely unknown territory to his favourites would encourage arrogance and waste by those who received the land. De Souza had a great deal of historical experience on which to base his judgement: since the beginning of the Portuguese "reconquest" of Portugal from the Moors in the twelfth century, the Portuguese king had used promises of generous land grants to recruit noble or would-be noble warriors to his cause (Sodero 1990: 25). Once ensconced on their land, the warrior knights made troublesome subjects, reluctant to pay taxes and inclined to burn or pillage each other's holdings in constant quarrels. The king found that new conquests yielding new revenues were the only practical way to maintain some independent legal and military authority, but he would also find again and again that new conquests brought the same old troubles from his nobility. The solution was to keep expanding, leading not only to the establishment of Portuguese territories all along the coasts of Africa and Asia, but also to the control of a large portion of the world's trade. With the acquisition of Brazil, a land mass whose shape and dimensions were only dimly understood, Portugal, a nation of not much more than one million people, became colonial ruler of a territory that in its modern form is larger than all of Europe to the Urals. The only way the monarchy knew how to assert rule in Brazil was once again to give land grants, along with extensive administrative authority, to friends of the Crown.

Tome de Souza's warning about the consequences was well justified by subsequent experience. The land barons were not only difficult to rule, but their vast land grants encouraged them to use land as though it were limitless. When native labour and Portuguese immigration failed to provide enough motive force to make it profitable, they reached out to Africa to import what would become the largest number of slaves of any country in the Americas. They used this labour to destroy most of the northern portion of the Atlantic Coast Rainforest (still one of the world's most diverse, though less than seven or eight per cent remains) by the end of the colonial period, and by the end of the nineteenth century they had destroyed most of the southern portion. Religious orders, most famously the Jesuits, tried in their own authoritarian fashion to protect indigenous Brazilians and use the land more conservatively, but, outside of the far south, their overall effect was negligible and was largely erased after the Jesuit expulsion in 1759.[3] Colonial officials tried to rein in the disorder and destruction at various points, particularly in the latter part of the eighteenth century, but the power of the landowners proved more decisive than that of reformist officials (Dean 1995). Some prominent Brazilians became convinced that the wasting of the land was a major cause of Brazil's economic stagnation. The Brazilian scholar (and director of Greenpeace Brazil in the 1990s) Jose Augusto Padua, has shown how widespread and important this critique would become (2002).

The man known as "the father of Brazilian independence," Jose Bonifacio da Silva, was convinced that Brazil's profligate misuse of land and labour was a major barrier to its economic and social progress. Throughout the nineteenth century, other major politicians and writers made the same case. Padua points out that what we might now call an "environmental critique" of Brazil's economy and society was held by a minority, but it was a minority who would come to be seen as the most important Brazilian thinkers of their time. Many were abolitionists who believed that the end of slavery would bring about more responsible land use. They were wrong in this belief. However, their more general point, that destructive land use was tied to larger social factors, and that it weakened and endangered Brazilian society, remains a view widely held by Brazilians today (Padua 2002: 10-23 and *passim*).

There are other deep traditions of thoughtful regard for the Brazilian land which, for the sake of brevity, can be mentioned only succinctly here. One, firmly falling within Buarque de Holanda's (1969) interest in "the vision of paradise," is represented by the hundreds of prints, drawings, and paintings of eighteenth- and nineteenth-century landscapes and social scenes. These works of art show profound interest in the unique beauty of the Brazilian land. Some portray the devastation of forests and ruined lands, in which poverty and oppression are pictured as consequences or the treatment of the land, a paradise lost. The works of the writers who had drawn the same association likely influenced many of these. The landscape art of nineteenth-century Brazil has been called an *iconografia*, used in billboard commercials as well as in serious works of social criticism. It is a kind of subliminal set of visual illustrations, held more or less consciously in the minds of most Brazilians, for the text of today's environmental critique of Brazilian development (Correa do Lago 2001).

More concrete expressions of twentieth-century interest in the Brazilian environment included the founding in Belem in 1866 of the first scientific research institute for the study of nature in the Amazon, the Goeldi Museum. Dom Joao VI, Portuguese regent in Brazil, founded the magnificent Botanical Garden in Rio de Janeiro in 1808 and Brazil's independent monarch, Dom Pedro II, generously supported its work to study and celebrate the nation's natural endowment. President Vargas founded the Institute for Research on the Amazon (INPA) in Manaus in 1952.

The "positivism" of the French writer Auguste Comte had an immense influence in most of Latin America, and especially in Mexico and Brazil. Considered by some the foundation of modern sociology, Comtian positivism sought to create a "science of society" that among other things required the study of physical and cultural geography as a starting point for understanding society. Positivism was especially attractive to authoritarian-minded reformers and most particularly to military officers. But while historians usually discuss positivism because it provided an ideology justifying rapid authoritarian technological modernization centred on building railroads, it had another aspect as well, epitomized by the former military officer and journalist Euclides da Cunha. In writing a magnificent chronicle of a war fought against a mystic millenarian leader of the poor in the rural northeast in the last decade of the nineteenth century, da Cunha spent many long chapters describing what would now be called the environmental history of that region (1944). This he considered essential in understanding the character of the rebels and their leader, as well as in understanding the formation of the Brazilian character more generally. Read by virtually every high school student in Brazil over the last century, da Cunha, like other positivists, engraved in the Brazilian collective psyche the idea that people are shaped by the land they inhabit and by the uses to which the land is put.

In the late nineteenth century and into the twentieth century, liberal novelists, poets, and essayists saw the export-plantation economy of Brazil as the great curse of Brazil, condemning rural workers to a life of drudgery, the land to destructive practices, and the national economy to dependence on volatile export markets (see Stein 1985). The twentieth-century Bahian novelist Jorge Amado, long one of the world's best-selling authors, described plantation development in the Atlantic Coast forest as a process of "derrumbando mata, matando gente," or "felling the forest, killing people"—a phrase all too accurate in describing events along the Amazon frontier in the twenty-first century (Amado 1943: 22). Throughout Brazil in the first half of the twentieth century, there were people of some influence who tried to argue for a more diverse and humane structure of agricultural production that would at once be more conservative of land and resources and more effective in providing the foundation for industrialization. For example, with the onset of severe problems with the monocrop cacao economy of southern Bahia, a group of planters from old families of the region argued in the 1930s for the adoption of a regional planning approach based on the book *American Regions* by Howard Odum, Sr.—father of ecologists Eugene and Howard, Jr. (Lavigne 1967). By the mid-twentieth century, Brazil had a long tradition of critical thinking about the nature of its economy and its ruthless exploitation of nature.

THEME TWO: THE MILITARY DICTATORSHIP

When the military overthrew the civilian government in 1964 it promoted an aggressive version of capitalist development theory laced with a conservative nationalism. The developmentalism espoused by the military was a direct counter to the increasingly influential socialist and communist ideas of national development that the military sought to repress and eliminate. The Brazilian military and its allies in the US State Department justified the coup by claiming (almost certainly erroneously) that without it a socialist revolution was about to occur in Brazil. Developmentalism of this era, whether on the left or the right of the political spectrum, had relatively little regard for nature or environmental protection, promising rapid economic advances through the maximum use of natural resources. Nonetheless, as the military pursued its program of authoritarian capitalist development, and as even more critiques of this program began to emerge in spite of rigorous political repression, we can see threads of the older traditions of regard for the Brazilian landscape reasserting themselves (Skidmore 1988).

The military was influenced by a fascistic kind of sentimental nationalist vision of land and people, as well as by a geopolitical vision of the nation's territory. When President Emilio Medici (1969–74) announced in 1970 that the government would launch a program for the rapid development of the Amazon, he declared that the Amazon was "a land without people for a people without land." On the one hand, this was a cynical attempt to defuse the insistent calls for land reform in Brazil, one of the ignition points for the crisis that led to the 1964 coup. It also cast out a lure to the desperate rural poor of Brazil to come to the Amazon, where they could be used to build roads and dams, fell timber, and mine ore, attracting foreign investment and providing a flow of resources for national development. On the other hand, the slogan was a more or less sincere expression of the belief that "developing the Amazon" represented a genuine solution to many of Brazil's problems, a view that had been largely shared in previous civilian administrations that had done less about it. Military planners were obsessed with the idea that the Amazon constituted a national security problem. They thought that there were too few people in the region and too little infrastructure to defend its borders and to guard against the kind of domestic insurgency that had arisen in relatively remote areas in other Latin American countries. Economic development would provide a structure for national defence (Foresta 1991).

The military also undertook what seemed to them an ambitious program of reserves to preserve species and unique ecosystems of the Amazon. They began this effort at a time when there was little domestic or foreign attention being paid to the Amazon, least of all to Amazon conservation. Their approach to the conservation of the Amazon was profoundly technocratic and authoritarian, but then again, so was the approach to conservation in much of the rest of the world. It was based on setting aside reserves serving specified purposes, among them, conservation of species and unique landscapes, research, protection of indigenous people, and tourism. Each reserve would be given a level and type of protection appropriate to its purpose. The selection of the number of reserves and their size and boundaries were mostly to be determined by scientific and technical advisers. The military enlisted an impressive group of ecologists, biogeographers, and other scientists, both Brazilian and foreign. The scientists argued from the most advanced ecological theory of the time based on island biogeography, a method that had been recently developed to study the rise and fall of species and of species diversity. It represented tremendous advances in ecological theory, but it was later seen to be inadequate and misleading in a number of important ways The generals proceeded to establish large reserves based mainly on the recommendations of the scientists, though somewhat adapted to what the planners saw as military or commercial necessity.

Time would demonstrate that the conservation planning process was deeply flawed for two reasons. First, much of the ecological theory turned out to be simply wrong. Second, the planners did not seriously consider the viewpoints, interests or survival strategies of those affected by the plans: indigenous people, peasant colonizers, ranchers, residents of small towns and cities of the region, and even, in some cases, significant commercial interests such as mineral prospectors. There was very little participation by such people; they were seen as the problem, as objects to be controlled. The reserves were not particularly well chosen for the goal of species preservation, they were often poorly delimited for the purpose of protecting indigenous people, and they proved to be very difficult to protect from intrusion by the many people and companies interested in exploitation of the land and resources. In direct contradiction of its conservation reserve plan, the regime's development plan was encouraging intrusion through provision of roads, electricity, and fiscal incentives (Foresta 1991).

Meanwhile, the region was beginning to break out in trouble spots that called increasing attention to the abuses of land barons, lumber and mining firms, large-scale cattle ranchers, miners, and the official planners of such large-scale projects as roads and hydroelectric dams. The official program to colonize large areas of the Amazon with small landowners was failing demonstrably, due to a combination of poor appreciation for the difficulties of settled agriculture under Amazon conditions, poor assessment of the potential of particular areas, lack of resources, incompetence, corruption, and malfeasance. Local resistance to these abuses and failures led to intimidation, beatings, and murders. These in turn inflamed further resistance and brought new organizations into the battle, particularly rural agricultural unions, Catholic religious orders, and even prominent members of the national Catholic hierarchy, including the National Conference of Brazilian Bishops (CNBB). In spite of fierce repression by local authorities and the military government, the official program of Amazon development launched in 1970 was creating international scandals by the mid-1970s with assassinations of rural leaders and foreign priests. In 1975, the CNBB founded the *Comissão Pastoral das*

Terras (CPT), or Bishops' Commission on Land, in specific response to growing conflict and violence in the Amazon. (The CPT continues today to play a significant role in promoting land reform and justice throughout rural Brazil.) By the end of the 1970s, the military's development program had created significant organized resistance, along with growing national and international awareness. The military's attempt to push rapid development of the region, counterbalanced with conservation reserves, was coming unglued at both ends (Wright and Wolford 2003: 181-206; Trecacanni 2001).

Similar unravelling occurred outside the Amazon. The military's program of rapid industrialization had changed Brazil irrevocably. While growing at a rate that doubled its size nearly every twenty years, the population became two-thirds urban rather than two-thirds rural in less than two decades. Industrialization urbanized the nation and made it a major world producer of ships, autos, and many other industrial goods. The Western Hemisphere's largest petrochemical complex was expanded on the edge of the bay in the city of Salvador da Bahia, one of Brazil's poorest large cities. The government offered promises of relative freedom from regulatory pressure along with fiscal incentives to industrial investors. Outrageous examples of industrial pollution were not only concentrated in places such as Cubatão, but were also scattered across Brazil's urban landscapes, polluting the water and air in ways that were often highly visible and obnoxious to everyone. Scientists who studied the problem were sometimes intimidated by government threats, even in cases where their studies had been financed by the government itself (Moreira Alves 1985; Gaspari 2002).

THEME THREE: INEQUALITY AND THE DEVELOPMENT OF THE BRAZILIAN ENVIRONMENTAL PERSPECTIVE

The size of the Gross National Product approximately trebled in the high growth years from 1960-75. During the same period, the enormous inequalities in the distribution of wealth and income became markedly worse. Overall growth certainly benefited the poor in many ways, but it also caused a great deal of distress among various sectors of society. Modernization of agriculture, particularly in Brazil's most fertile regions, led to even greater concentration of land in the hands of a few. People were forced out of the countryside and into cities with few employment opportunities and extremely poor living conditions. Cities with millions of inhabitants provided sewage drainage for only a minority, and the sewage of even the minority typically went into the ocean or waterways with little or no treatment. Potable water was a luxury item. Health care for the poor was nearly non-existent in many areas, while the poor were disproportionately exposed to the air and water pollution that was endemic to the industrial and urban growth. While some diseases came under much better control, diseases that had been rare previously, such as anacephalia and spina bifida caused by industrial pollution, appeared in some poor populations at unprecedented rates.

The social networks on which people had depended for survival were difficult and sometimes impossible to reconstruct in new urban neighbourhoods. The poor were often forced to live far away from their workplaces, with poor public transport. These problems were often most notorious in neighbourhoods in which the military had provided what they saw as hygienic housing, housing that met relatively high technocratic criteria for "livability." This type of housing was the kind featured in the recent Brazilian movie on drug gangs, *City of God*. City people of all classes began to be severely preoccupied with fear of rapidly growing rates of violent crime. While economic growth unquestionably provided enormous benefits, the price of such growth became increasingly evident to many Brazilians. When the country's economic growth slowed significantly in the mid 1970s, this became doubly true.

THEME FOUR: REBUILDING CIVIL SOCIETY AND CITIZENSHIP

Despite strong repression, new forms of resistance to the military regime were gaining strength all over Brazil by the late 1970s. The extremely fast economic growth of the late 1960s and early 1970s (as high as 11 per cent per year) was beginning to collapse under the burden of too many poorly planned investments and stagnation in the international economy. As a consequence, public and private debt increased quite dramatically. As growth slowed it became more difficult to bear the environmental and social costs, and the government was forced into austerity budgets. President Ernesto Geisel (1974-79) began to recognize the severity of the problem and announced an official *abertura* (opening) of the repressive valve controlling political expression. However, as unexpected waves of mobilization emerged, a new round of repression served mainly to arouse greater political activity in opposition to the regime (Moreira Alves 1985; Gaspari 2002).

Labour unrest in the city and state of São Paulo posed the most serious challenge to the regime. From the labour struggle would emerge a whole new political class of capable and dedicated people, most famously Luis Inacio da Silva (Lula), elected president of Brazil in 2002. His extraordinary leadership abilities and his background as a poor kid from the rural northeast who started as a shoe-shine boy and later qualified as a skilled drill-press operator combined to make him a very attractive symbol for a new kind of Brazilian politics. The success of the labour unions of São Paulo's ABC industrial complex, centred on automobile production, was the most important single element in a broad wave of resistance. Such resistance was remarkably achieved against determined repression, including the jailing of Lula and others.

The labour struggles in São Paulo, however, did not occur in isolation. In contrast with the necessarily clandestine operations of relatively isolated armed resistance movements, the battles waged by labour unions were highly publicized and enjoyed broad public support. They occurred within a national context of a flowering of numerous new organizations that were formed to address a wide variety of social grievances. These included organizations campaigning for basic civil rights and against political exile, torture, and murder of dissidents. In effect, such was the social support for these new organizations that some of them enjoyed support from powerful leaders within the élite business and financial community. Others fought the corruption, violence, and incompetence of local and state officials. While some were based on protests against grotesque examples of industrial pollution, such as in Cubatão on the coastline of the state of São Paulo, others were the organized reassertion of the demand for land reform. The most notable of these would come to be called the Landless Peasant Movement, or *O Movimento dos Trabalhadores Rurais Sem-Terra* (MST) (Wright and Wolford 2003).

The great variety of new organizations and campaigns that arose in the late 1970s and 1980s were not a unified force, however, as their interests and strategies varied. Moreover, the military regime did not automatically collapse, even though it did lose its informal and formal authority over a period of nearly a decade. During this time, and long after, an intoxicating sense of new possibilities continued to flourish, but it was tempered by the difficulty of building and maintaining organizations under difficult conditions. This difficulty created a degree of maturity and realism among those mobilized that might have been more fragile if the regime had given way in a single day, month, or year. However, in the early years, the sense of euphoria created by successful opposition to what had seemed like an implacable dictatorial regime was widely shared and important. Both in Brazil and internationally, many of the ideas and approaches of the older socialist and communist movements had been largely abandoned, and people were searching for new approaches to social change. Violence was nearly universally rejected, but aggressive civil disobedience was given new credibility by the São Paulo union movement and others. People were more open than before to listening to and supporting the ideas of others, primarily because they all wanted to contribute to replacing the old with something new.

Starting in the mid- to late-1970s, the nascent environment movement struggled to find its place amidst widespread social mobilization inspired by a dizzying variety of grievances and hopes. Fortunately for environmentalists, there was a great deal of sympathy and support across movements. Labour unions, and Lula specifically, lent significant and highly public support to the movement for land reform and for reform of housing and urban services. The battle for land reform was also a battle against landlords and companies who were often the chief despoilers of the environment. Particularly in the Amazon, people placed great effort and creativity into fashioning alliances between the poorest rural people, indigenous people, and environmentalists. These alliances were sometimes opportunistic and fragile, although they did endure in some cases. Chico Mendes, who had been educated by an aging socialist in exile in the Amazon, and who grew up in the socialist rural union movement, became the chief proponent of a model of Amazonian development meant to preserve the forest while enhancing livelihoods based on harvest of forest products and sustainable cultivation of others. Mendes was determined to emphasize the forest conservation side of his argument by environmentalists working in the Amazon, who needed him to strengthen their own cause against those who asserted that environmentalists worked against the interests of the poor. Mendes needed the environmentalists for their ability to organize urban support in Brazil and abroad, including organizations and foundations that could provide political leverage and funds. Mendes's assassination in 1988, perhaps one of the most famous examples of a death foretold, became an immediate international *cause célèbre* because of this alliance. The new civilian governments have felt impelled to adopt the alliance's proposal for demarcating *extractive reserves* of the kind proposed by Mendes (Revkin 2004).

Another example of overlapping interests comes from the history of the land-reform movement. The MST, the most militant and effective among land-reform organizations, opposed the government's program of expansion into the Amazon from its earliest days because it was in direct conflict with their own program. The MST's major weapon was land occupation, during which poor people occupied land and demanded its redistribution. When the government sought to destroy the MST and its predecessors, one of its chief methods was to try to lure away those who had occupied land by offering them free passage to a Promised Land in government colonization projects in the Amazon. For a variety of reasons, the MST was convinced that the promise was illusory, and it was also convinced that it had to break the power structure that governed land ownership, not simply accept frontier clearings. It was essential to the MST, then, to support campaigns against the government's Amazon program. Slowly, environmentalist organizations began to see that it also made sense for them to support land reform as an alternative to Amazon frontier development (Wright and Wolford 2003).

In its early years, the MST favoured a perspective that envisioned cooperative farms using highly mechanized, chemical intensive agriculture. Technologically, it was akin to the methods used by the richest, most successful Brazilian farmers, by US agriculture, and by farms in Cuba and the Soviet Union. As a result of disappointments with this model, and strongly influenced by international and national ecology-based movements, the MST became the champion of an ambitious form of agro-ecology.[4] The MST also worked with environmental organizations and individual field biologists to create projects that would support land-reform recipients while enlisting them to work in forest conservation activities. With time, environmentalists and land-reform advocates have had a tendency to learn from each other and be transformed by the experience (Wright and Wolford 2003).

Indigenous groups working to protect their territories and cultural integrity also found that environmentalists often made effective allies. This was particularly true in the campaigns against large mining and infrastructure projects of the sort financed by the national government, the World Bank, and the Inter-American Development Bank (IDB). Environmentalists in Europe and the United States were able to offer native peoples and Brazilian environmentalists concrete help in the form of political leverage applied in the countries providing financing. As was the case in the Mendes affair, foreign environmentalists were usually strongly attracted to campaigns that tied their concern for forest preservation to the interests of people seen as oppressed or excluded. This was also the case with the newly identified, though historically venerable, cause of the *quilombo* communities scattered throughout remote regions of Brazil and composed of descendants of runaway slaves. In the case of proposed hydroelectric dams on the Trombetas, a major Amazon tributary, environmentalists joined *quilombos* and indigenous groups in a successful international effort to cancel the project. These alliances were to become more effective when both indigenous people and *quilombo* communities were given strengthened territorial and political rights under the 1988 Constitution and implementing legislation.

Brazilian environmentalists concerned with urban issues were far more focused than their counterparts in the United States and Europe on issues of sanitation, housing, and palpable levels of pollution. These issues made for natural alliances with everything from small-neighbourhood defence committees to labour unions, public health groups, housing rights groups, fishing cooperatives, and groups militating for improved public transport for the poor. Many teachers and teachers' unions found environmental issues to be natural entry points for methods of education long popular in Brazil. Brazilian feminists tended to take environmentalism for granted and adhered to the idea that patriarchy is at least partially to blame for the world's environmental crisis. The very active and effective Brazilian anti-AIDS movement and the movements for gay rights also tended to associate themselves with environmental goals. Perhaps this was because inattention to pressing public health needs and homophobia were inevitably associated with the intensely conservative attitudes characteristic of supporters of the dictatorship, who are also seen as enemies of the environment. In any case, the strong tendency to see underlying commonalities among a broad range of social causes is a clear heritage of the history of resistance to the military regime.

Before the rather more limited term "environmental justice" came into widespread use in the United States, Brazilians were talking about *socio-ambientalismo*, or socio-environmentalism, "which stressed the need to address environmental ills and development strategies [in ways] that would benefit the poor at the same time" (Hochstetler and Keck, 2007: 206-07). This emphasis was made the theme of the 1992 Rio conference on Environment and Development, and it tends to be taken as a given in most environmental discourse in Brazil. Socio-environmentalism may be hard to define and even harder to put into practice in many particular cases, but conceptually it makes far more sense in Brazil and in much of the world than does a categorization that would separate efforts to reduce poverty from environmental protection.

As in the US, some older conservation organizations in Brazil that focused on creating and maintaining parks and reserves feared this inclusive interpretation of environmentalism. They worried both because socio-environmentalism might dilute what they saw as immediate and compelling efforts for landscape and species conservation, and because many such organizations were made up of middle- and upper-class people who had little understanding of poor people and organizations representing them. Members and staff of some organizations in Brazil, and some based in the US, insisted in describing Brazil's environmental problem as one created by "the rapists and the ants," that is, the landlords and corporations who raped the land and forests, and the poor who relentlessly nibbled the forests and land to death. In this view, making alliance with the ants seemed neither politically fruitful nor in the interests of the environment.

This point of view (while it can still be found in São Paulo, Brasilia, Washington, and San Francisco) did not flourish in the particular context of the end of the military regime and the building of a new civilian regime and new civil society. It was easily associated with the mentality of the dictators and their supporters. Most important, it stood squarely against what most people thought was the central task: to rebuild democracy, civil society, and a broad, shared sense of citizenship.

However, some of the most prominent international environmental organizations carried on projects in Brazil in the 1980s and 1990s that actively deepened inequities while attempting to protect nature. Forest and biodiversity preservation projects, for example, sometimes attempted to cut through the morass of conflicting claims to ownership of land in forested areas by simply compensating the most powerful land claimants for economic losses. Compensation was granted on the condition that poorer counter-claimants be removed from the area, without worrying about the injustices or violence such policies engendered (Wright 1992). In other cases, projects simply did little to address broader social and environmental concerns. As a result of intense Brazilian and international debate over these failings, by the mid-1990s virtually all conservation organizations recognized the necessity at least to be seen to address social issues and the need for broad participation. They dressed up their projects with elements of local participation and social equity, but often with little substance. The quality and character of such efforts were and remain central elements of public and private debate about the role of the various major conservation organizations, both Brazilian and international.

The bulk of private funding for conservation initiatives comes from international conservation organizations and the distribution of public funding through institutions such as the IDB and the World Bank. The Brazilian government itself is often heavily influenced by these international institutions. Brazilian environmentalists who seek such funding must either work directly for these organizations or form new non-governmental organizations (NGOs). Indeed, scores if not hundreds of NGOs have been founded primarily to obtain international funding for environmental work. The funders must be convinced that the NGO receiving funds is consistent with their goals and that they be competent and trustworthy. As a result, a great deal of the internal politics of Brazilian environmentalism is inevitably concerned with the sometimes challenging task of pleasing the international conservation organizations, while working in ways that fit within the often much broader agendas as understood by Brazilians. It has also meant an inevitable trend toward the professionalization of Brazilian organizations, with many Brazilian activists convinced that this diluted the participatory and creative character of the environmental movement.

This can be very complicated as, for instance, it is not always the Brazilians who are arguing for a broader agenda. Moreover, in some cases the argument for addressing broader concerns is a deliberate or inadvertent way to undermine legitimate conservation efforts; this is sometimes the case with money for supposed conservation purposes coming from the IDB and World Bank that is really meant more to promote economic growth. The politics and discourse in this whole area have been delicate and are often highly perilous for the Brazilian environmentalists engaged in them.

While this set of concerns would continue to develop over the decades during and after the collapse of the military regime, in the 1980s Brazilians were strongly focused on what could be done through the construction of a new body of laws and the formation of new political institutions. The intense criticism of environmentally destructive growth and the economic costs of repairing such notorious situations as the massive industrial pollution of Cubitao led even the military to adopt new environmental measures. In August 1981, the regime approved a National Environmental Policy and set up a system for environmental monitoring and planning at federal, state, and local levels (Law 6938). This measure created the National Council for the Environment (CONAMA), made up of representatives of industry, commerce, agriculture, government, labour unions, and NGOs. CONAMA was invested with legislative authority. The law established permit systems for polluting activities, a requirement for environmental impact statements and zoning requirements. Many of the measures were modeled on somewhat earlier efforts of the same sort in the key industrial states.

The build-up to the convention that would draft the new civilian constitution of 1988 involved intense campaigning by every interest group in Brazil. There was relatively little confidence in the effectiveness of the military's environmental legislation and it was hoped that, with the entrenchment of strong environmental provisions in the Constitution, it would be possible to gain more uniform and disinterested enforcement of environmental law. Environmentalists were disappointed that they did not succeed in gaining strong representation in the convention, however. They were delighted when the drafters of the Constitution nonetheless included strong and detailed environmental provisions, demonstrating concretely the breadth of appeal of the environmental message. Based on the provisions of the Constitution, the Brazilian Congress would pass a thorough body of legislation to implement its environmental provisions. While there was additional impetus to enact such laws from the 1992 Rio Conference on Environment and Development, the Constitution itself and the general sympathy for environmental goals made its passage quite likely in any case. By the end of the century the nation boasted a formal legal structure similar to that of the United States and most nations of Europe, and in principle as strong. Nevertheless, it would become progressively clearer than before that the key issues were the lack of enforcement of the laws rather than their insufficiency (Brazil 1988/1996; Hochstetler and Keck, 2007).

Through much of the 1980s and 1990s, environmentalists and everyone else interested in politics put enormous energy into building new political parties. A small Green Party (*O Partido Verde*) was founded in 1986 on the European model and was largely led by people who had returned from long political exile in Europe, and it had a brief series of successes. Fabio Feldmann, one of Brazil's most prominent environmentalists, worried that the Green Party would isolate environmental concerns from other issues—precisely what socio-environmentalism sought to avoid. Feldmann and most other environmentalists hoped to imbue as many parties as possible with an environmental perspective. Most of the numerous political parties expressed at least nominal sympathy with environmental goals, and no significant national party (as opposed to some state governors) took openly anti-environmental stands, though some favoured projects and policies that would do environmental harm (Hochstetler and Keck, forthcoming: 111-17).

The most important development in the complex history of party-building was the growth of the Workers' Party (*Partido dos Trabalhadores*), or PT, led by Lula. The PT enjoyed a great deal of prestige because it was formed by the people who were widely understood to be those who had done the most to bring about the end of the regime. The party attracted a broad array of people with overlapping as well as many contradictory goals. Through the 1990s, the party came to control many large municipal governments and several state governments. The 1988 constitution had given new powers to municipalities, and the PT worked with dedication and imagination to show what could be done to improve conditions in the cities and address long-standing inequities. Many improvements involved strong environmental elements, such as urban sanitation, water supply, and other public health measures. Land-reform and environmental activists throughout Brazil were also either PT organizers or worked closely with PT offices. Through such programs as "participatory budgeting" in municipalities, the PT had also demonstrated that the exercise of citizenship could be both exhilarating and effective. Lula's repeated failed attempts to win the national presidency were disappointing to party activists, but the campaigns did much to build the party and force it to prove itself to voters. In the 2002 presidential elections, one of Lula's greatest strengths was the reputation of the PT for integrity and creative pragmatism in solving long-standing problems at the local and state levels.

The balance sheet running up to the 2002 election was long and complicated with respect to environmental issues. The previous civilian administrations could claim credit for ending fiscal incentives for ranching in the Amazon in 1988, at the time thought to be one of the primary causes of deforestation. There were new laws requiring landowners in forested regions to maintain most of the land in forest; in many circumstances, the law required the retention of 80 per cent of the land in forest. These were backed up by new laws and new technical means for spotting and enforcing illegal logging and land clearing (Laurance et al. 2001). The administration of Fernando Henrique Cardoso (1995-2003) had shut down some of the most damaging small- and large-scale mining operations in the Amazon, and delayed or abandoned plans for some of the additional dams that had been projected in the region. It had newly demarcated reserves for conservation that, taken together with earlier reserves, by 2002 totalled three times the size of France and had indigenous protection over areas comparable to twice the area of France (Earthscan 2003; De Carvalho and de Mello 2004). They had begun a much-needed national reform of water law and had exposed massive fraud in the holding of land, especially in the Amazon.

In natural areas outside the Amazon, the civilian governments, including Cardoso's, had created new protected areas. In the Pantanal, the great marsh that supported the Western Hemisphere's most abundant populations of birds and other animals, environmental organizations and government had planned a more or less traditional conservation reserve program. Local people, mostly ranchers, objected strongly and insisted on a voice and an ongoing role in the management of the area. The result was a participatory process that resulted in a program based on ranchers' commitments to conservation management, largely undertaken by the landowners themselves, attracted by the prospect of tourist revenues funnelled to them. A hugely popular dramatic series, or *telenovela*, produced by the largest national television network, *A Globo*, ensured popular support and abundant tourist traffic to the Pantanal.

With government support, universities and schools established strong training programs for environmental professionals and technicians, raising the technical proficiency of most environmental agencies substantially. Working with state governments, the federal administration dramatically reduced industrial pollution in most regions and had taken significant steps to reduce automobile pollution, particularly in the city of São Paulo. Active public prosecutors, particularly in the state of São Paulo, used the new structure of environmental law, including a new federal law on environmental crimes, and older principles of public authority over land and resources to pursue cases aggressively, resulting in substantial improvements. In the face of organized opposition among Brazil's scientists and the public, the government had settled for a small nuclear energy program, while the military government had planned a much larger commitment. Most people still had no access to potable water and were not hooked up to sewage treatment facilities, but there were improvements and plans for more. Although the MST and other land-reform organizations decried the lack of attention to land reform, by 2002 there were more than half a million families settled on more than a million and a half hectares of land redistributed by the government—at most a few percentage points of the national agricultural land, but a beginning.

It was a fair judgment on the part of environmentalists that the accomplishments of government between the end of the military regime and the end of the twentieth century could be counted as the achievements of the countless individuals and hundreds of environmental organizations who had pressured for them. None of the administrations or parties in power in the period had any serious political or ideological commitments to environmental protection. Because environmentalists had succeeded in making their cause popular, politicians thought that taking environmental action made good press for voters. It was also a measure of the success of environmentalists that a large number of individual government officials in the federal government ministries had become convinced of the importance of environmental issues and were willing to work hard to implement environmental measures. In many cases, environmentalists had simply won in confrontations with the government on a number of environmentally damaging projects, such as the *Hydrovia* canal on the Parana River, that would have made it cheaper to ship soybeans but likely would have done irreparable ecological damage to the Pantanal.

Environmentalists worried that the most active among them were aging survivors of the military era, and new recruits to the cause seemed fewer than before. Many people interested in environmental issues chose to become professionals in the field and distance themselves from the movement-based environmental politics that was seen as critical to the older generation. Environmentalists could nonetheless pat themselves on the back for a great deal of work well done. It seemed clear that this was above all due precisely to the strengthening of civil society and citizenship (Hochstetler and Keck 2007).

Of course, there were many important things for which the environmental movement could claim no credit. Perhaps as important as any specific environmental accomplishments, under the leadership of President Cardoso, who was finance minister before assuming office, the nation had brought under control its staggeringly high and chronic inflation problem. This made it possible for both government and business to plan more rationally and effectively in all areas, including the environment, as well as to increase the confidence of investors in the economy.

THEME FIVE: THE CREDIBILITY CRISIS

However, Brazilians were far from satisfied with these accomplishments. The Amazon is one of the best examples. In spite of all the reforms, the rate of deforestation failed to decline significantly, and in some years it in fact spiked dramatically upwards. The use of remote sensing technology, new legal tools, and strengthened enforcement agencies failed to produce expected results in slowing deforestation. Soybean farmers—once a major concern of environmentalists because they were bringing devastation to the unique ecosystem and species of the savanna known as the *cerrado*—were beginning to move into the humid forests of the Amazon. A powerful state governor, proud of his title as The King of Soy, led this assault on the forest, while gaining support from the national government and international institutions for signing "best practice" agreements for his operations (Laurance et al. 2001).

Elsewhere, continuing air and water pollution tended to obscure the substantial progress that had been made. According to official statistics, for example, far less pollution than before entered Rio de Janeiro's Guanabara Bay, but the upper reaches of the Bay still stink and glisten with massive amounts of spilled oil, sewage, and industrial pollution. Industrial pollution in both the city and the state of São Paulo, according to the statistics, decreased markedly in the 1990s, and legislation required that new cars would soon have to meet pollution criteria comparable to standards in the US and Europe. Nonetheless, outrageous examples of pollution continued to flow into São Paulo's main canals and rivers from time to time, and the automobile pollution levels from notoriously gridlocked traffic sometimes competed with the worst in the world. In both the state of São Paulo and the national government, aggressive public prosecutors had pursued countless, and at times spectacular, cases of violation of environmental law. However, well-publicized examples of firms getting away with violating the law often seemed more important to people than those cases of effective enforcement whose consequences are difficult to perceive. When landowners literally got away with murder in repressing rural workers and environmentalists—as happened in hundreds of cases in the Amazon, southern Bahia, western São Paulo, and other regions during the 1990s—it seemed as though justice was hard to find, environmental or otherwise.

In the new century, there has been relatively little enthusiasm for simply adding new laws in the hope of improving environmental performance. The problem was the lack of implementation and enforcement. The deeper problem was the age-old problem of a state founded on deep inequalities of power and money. This was clearly seen in the financial and influence-peddling scandals that have plagued government at every level. One president, Fernando Collor de Melo (1990–92), was successfully impeached for stealing millions of dollars, and no president's administration has emerged untouched by serious scandal. Innumerable cases of local and state corruption that permitted forest destruction, rampant ongoing pollution, bankrupting of public agencies entrusted with building sewage and water systems, and other environmental outrages remained a large part of the daily news. That political leaders and administrators seemed nearly incapable of operating without indulging in avaricious and craven behaviour led many to conclude that things would never really get better.

The activists who supported the PT and the election of Lula in 2002 replied to this sad and cynical conclusion with the slogan, "A Better World Is Possible." This faith was based on two ideas. One was that the rebirth of civil society provided a broad foundation for society's renewal, through the continued work of the thousands of organizations that had sprung up for citizen action throughout Brazil. The second was that the PT in particular, with its growing reputation for honesty, creativity, and effectiveness gained through its management of state and local governments, could renew both faith in government and the creaky machinery of government itself. The voters had only to bring it to national power by electing Lula.

The voters responded, electing Lula by nearly a two-thirds margin in the run-off election for president. True, the PT did not win enough seats to control Congress, but the administration had prospects for forging alliances to get its essential program passed. The sense of elation and renewed hope in Brazil was powerful and palpable on inauguration day, January 1, 2003. Even a large share of those who had voted against Lula confessed to be caught up in the hope and enthusiasm. There was a sense that with the election Brazil had succeeded in rebuilding civil society and citizenship, for no one represented these things more than Lula himself.

Many people from environmental organizations had been PT activists and were recruited into ministries and staff positions in Congress. Lula chose a senator from the Amazon state of Acre, Marina Silva, as minister of the environment. Silva enjoyed enormous respect among environmentalists: she is a charismatic woman from a poor family in the Amazon and was a close associate of Chico Mendes in building the alliance for rainforest preservation. She has proven to be imaginative and dedicated, particularly with regard to the Amazon. She appointed Maria Allegretti, an anthropologist and environmentalist known for her work with Mendes and others in the Amazon, as special secretary for the Amazon. The Ministry initiated new, technically sophisticated get-tough programs of surveillance and enforcement to prevent deforestation. The government withdrew logging and mining permits that were either fraudulently obtained or abused (Hochstetler and Keck 2007). Lula made land reform one of the key themes of his inaugural speech and chose old land-reform advocates to direct the land-reform efforts.

However, the performance of the Lula administration has been largely disappointing to environmentalists. In spite of the new programs, rates of deforestation have climbed (Asner 2005). Soybean farming has accelerated its march into the forests. Ranchers have discovered that they can turn handsome profits without government subsidies, and they continue to expand their holdings into the Amazon. While the government has long since abandoned large-scale colonization in the Amazon by poor settlers, the settlers keep coming on their own initiative. The Lula administration continues to push roads further into the frontier, inevitably promoting timber operations, settlement, and other destructive activities. Violence in the Amazon and elsewhere in rural Brazil continues, punctuated in 2005 by the brutal murder of the American nun, land reform activist, and environmentalist Dorothy Stang. As usual, the culprits were gunmen hired by ranchers.

The Lula administration has also opened the door to genetically modified crops in Brazil, disappointing environmentalists who had fought successfully against GMOs (genetically modified organisms) for more than a decade and thought Lula would support them. Lula has taken steps to revive Brazil's civilian nuclear energy program, and he has also pushed hard for a diversion project on the São Francisco River that many in the region, including a Catholic priest who went on an extended hunger strike in protest, and most Brazilian environmentalists fiercely opposed. The pace of land reform has been speeded up considerably, but far short of the various moving targets the administration has set for it, and even shorter of the expectations of the land-reform movement. One year into his administration, more than 500 environmental organizations and prominent activists sent a public letter of protest to Lula regarding the poor environmental performance of his regime, and the next year, Allegretti left her post in frustration (Hochstetler and Keck 2007).

Lula's government has kept the promise Lula made before his election to nervous financiers and political allies outside the PT to practise fiscal austerity and pay down Brazil's debt. The financial community that once feared Lula to the point of near hysteria now praises him for his fiscal policy, as do the IMF and the US State Department. Economic growth has been on average much stronger than in the previous administration of Cardoso. Lula has undertaken a number of foreign-policy initiatives that have put Brazil among the leaders of countries working for international trade, finance, and intellectual property agreements better designed to serve the interests of the world's poorer countries. Various programs in health, education, and culture have enjoyed modest success. Social-service and welfare programs have improved the income of the poorest Brazilians, and it is striking that as support for Lula has been drying up among the middle class, his support remains highest among the poorest people and in the poorest regions. In spite of environmentalists' frustrations, Lula enjoyed high approval ratings into 2005.

In that year, the country was in for another rude shock. The Lula regime was hit by a complicated set of bribery and corruption scandals that nearly brought government to a halt for some months. Lula's chief advisers and the highest officials in the PT were forced to resign. Scandal after scandal revealed that the PT no longer deserved its reputation for integrity. Even though the most serious of the corruption schemes was initiated by the party leadership to advance its broad legislative program and not for the personal enrichment of the party officials themselves, the behaviour was clearly unethical and much of it almost certainly illegal. At the time of this writing, Lula has not been proven to be personally involved in the scandal, though reasonable deduction would make it hard to believe he did not know of the operations. His administration has been seriously shaken and must move forward without many of its most capable leaders.

CONCLUSION

So, once again, Brazilians have been disillusioned by their government. Those who had earned the greatest confidence through their decades of long and courageous work to bring down a dictatorship and rebuild a society of active citizenship and democracy were among those who played the most prominent role in the scandals. While the government has proven competent and effective in leading economic growth, it has had no spectacular successes in its agenda for the reduction of poverty, and its environmental record has been lacklustre at best.

Lula has repeatedly and publicly warned the country that it should expect no miracles from his government. The task of creating a genuinely democratic and reasonably egalitarian society, he has said, is a job that will take at least fifty years. Indeed, if environmentalists and other social activists are to take a lesson from the experience of the last few years, it is that the construction of citizenship and civil society cannot depend on heroic figures or on particular governments, but must be built brick by brick, day by day, perhaps over generations. As a prominent Brazilian lawyer and urban planner concluded in the summer of 2006, "in the last analysis, the future of environmental policy will depend on society's renewed political mobilisation within and without the state apparatus" (Fernandes 2006). The great challenge is to work fast enough to avoid unacceptable levels of irreparable environmental harm in the meantime.

Notes

1 I am strongly indebted to Kathryn Hochstetler and Margaret Keck for allowing me to consult the unfinished manuscript prior to its publication, *Greening Brazil: Environmental Activism in State and Society*. I have relied on it heavily here. While our interpretations are compatible, they are not identical. When published, their book will no doubt become the definitive text on the Brazilian environmental movement. I also want to give special thanks for the insights of Jose Augusto Padua.

2 "Dependent capitalist development" is a term deriving from an analysis that attributes many of the problems of relatively poorer nations to their subordinate place in an international economic and political structure. Dependent capitalist development is characterized by some mix of the following: 1) a disproportionately large share of national production owned or controlled by foreigners and/or a particularly small and politically powerful domestic class; 2) a relatively large share of production being for export to international markets; and (3) political arrangements that make it difficult for workers to make gains in wages and social benefits, usually because those who own and control production are able to restrict broad effective local political participation. Production tends to be centred on products requiring relatively low levels of education and skill. Owners have more incentive to sell their products to richer foreign markets than to increase the limited purchasing power of the local population. When national governments and political movements threaten to change this structure, private corporations and financial institutions can combine their own power with that of such international institutions as the World Bank and the International Monetary Fund and the intelligence and military establishments of the more powerful nations to undermine or destroy the initiatives toward change. The structure supporting these arrangements was in most nations constructed under formal colonialism and maintained and sometimes reinforced in the post-colonial era. Various forms of economic nationalism have been directed at trying to reduce the ill-effects of this structure, as in Brazil in most of the years from 1930 to 1964, or destroy it, as in Cuba under Fidel Castro. The literature on the topic is vast and the complications

enormous. Interestingly, Brazil's president Fernando Henrique Cardoso (1995-2002) was long recognized internationally as a pioneer in the development of theory on the topic, but when he came to power he governed through policies that most observers (including, he confessed, himself) considered to be a rejection of the theory he had been so important in formulating.

3 The Jesuits were expelled from both Brazil and Spanish America during this period. The primary reasons were probably that the colonial governments saw them as a threat to government authority, and they tied up land and labour that private entrepreneurs were anxious to exploit. While they were successful agricultural producers, for several reasons they did not contribute much to tax revenues. They were relatively immune to political pressure from governmental authority and powerful landowners. The Jesuits also aroused suspicion and anger among other Roman Catholic religious orders. Their effects and influence on indigenous people and on the land remain highly controversial. There seems little doubt, however, that while they were certainly paternalistic and often brutal, their record as exploiters of land and labor was on balance at least less shameful than that of private landowners.

4 Agro-ecology, perhaps best defined by the widely used text of Miguel Altieri, a Chilean who now works directly with the MST, seeks to go beyond the concept of organic agriculture. Instead of defining itself largely by what it does not do—organic agriculture rejects the use of most synthetic fertilizers and pesticides—agro-ecology defines itself as a more ambitious and comprehensive approach to understanding and practising agriculture. It sees agriculture within the terms of ecological science, with the purpose of providing stable livelihoods and food production while providing ecological benefits and minimizing ecological damage (Altieri 1995).

References

Amado, Jorge. 1943. *Terras do Sem Fim.* São Paulo: Editora Martins.

Asner, Greg. 2005. "Satellite Study Doubles Forest Disturbance Estimates in Brazil—Impacts Widespread." Washington, DC: Carnegie Institution.

Brazil. 1988/1996. Constitution 1988 with 1996 Reforms. Title VI. Environmental Protection.

Correa do Lago, Pedro. 2001. *Iconografia Brasileira.* São Paulo: Itau Cultural.

da Cunha, Euclides. 1944. *Rebellion in the Backlands*. Trans. Samuel Putnam. Chicago: University of Chicago Press.

Dean, Warren. 1995. *With Broadax and Firebrand: The Destruction of the Brazilian Atlantic Forest*. Berkeley: University of California Press.

De Carvalho, Karla Bento, and Milton Thiago de Mello. 2004. "Indian Reserves and Protected Areas in the Brazilian Amazon." Paper presented at the 3rd IUCN World Conservation Congress, Bangkok.

Earthscan. 2003. "Country Profiles: Biodiversity and Protected Areas—Brazil." http://earthtrends.wri.org.

Fernandes, Edesio. 2006. "Sustainable Development and Environmental Policy in Brazil: Confronting the Urban Question." Brazil Network. www.brazilnetwork.org.

Foresta, Ronald A. 1991. *Amazon Conservation in the Age of Development*. Gainesville: University of Florida Press.

Gaspari, Eloy. 2002. *As Ilusoes Armadas: A Ditadura Envergonhada, vol. 1, A Ditadura Escancarada, vol 2.* São Paulo: A Companhia das Letras.

Hochstetler, Kathryn, and Margaret E. Keck. 2007. *Greening Brazil: Environmental Activism in State and Society.* Durham, NC and London: Duke University Press.

Holanda, Sergio Buarque de. 1969. *O Visão do Paraiso: Os motivos edenicos no descobrimento e colonizacao do Brazil*. 2nd ed. São Paulo: Cia Editora Nacional.

Laurance, William, Ana K.M. Albernaz, and Carlos da Costa. 2001. "Is Deforestation Accelerating in the Brazilian Amazon?" *Environmental Conservation* 28: 305-311.

Lavigne, Eusinio. 1967. *Cultura e regionalism cacaueiro*. 2nd ed. Recife: Editora Cultura Brasileira.

Moreira Alves, Helena. 1985. *State and Opposition in Military Brazil*. Austin: University of Texas Press.

Padua, Jose Augusto. 2002. *Um sopro de destruicao: pensamento politico e critica Ambiental no Brasil escravista*. Rio de Janeiro: Jorge Zahar Editora.

Revkin, Andrew. 2004. *The Burning Season: The Murder of Chico Mendes and the Fight For the Amazon Rain Forest*. 2nd ed. Washington, DC: Island Press.

Skidmore, Thomas. 1988. *The Politics of Military Rule in Brazil, 1964-1985*. New York: Oxford University Press.

Sodero, Fernando Pereira. 1990. *Esboco historico da formacao do direito agrario no Brasil*. Rio de Janeiro: FASE.

Stein, Stanley. 1985. *Vassouras: A Brazilian Coffee County, 1850-1900. 2nd ed*. Princeton, NJ: Princeton University Press.

Trecacanni, Girolamo Domenico. 2001. *Violencia e grilagem: instrumentos de acquisacao da propriedade da terra no Para*. Belem: UFPA, ITERPA.

United States State Department: Bureau of Western Hemisphere Affairs. 2006.

"Background Note: Brazil." http://www.state.gov/r/pa/ei/bgn/35640.htm.

Wright, Angus. 1992. "Land Tenure, Agrarian Policy, and Forest Conservation in Southern Bahia, Brazil—A Century of Experience with Deforestation and Conflict Over Land." Paper presented at the Latin American Studies Association Meeting, Los Angeles.

Wright, Angus, and Wendy Wolford. 2003. *To Inherit the Earth: The Landless Movement and the Struggle for a New Brazil*. Oakland, CA: FoodFirst Books.

CONCLUSION

Jordi Díez

In the Introduction to this collection, O.P. Dwivedi and I argued that the experience of developing countries with environmental management must be looked at in relation to the process of globalization, the international environmental agenda, and the strengthening of civil society actors. In this book we have presented eleven case studies from across the Global South with the intention of learning more about the effects of international actors, forces, and institutions on the environmental politics and policies of developing countries within the larger process of globalization. The cases provide invaluable insights into the challenges and opportunities that developing countries have faced in instituting and implementing the necessary environmental management tools to foster more sustainable development practices. They also provide us with a truly great opportunity to draw general cross-national lessons on these national experiences. In this short conclusion, I therefore present very succinctly some of the most salient general lessons that can be drawn from our cases.

Chapter 1 argued that the latest phase of the process of globalization, which has been characterized by its scope and speed, has brought about numerous challenges to the natural environments of countries of the Global South. The case studies of this volume seem to confirm this argument. Most countries studied in this book have undergone important processes of economic reform. Two of them, Chile and Lebanon, began these reforms in the 1970s, but by the turn of the century most had adopted market-friendly economic policies based on neo-liberal tenets. In some countries, the models have not been completely based on strict neo-liberal models, since parts of their economies have been left protected, such as Thailand and Indonesia. In the case of China, economic reform has been introduced in some economic sectors, while reform in others has not taken place. However, the economies of all the case studies presented have been almost completely reoriented toward export markets. As a result, their economies have become increasingly linked with the forces of global capitalism. The integration of these economies into the international economic and trade system appears to have been accompanied by numerous environmental challenges. To be sure, in most cases signs of environmental degradation were present before the acceleration of global economic integration. But the adoption of economic policies based on exports points to a deepening of environmental degradation. Because the comparative advantage of these economies rests on the export of commodities and agricultural goods, the increase of exports of these products has consequently placed more pressure on natural environments. Whether it is aquaculture farming in Indonesia, soy cultivation in Brazil, orange production in Chile, avocado farming in Mexico or timber extraction in Thailand, in all of the countries studied in this book, increased demand for these products in international markets has resulted in increased demands for natural resources and a subsequent expansion of agricultural land. The consequence of this phenomenon has been continuously high levels of deforestation, soil erosion, and desertification. The majority of the countries possess vast forest covers. Indeed, the countries included here possess more than half of the world's tropical and temperate forests. Yet none of the contributors identified any tangible decline in deforestation rates and, consequently, biodiversity loss, soil erosion, and the increase in greenhouse gases continue apace.

More importantly, however, poverty has not been reduced in any significant way. In many of the countries under study, the restructuring of economic reform has brought about tangible economic growth. This is the case for our Asian case studies and Chile, and to a lesser extent Brazil and Mexico. But in the other cases, it appears that poverty, the main cause of environmental degradation in the developing world, has not been reduced and in certain instances it has in fact worsened. A manner in which continued poverty has manifested itself is through the significant migration that has taken place from rural to urban areas. Because of the inability of the rural poor to find employment in rural areas, and because urban areas offer more employment opportunities, in most countries, especially the larger ones, urban populations have increased dramatically. This is even the case for China, whose economy has grown at an astounding pace over the last two decades; the integration of its eastern coastal regions with global markets has translated into significant migration into its eastern cities. Such migration has applied increased pressure on local governments to provide basic services such as sewage and potable water. Given that in the Global South municipal governments generally tend to be underfunded, as several contributors have made clear, they have not been able to deliver these services, and the result has been an increase in urban waste pollution and an array of related health problems for the urban poor. In many cases, these problems have been compounded by the privatization of some services, such as the provision of water, which has increased its price and made it even less accessible. As Rebecca Tiessen succinctly put it in her chapter, "Neo-liberal economic reforms therefore exacerbate social and environmental challenges by reducing access for low-income households to basic necessities such as water and wood for cooking" (209).

These pressures on the environment in the Global South have at times been exacerbated by the role played by certain international institutions that have become key international players within the new globalized world. The World Bank is perhaps the most salient example; it is the tutor and sponsor of economic neo-liberal reforms for developing nations, and its influence on their economic policy has increased dramatically. While the World Bank has increasingly paid attention to environmental concerns, its primary aim has been to foster export-led economic growth in the Global South. As a result, many of the policies and programs that it recommends support primarily economic growth and do not have environmental concerns as a priority. Indeed, in many cases, the Bank has directly financed, either partially or wholly, projects that have had disastrous environmental consequences, such as the Sardar Sarovar dam in India, the Bujagali dam in Uganda, and the Pangue dam in Chile.

Clearly, as these contributions demonstrate, the manner and degree in which globalization has brought about environmental challenges to the Global South has varied across our case studies. But it does appear, based on our eleven cases, that globalization has indeed brought about numerous challenges.

As also argued in this book, the process of globalization has not been limited to economic integration; increased global interconnectedness has facilitated communication among people around the world, and this has contributed to both the strengthening of transnational links and interaction among non-state actors, which has allowed for the transnational organization and mobilization of various groups. As we saw, the environmental movement has been at the core of this phenomenon; the number of environmental non-governmental organizations (ENGOs) has increased dramatically over the last two decades, as have their international visibility and influence. The growth in the number of international ENGOs has been characterized by an increase in their operations worldwide, especially in developing countries. What general lessons can we draw from the various contributions in regard to the role played by these organizations in the environmental management of the Global South?

Two general lessons can be drawn. First, in all of the countries presented here, it appears that, similar to international trends, there has been an increase in the number of ENGOs. In most cases, the increase in numbers occurred during the 1980s and accelerated in the 1990s. The emergence of these organizations has been partly the result of political opportunities that have emerged in developing countries as they have experienced processes of political liberalization and democratization. These democratic openings have allowed for the mobilization of certain groups, and the environmental movement has clearly been one of them. Domestic political conditions seem therefore to contribute to the emergence of environmentalism in the Global South, and in most cases it has occurred within a larger re-ordering of a relationship between states and civil society. This has certainly been the case for Thailand, Mexico, South Africa, Brazil, and Chile. The one case that clearly stands out from this collection is China: although the country has not been part of the so-called Third Wave of democratization, it has also experienced the emergence of ENGOs and, as Stephen Ma shows, they have begun to serve as "conduits" between government officials and citizens and have had an influence on raising environmental awareness and prompting further citizen mobilization. While domestic political conditions have been determinant in the strengthening of environmental mobilization, it has also occurred as a result of the close associations formed between international and domestic ENGOs through the administration and management of joint environmental projects. This has clearly been the case for Mexico, Indonesia, South Africa, and Chile. In all cases, the emergence and proliferation of ENGOs has been the result of increased funding from international organizations and donors. Indeed, in some countries, the funding received by ENGOs comes entirely from foreign sources, and they have dramatically been empowered by funding from international donors.

Such increased interaction between southern environmentalists and their northern counterparts, as well as increased funding, has resulted in the proliferation of ENGOs in the Global South, but it does not necessarily mean that environmentalism is new in developing countries or that it is strictly the result of international influence. In fact, as many of the contributors to this collection have shown, in several countries environmental awareness and mobilization have preceded the latest international environmental wave. As Angus Wright shows, for example, Brazilians' environmental awareness has been an important part of the country's popular consciousness for centuries; environmental movements in Brazil are old and not foreign imports. In India and Indonesia, environmental mobilization has emerged from local indigenous cultures and traditions, making them quite distinct from contemporary international environmental movements. Moreover, growing cooperation between northern and southern environmentalists does not mean that they share environmental priorities and norms; in some countries, frictions have characterized their relationships and in others local environmental movements have in fact challenged the activities of northern-based ENGOs as they oppose the international norms they pursue.

The dependence of some environmentalists in developing countries on international funds has meant that, in some cases, they have undergone a "professionalization" of their activities as they have become highly institutionalized through well-established and well-funded ENGOs. In countries such as Mexico and Brazil, this has meant that these ENGOs are mostly operated by individuals from middle- and upper-class backgrounds, which has in turn alienated groups that have a more grassroots orientation. Moreover, this dependence has also resulted in competition for scarce resources among environmentalists in the Global South, something that has contributed to the fragmentation of their movements.

Second, ENGOs have not only increased in number, but they have also become more visible and in many cases active and influential actors in environmental management. As several of our cases show, ENGOs have been very influential in pushing environmental issues to the political agendas in many developing countries, thus contributing to raising environmental awareness and equipping citizens to demand better environmental protection from their governments. Beyond agenda-setting, ENGOs appear to have been successful in gaining participatory spaces in the formulation of environmental policies in the countries under study. In part, the opening of such spaces was the result of commitments made by countries at the Rio Summit of 1992, at which, as we may recall, signatory countries agreed to allow civil society actors a say in environmental management. As a consequence, ENGOs do appear to have become participants in environmental decision- and policy-making processes in the Global South. As Jorge Nef and Christopher Gore show in their contributions, participation has not always translated into influence in both Chile and Uganda. However, environmental mobilization has often resulted in some rather important policy triumphs, such as the cancellation of some projects that would have had important environmental repercussions (Mexico, Thailand, and India) and the modification and execution of others (China). In effect, in the cases of Brazil, Mexico, and Nigeria, the emergence of environmentalism has been accompanied by the establishment of green political parties and the appointment of renowned environmentalists as environment ministers.

Moreover, in some cases, they have had a direct and tangible impact on environmental policy-making and management. This has clearly been the case for the Indonesian, Mexican, and South African cases; in the three countries, ENGO influence has been substantial. Indeed, in many of the countries analyzed in this volume, the enactment and reform of environmental policies were characterized by highly consultative policy-making processes. Importantly, ENGOs have managed to secure permanent spaces of influence in environmental decision- and policy-making, partly as a result of commitments made at the Earth Summit, which means that further opportunities to influence environmental management have become institutionalized. ENGOs have also contributed to making governments more accountable (South Africa, Mexico, and Thailand), proposed alternative solutions (Thailand and South Africa), pressured governments to force corporations to release information on industrial pollution (South Africa and Mexico), and contributed to environmental education (Chile, China, South Africa). In fact, in some countries, ENGOs are more effective than governments in monitoring and implementing some environmental projects and programs (China and Indonesia). In many cases growing influence by ENGOs has been the result of the increased recognition by governments of the important role that ENGOs play as they have become important political players in their own right. The inability of government institutions to tackle environmental problems has provided an opportunity to ENGOs to fill the vacuum. In some cases, environmental projects that involve ENGOs appear to be more successful (Indonesia and China).

Nevertheless, the increased influence of civil society actors in environmental politics does not automatically translate into positive outcomes in environmental management nor does it mean that relations between governments and ENGOs are always without quandaries. In effect, as Christopher Gore and Paul Kingston show in their contributions, the increased influence of ENGOs has been problematic. Christopher Gore argues that, in Uganda, increased influence by civil society actors in environmental issues has created deep tensions between them and the government. As he argues, donors claim that civil society should assist in managing environmental resources, holding governments accountable while national governments, who are implementing donor-designed reforms, are often antagonistic to civil society organizations that challenge their policies, labelling them as anti-development, "enemies of the state" or "economic saboteurs" (157). The execution of environmental policies, programs, and projects therefore becomes more difficult. Gore argues that, as a consequence, the already "messy" processes and practices of development and the environment become messier through the presence and dominance of actors that are agents of global norms. He suggests, then, that "it is important to understand how global actors work through and in domestic decision-making, and how the actors and ideas inherent to globalization become embedded in the domestic processes where decisions over resource extraction, policy, law, protection, conservation, and management are scrutinized" (157). In the case of Lebanon, Kingston argues that NGOs have formed strong linkages with international developmental agencies that fund their environmental activities, something that has allowed them to intervene deeply into political ecologies of developing countries, thereby serving to legitimize their right to intervene in the environmental affairs of the Global South. As Kingston shows, this intervention has in turn set off processes that have led to significant and often counterproductive restructurings of local environmental governance; such intervention has empowered local elites, who have created and controlled "environmental monopolies." As a result, foreign donor intervention has had negative repercussions for civil society organizations involved in environmental advocacy, pre-empting and disempowering some local, progressive actors. The growing presence and influence of ENGOs in environmental management in the Global South have not always had positive effects, as the Lebanese case demonstrates.

The process of globalization and the strengthening of transnational civil society actors have unfolded against the backdrop of collective efforts at the international level to deal with environmental problems, as we have witnessed increased environmental awareness across the world and the establishment of an international environmental regime. What have been the responses in developing countries to these international factors and how effective have these management tools been?

All of the countries analyzed in this volume have made serious attempts to adopt environmental management tools to conform to the commitments upon which they agreed at international fora. Based on the contributions to this volume, it is apparent that all of these countries have adopted a truly unprecedented array of environmental institutions and laws over the last twenty years. Many countries adopted some pieces of environmental legislation in the early 1970s, such as China, Mexico, and India, following the 1972 Stockholm conference. However, the Rio Summit of 1992 appears to have given new impetus to the adoption of environmental management tools; most environmental legislation that exists in these countries was enacted during the 1990s. Moreover, in several countries analyzed in the volume (e.g., India, Mexico, Thailand, Uganda, South Africa, and Brazil), environmental provisions have in fact been entrenched into the countries' constitutions.

Ministries and other environmental institutions were also established during the 1990s and, in countries in which some environmental institutions already existed, they were significantly reformed. The various conferences have therefore been highly influential in the establishment and reform of environmental institutions. International financial institutions have also been influential. As the African cases demonstrate, during the early 1990s the Bank initiated a drive for African countries to develop National Environmental Action Plans (NEAPS). These plans required the establishment of environmental institutions and administrative units. The creation of new environmental policies and institutions in some countries has therefore been largely due to requirements made by the World Bank. The institutionalization of environmental management has also been the result of the desire of developing countries to secure international funding for environmental protection. As we may recall, the Rio summit was held soon after the establishment of the UN's Global Environmental Facility (GEF), an institution tasked to channel resources to developing countries for the implementation of environmental protection measures. The establishment of environment ministries and institutions was thus needed to secure these funds. In some cases, these institutions have become highly dependent on international funding. As Gore shows in Chapter 7, Uganda's environment ministry is wholly funded by the World Bank.

Despite the importance of the establishment of these environmental management tools, our case studies point to numerous obstacles to the effective functioning of these environmental institutions and legislation. In almost all cases, the contributors point to significant implementation problems. Often this has been the result of the limited resources that have been assigned to the various environmental institutions. Inadequate resources in turn result in limited expertise by government officials on environmental issues (India and Mexico) or inadequate administrative capacity (Mexico, India, Chile, Nigeria, and Brazil).

The creation of environmental institutions and the enactment of legislation took place mostly at the beginning of the 1990s, when the majority of developing countries had just begun to recover from the Debt Crisis and when many were in the midst of implementing economic reform programs. Both of these developments meant a considerable reduction in government spending across all policy areas, including the environment. However, weak policy implementation has also been partly the result of a lack of political will on the part of governments to enforce environmental regulation. As several contributors to this volume have noted, frequently weak environmental enforcement is intentionally promoted not to discourage foreign investment. In some countries, weak implementation is a result of the various problems that arise from the overlapping of responsibilities across levels of government and administrative units, given the highly fragmented composition of their bureaucracies, a lack of inter-governmental coordination, and the manner in which decentralization processes have unfolded. As several authors note, policy implementation responsibilities have in many instances been decentralized to municipal governments without the required funds to carry them out. In some countries, environmental legislation has often been enacted with vague language, making it difficult to implement.

Weak environmental compliance and policy implementation in several countries is also the result of the strong influence exerted by corporate and business groups. In some countries, such as India and Mexico, the influence of the private sector has resulted in the adoption of voluntary compliance schemes whereby certain businesses possess high levels of discretion with respect to their compliance with environmental regulations and standards. In other countries, such as Chile and China, the influence exerted by businesses means that governments deliberately chose not to enforce environmental regulations. Finally, partly due to the inadequate resources assigned to environmental institutions, in several countries environmental officials engage in corruption activities that impede full enforcement.

It appears, based on the cases presented here, that analyses of the experience of environmental management in the Global South must take into account an international dimension. The process of globalization has brought about numerous environmental challenges as developing countries have re-oriented their economies toward global markets. But globalization has also allowed civil society actors and governments to tackle these challenges collectively. The cases presented here show that, partly due to stronger civil societies, and partly due to the influence exerted by the international environmental agenda, developing countries have not only paid more attention to environmental issues, but they have also acted by instituting numerous environmental management tools. However, as we have seen, environmental institutions and laws will not be effective without a healthy dose of political will, resources, and administrative efficiency. As it stands, it would appear that such a dose is not forthcoming. Since the repercussion of environmental degradation is global in nature, the responsibility to make it happen is collectively shared by countries of the Global North and South and by civil society actors. As our environment continues to deteriorate, we have a mammoth challenge ahead of us in this century.

NOTES ON CONTRIBUTORS

JORDI DÍEZ

Jordi Díez is assistant professor of political science at the University of Guelph, Ontario, Canada. He has conducted research on environmental politics and policy in Latin America (with a special emphasis on forestry policy), North American security, and civil-military relations. He is the author of *Political Change and Environmental Policymaking in Mexico* (Routledge, 2006).

O.P. DWIVEDI

O.P. Dwivedi is University Professor Emeritus in the Department of Political Science at the University of Guelph, Ontario, Canada. He is a past president of the Canadian Political Science Association, president of the Canadian Asian Studies Association, and a former vice-president of the International Association of Schools and Institutes of Administration. He has authored, co-authored, and edited 34 books and over 120 articles and chapters in books and scholarly journals. He has been consultant to the World Bank, WHO, UNESCO, UN-ESCAP, UNO, and CIDA. In 1999, he was elected Fellow of the Royal Society of Canada. Professor Dwivedi was bestowed the Order of Canada in 2005.

CHRISTOPHER GORE

Christopher Gore is assistant professor in the Department of Politics and Public Administration at Ryerson University in Toronto, Ontario, Canada. He has conducted field work in East Africa and has a background in environmental studies and comparative public policy, administration, and politics. His most recent publications include book chapters and journal articles on energy-sector reform and environmental policy in East Africa, as well as cities and climate change in North America. His research focuses broadly on how multi-level interests converge and influence urban and environmental policy prescriptions in East Africa and Canada. Future research will focus on food and energy security, climate change policy, and the relationship between infrastructure and development.

CRAIG A. JOHNSON

Craig Johnson is associate professor of political science at the University of Guelph, Ontario, Canada. He has published widely in the field of Asian politics and international development, primarily on issues relating to governance, decentralization, vulnerability, and epistemology in development studies. His most recent publication (co-edited with John Farrington, Priya Deshingkar, and Daniel Start) is *Policy Windows and Livelihood Futures: Prospects for Poverty Reduction in Rural India* (Oxford University Press, 2006).

PAUL KINGSTON

Paul Kingston is associate professor of political science and international development studies at the University of Toronto, Scarborough, Ontario, Canada. His work focuses on the history and political economy of development in the contemporary Middle East, and he is presently completing a book on civil society, non-governmental organizations, and advocacy politics in post-war Lebanon.

STEPHEN K. MA

Stephen K. Ma, a Fulbright Senior Specialist in Public Administration and in Political Science, is professor of political science and director of the Institute for Executive Leadership at California State University, Los Angeles, California. He has authored or co-authored and edited or co-edited eight books, in English and Chinese. His research articles have appeared in *Pacific Affairs*, *Asian Survey*, *Journal of Contemporary China*, *Chinese Public Administration*, *Asian Journal of Political Science*, *International Journal of Public Administration*, and *Policy Studies Review*. He has contributed to the *Handbook of Economic Development* (1998), *Administrative Reform and National Economic Development* (2000), *Where Corruption Lives* (2001), *Administrative Culture in a Global Context* (2005), and *Public Administration in Transition: A Fifty-Year Trajectory Worldwide* (2007).

BRUCE MITCHELL

Bruce Mitchell is professor of geography and serves as associate provost, academic and student affairs, at the University of Waterloo, Ontario, Canada. His research has focused on integrated water resource management, with particular attention to policy, institutions, and governance. He was a team member and leader in two CIDA-funded projects in Indonesia that extended over 15 years, one in Bali and one in Sulawesi. He is a Fellow of the Royal Society of Canada, and a Fellow of the International Water Resources Association.

JORGE NEF

Jorge Nef is professor of rural extension studies and international development at the University of Guelph, Ontario, Canada, and is director of the Institute for the Study of Latin America and the Caribbean (ISLAC) of the University of South Florida. He has field experience in Latin America and the Pacific and has been a visiting professor and researcher in universities in Canada, the US, and overseas. His areas of research and interest are human security, political economy, inter-American relations, global studies, public administration/management and public policy, research methodology and epistemology, and higher education, He has written or edited 14 books and monographs and produced over 120 articles in refereed journals and books. A past president of the Canadian Association of Latin American and Caribbean Studies, he is the recipient of many teaching awards and recognitions throughout his career (such as the lifetime OCUFA Award for Teaching

Excellence), and he has been elected to be a member of the World Academy of Art and Science. Currently, he is the president of the Canadian Association for the Study of International Development (CASID). He is also a poet, having published one book of poems while numerous of his works have appeared in international anthologies.

GODWIN ONU

Godwin Onu is professor in the Department of Political Science of Nnamdi Azikiwe University, in Awka, Nigeria. He teaches in the fields of international economic relations, comparative politics, and public administration, among others. His areas of research interest include public administration and governance, environmental politics and governance, ethnic politics, and public policy, among others. Professor Onu was former director of research and president of the Nigerian Political Science Association and currently serves as executive member of two of the International Political Science Association's research committees. He is also on the editorial board of the *International Journal of Electronic Governance* and has published in the International Encyclopedia of Digital Governance.

REBECCA TIESSEN

Rebecca Tiessen is associate professor of global and international development studies at the Royal Military College in Kingston, Ontario, Canada. Her field research has taken her to Indonesia, Sri Lanka, Zimbabwe, Malawi, and Kenya. Her research areas include gender mainstreaming, gender inequality, environment and sustainable development in Africa, HIV/AIDS and human security, Canadian foreign aid to Africa, youth work/study abroad programs, and global citizenship. She has recently published *Everywhere/Nowhere: Gender Mainstreaming in Development Agencies* (Kumarian, 2007).

ANGUS WRIGHT

Angus Wright is professor emeritus of environmental studies at California State University, Sacramento, California. He is the author of *The Death of Ramon Gonzalez: The Modern Agricultural Dilemma* (University of Texas Press, 2nd ed., 2005) and co-author, with Wendy Wolford, of *To Inherit the Earth: The Landless Movement in the Struggle for a New Brazil* (FoodFirst Books, 2003). He has served on the boards of the Pesticide Action Network North America, Food First, and the Land Institute, as well as on various task forces and commissions.

INDEX

Recycled
Supporting responsible use
of forest resources
FSC
www.fsc.org Cert no. SGS-COC-003153
© 1996 Forest Stewardship Council
100%